Electrical Systems Engineering

Electrical Systems Engineering

Editor: Maurice Willis

NY RESEARCH PRESS

New York

Published by NY Research Press
118-35 Queens Blvd., Suite 400,
Forest Hills, NY 11375, USA
www.nyresearchpress.com

Electrical Systems Engineering
Edited by Maurice Willis

International Standard Book Number: 978-1-63238-651-9 (Hardback)

Cataloging-in-Publication Data

Electrical systems engineering / edited by Maurice Willis.
    p. cm.
Includes bibliographical references and index.
ISBN 978-1-63238-651-9
1. Electrical engineering. 2. Systems engineering. I. Willis, Maurice.
TK145 .E44 2019
621.3--dc23

Contents

Contents

Preface

The world is advancing at a fast pace like never before. Therefore, the need is to keep up with the latest developments. This book was an idea that came to fruition when the specialists in the area realized the need to coordinate together and document essential themes in the subject. That's when I was requested to be the editor. Editing this book has been an honour as it brings together diverse authors researching on different streams of the field. The book collates essential materials contributed by veterans in the area which can be utilized by students and researchers alike.

Electrical systems engineering is concerned with the development of electrical systems through the application of the principles of electrical engineering. The diverse aspects of generation, transmission, distribution and utilization of electric power are studied under this field. It further delves into the design of a range of electrical devices related to the generation of electricity, like generators, transformers, motors, etc. The applications of this field are in grid systems as well as off-grid systems. Specialized power systems are used in aircraft, electric railway networks, etc. The ever growing need for advanced technology is the reason that has fueled the research in the field of electrical systems engineering in recent times. This book attempts to understand the multiple branches that fall under this discipline. It elucidates new techniques and their applications in a multidisciplinary manner. Coherent flow of topics, student-friendly language and extensive use of examples make this book an invaluable source of knowledge.

Each chapter is a sole-standing publication that reflects each author's interpretation. Thus, the book displays a multi-facetted picture of our current understanding of application, resources and aspects of the field. I would like to thank the contributors of this book and my family for their endless support.

Editor

Inverse Filtering and Principal Component Analysis Techniques for Speech Dereverberation

Mohamed Anouar BEN MESSAOUD, Aicha BOUZID

Department of Electrical Engineering, National School of Engineers, University of Tunis El Manar,
BP 37 Belvedere, Tunis, Tunisia

anouar.benmessaoud@yahoo.fr, bouzidacha@yahoo.fr

Abstract. *In this work, we present a single channel approach for early and late reverberation suppression. This approach can be decomposed into two stages. The first stage employs the inverse filter to augment the signal-to-reverberant energy ratio. The second stage uses the kernel PCA algorithm to enhance the obtained dereverberant signal. It consists in extracting the main non-linear features from the speech signal after inverse filtering. Our approach appears to be efficient mainly in far field conditions and in highly reverberant environments.*

Keywords

Inverse filtering, kernel algorithm, principal components analysis, reverberant speech.

1. Introduction

It is well known that the intelligibility of speech in natural environments is influenced by room reverberation [1], [2]. The presence of reverberation alone can degrade speech perception. Several factors contribute to the influence of reverberation on intelligibility including: source distance, room reverberation time, and the characteristics of both late reverberation and early reflections. Speech dereverberation is an active area of research designed to mitigate the influence of reverberation on speech intelligibility under monaural presentation.

Several speech dereverberation approaches have been proposed in literature. As the linear prediction residual signal includes the effects of the reverberant channel, Brandstein et al. [3] propose to use the residual and the all-pole filter resulting from prediction analysis of the reverberant speech. Allen [4] aim to suppress the effects of reverberation by identifying the Linear Predictive Coding (LPC) parameters.

Yegnanarayana et al. [5] apply the Hilbert envelopes to indicate the strength of the peaks in the Linear Prediction (LP) residuals. Gillespie et al. [6] propose to maximize the kurtosis of the residual by using an adaptive filter for multiple microphones. In [7], [8], the authors use a spatial averaging approach. Then the prediction residual is improved by applying a temporal averaging of larynx cycles. In [9], the authors identify the magnitude and phases of sinusoidal speech model by estimation of the pitch. Consequently, the speech dereverberation is applied to deduce an equivalent equalization filter. Habets et al. [10], Lebart et al. [11] have proposed to use the spectral subtraction technique to dereverberation. They used a statistical model of the room impulse response including Gaussian noise modulated by a decay rate of an exponential function, and then the power spectral density of the impulse response can be suppressed by spectral subtraction. A similar method proposed in [12] estimates the late reflection by using a multistep LP. In [13], the authors use the subspace method to determine the inverse filter independent of the source characteristics. Nakatani et al. [14] proposed to formulate the dereverberation problem as a maximum likelihood using a hill-climbing technique. In [15], the authors optimize the scaling factors by using the minimum mean square error criterion to eliminate the late reflection components.

Principal Component Analysis (PCA) is a subcategory of subspace approaches. It is an important tool for data analysis designed to reduce multidimensional signal. The essential object is to obtain a set of orthogonal factors that describe the variance of the observations and track the new factors considered to determine the necessary features. Speech processing by PCA [16] is extensively applied as a classical multivariate speech processing tool. PCA has been used in different fields of science to extract relevant information from complex matrix data [17]. For speech separation, a robust extension of classical PCA by generalizing eigenvalue decomposition of a pair of covariance matrices has been proposed by [18]. It may be used in speech identification [19], speech recognition [20], speech enhancement [21], [22], and [23] and speech dereverberation [24], [25], and [26].

In this work, we propose an approach of dereverberation in the field of single-channel speech enhancement which is appropriate for real-world applications. For suppressing the late and early reverberations, we apply a single-microphone two-step approach. In the first stage, the inverse filter is applied to enhance the speech to reverberation ratio and reduce the coloration caused by the early reflections and late reverberation. The second stage consists in using the kernel principal component analysis algorithm to enhance the obtained dereverberation signal.

Our approach is tested on a corpus of speech utterances from thirty two speakers. The evaluation shows that our approach is able to reduce early reflections and late reverberation in highly reverberant environments.

The rest of the paper is organized as follows. In Section 2. we present the first stage based on the inverse filter. Then, we integrate the kernel PCA into our approach. Next, in Section 3. we provide the experimental results of our approach.

2.　　Proposed Approach

We propose an approach of speech dereverberation employing monaural recordings. It can be decomposed into two essential stages, as shown in Fig. 1.

In the first stage of our approach, we apply a pre-processing procedure to reduce the short-term correlation of the speech, and identify the delayed late reverberations. Then, we determine the inverse filter by maximization of kurtosis. The second stage consists of using the kernel principal component analysis algorithm to reduce the effects of the long-term reverberation.

2.1.　　Inverse Filter Stage

1)　　Pre-Processing Procedure

The pre-processing procedure consists in adjusting the reverberant speech signal by the right choice of the short order linear prediction.

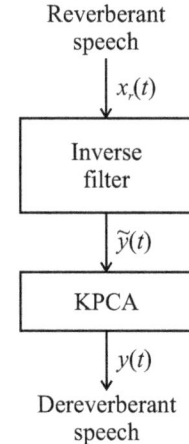

Fig. 1: Block diagram of our proposed approach.

We consider the reverberant speech signal $x_r(n)$ as the result of the convolution between the clean speech and the room impulse response.

$$
\begin{aligned}
x_r(n) &= \sum_{k=0}^{N} g(k)x(n-k), \\
&= \sum_{i=0}^{Q-1} g(i)u(n-i),
\end{aligned}
\tag{1}
$$

where is the response impulse filter determined by combining the effects of the human speech production and the room impulse response, and is the clean speech signal.

The finite impulse response filter is described as follow:

$$
g(n) = \sum_{j=0}^{M} a(j)b(n-j).
\tag{2}
$$

Such a filter would produce the speech signal from the white noise $u(n)$. The reverberant speech signal can be written in vector form:

$$
x_r(n) = Gu(n),
\tag{3}
$$

where:

$$
\begin{bmatrix}
g(0) & \cdots & g(N-1) & 0 & \cdots & 0 \\
0 & g(0) & \cdots & g(N-1) & \cdots & 0 \\
\vdots & \ddots & \ddots & & \ddots & \vdots \\
0 & \cdots & 0 & g(0) & \cdots & g(N-1)
\end{bmatrix}
$$

is a matrix G of size $(N-1) \times (Q+N)$ with Q and N are the length of the vector g and $x_r(n)$, respectively. With

$$
g = [g(0),\ldots,g(N-1)],
$$
$$
x_r(n) = [x_r(n),\ldots,x_r(n-Q)]^T.
$$

A bias provoked by the transfer function of the human speech production exists in estimated late elements of $G(z)$. So, we implement a linear prediction with small-order. In [27], the authors have proposed to use an order of pre-whitening equal to 20 taps. However, in our approach, the order is arranged according to the length of the RIR. This pre-processing compensates for the bias provoked by the transfer function of the human speech production bearing in mind its convolution by the RIR. Therefore, the resulting pre-processing will be more adapted to the reverberant signal. In experiment, we have chosen for the room impulse response with reverberation time equal to 0.5 s and 0.7 s, a short linear prediction order equal to 6 and 20 taps, respectively.

After processing the short-order linear prediction, we reduce the effects of the long-term reverberation.

The long-term linear prediction is described by the following equation:

$$x_r(n) = \sum_{l=0}^{L} f(l)x_r(n-l-D) + er(n), \qquad (4)$$

where $x_r(n)$ is the reverberant speech signal, $f(l)$ are the coefficients of filter, L is the number of the coefficients, D is the delay and $er(n)$ is the error of the speech signal.

In this step, we apply the Levinson-Durbin method to minimize the energy of the error $er(n)$.

Based on the method proposed by Kinoshita et al. [27], we apply the Wiener-Hopf equation specialized for delayed LP.

$$f = \left(E\{x_r(n-D)x_r^T(n-D)\}\right)^{-1} \cdot \\ \cdot E\{x_r(n-D)x_r(n)\}, \qquad (5)$$

with

$$E\{x_r(n-D)x_r^T(n-D)\} = v_x^2 GG^T,$$

$$E\{x_r(n-D)x_r(n)\} = v_x^2 Gg_{late},$$

where v_x^2 is the variance of white noise and

$$g_{late} = [g(D),\ldots,g(L-1),0,\ldots,0]^T.$$

As a result, we are having:

$$f = \left(GG^T\right)^{-1} Gg_{late}. \qquad (6)$$

The estimated power of late reverberation is:

$$E\left\{(x^T f)^2\right\} = ||v_x g_{late}||^2. \qquad (7)$$

The linear prediction filter order, L, is a large number. Consequently, the residual speech signal each time is calculated based on samples $L+D$. Hence, the linear prediction residual speech signal is capable of providing the long-term reverberation of the signal.

2)   Inverse Filtering

In order to suppress the long reverberations, we apply an inverse filtering by maximizing the kurtosis of linear prediction residual speech. As the reverberant speech signal has a lower kurtosis than that of linear prediction residual of clean speech signal, we suggest to estimate the inverse filter by using the kurtosis maximization.

Figure 2 shows the diagram for inverse filtering. We approximate the LP residual of the processed speech by the inverse-filtered LP residual of the reverberant speech.

The inverse-filtered speech signal is:

$$\widetilde{y}(t) = h\widehat{r}(t), \qquad (8)$$

where $\widehat{r}(t) = [r(t-K+1),\ldots,r(t)]^T$ is the residual signal of the reverberant speech after the application of the pre-processing procedure, h is the inverse filter, and $\widetilde{y}(t)$ is the inverse filtered speech signal.

The kurtosis of $\widetilde{y}(t)$ is described as follow:

$$I(t) = \frac{E\{y^4(t)\}}{E^2\{y^2(t)\}} - 3. \qquad (9)$$

According to the work of Gillespie et al. [6], the gradient of the kurtosis is described by the equation in below:

$$\frac{\partial I(t)}{\partial h(t)} = \qquad (10)$$

$$= \frac{4\left(E\{\widetilde{y}^2(t)\}E\{\widetilde{y}^3(t)\widehat{r}(t)\} - E\{\widetilde{y}^4(t)\}E\{\widetilde{y}(t)\widehat{r}(t)\}\right)}{E^3\{\widetilde{y}^2(t)\}}.$$

Then, we approximate the gradient by the following equation:

$$\frac{\partial I(t)}{\partial h(t)} \approx \frac{4\left(E\{\widetilde{y}^2(t)\}\widetilde{y}^2(t) - E\{\widetilde{y}^4(t)\}\widetilde{y}(t)\right)}{E^3\{\widetilde{y}^2(t)\}}\widehat{r}(t) =$$

$$= f(t)\widehat{r}(t). \qquad (11)$$

Then, we calculate the $E\{\widetilde{y}^4(t)\}$ and $E\{\widetilde{y}^2(t)\}$ to obtain the adaptive inverse filter.

In the time domain, the inverse adaptive filter can be described by the following equation:

$$h(t+1) = h(t) + \beta f(t)\widehat{r}(t), \qquad (12)$$

where β adapts the learning rate, and $f(t)$ is the function of the kurtosis.

However, on account of the several numbers of taps for RIR, signal is operated frame by frame into the frequency domain by using the Fast Fourier Transform (FFT) consistent with the work of Wu and Wang [28].

In the frequency domain, the Eq. (12) becomes:

$$H(n+1) = H(n) + \frac{\beta}{P}\sum_{p=1}^{P} F(p)\widehat{R}^*(p), \quad (13)$$

where $H(n)$ is the nth iteration of h, $F(p)$, $\widehat{R}^*(p)$ denote, respectively, the FFT of $f(t)$, $\widehat{r}(t)$, with p is the pth frame, and P is the number of frames.

Then, we obtain the inverse-filtered speech by convolving the inverse filter with the reverberant signal.

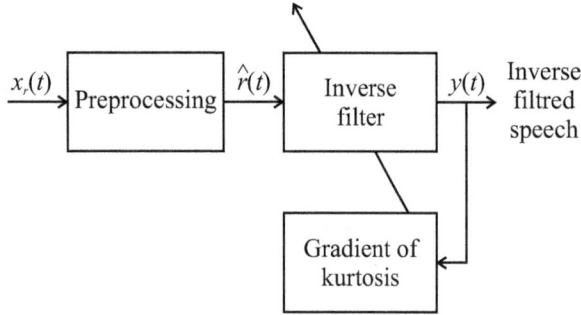

Fig. 2: Block diagram of the first stage.

In the next sub-section, we add the kernel PCA technique to enhance the processed signal.

2.2. Kernel PCA Stage

In Fig. 3, we illustrate the kernel PCA algorithm for speech dereverberation. In the KPCA technique, the speech components will be projected onto low-order

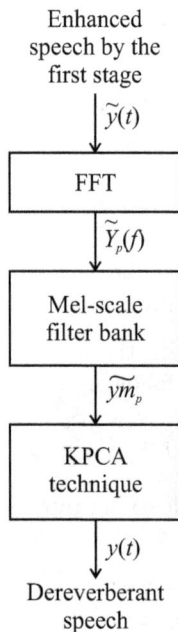

Fig. 3: Block diagram of the first stage.

while reverberant components will be projected onto high-order.

We compute the Fast Fourier Transform (FFT) of the obtained speech signal by the first stage, $\widetilde{y}(t)$. We obtain the short spectrum for the clean speech and the transfer function of the frequency f at the pth frame. Our goal is to determine the estimated clean speech $y(t)$.

The FFT of $\widetilde{y}(t)$ can be expressed as:

$$\widetilde{Y}_p(f) = Y_p(f)\cdot\widetilde{H}_p(f). \quad (14)$$

Then, we calculate the log-spectrum of $\widetilde{Y}_p(f)$. We obtain the following equation:

$$\widetilde{Y}_{\log-p}(f) = \log(Y_p(f)) + \log\left(\widetilde{H}_p(f)\right), \quad (15)$$

where $\widetilde{Y}_{\log-p}(f)$, $\log\left(\widetilde{H}_p(f)\right)$, and $\log(X_i(f))$ are the logarithm spectra for the first stage obtained speech signal by inverse filtering, estimated clean speech signal, and transfer function, respectively.

So that, we extract the clean speech based on KPCA,

$$Y = VY_{\log}. \quad (16)$$

The eigen-vector matrix V, is derived by the eigen-value decomposition, in which the filter $V = [v^{(1)}, \dots, v^{(p)}]$ consists of the eigen-vectors corresponding to the Q dominant eigen-values. The component of the convolution reverberant signal $[v^{(p+1)}, \dots, v^{(M)}]$ is eliminated by the filtering operation in Eq. (16).

According to the Eq. (16), the $\widetilde{y}(t)$ signal is represented under the assumption of non-correlation between the reverberant signal and the clean speech. Then, we apply a non-linear PCA called kernel principal component analysis (KPCA) to extract the clean speech.

Now, we detail the KPCA technique in the following subsection.

1) KPCA Technique

KPCA transforms the input logarithm spectra with nonlinear structure into a higher-dimensional feature space with linear structure, and then computes linear PCA on the mapped data. We can assume that reverberant signal will be eliminated in the high-dimensional space.

Consider the output of the mel-scale filter bank $\widetilde{ym}_p$ at p^{th} frame, we define the covariance matrix as follow:

$$C = \frac{1}{P}\sum_{p=1}^{P} \overline{\Psi}(\widetilde{ym}_p)\overline{\Psi}(\widetilde{ym}_p)^T, \quad (17)$$

where Ψ is the nonlinear map, and P denotes the number of samples.

Now, we must solve the eigen-vectors v and eigen-values α.

We have:

$$\begin{aligned} \alpha V &= CV, \\ \alpha\overline{\Psi}(\widetilde{ym}_k)^T V &= \overline{\Psi}(\widetilde{ym}_k)^T CV \quad \text{for all } k, \end{aligned} \tag{18}$$

where $k = 1, \ldots, P$.

Such as the eigen-vectors:

$$V = \sum_{j=1}^{P} \beta_j \overline{\Psi}(\widetilde{ym}_j). \tag{19}$$

Replacing the Eq. (17) and Eq. (19) into Eq. (18) leads to:

$$\alpha \sum_{j=1}^{P} \beta_j \overline{\Psi}(\widetilde{ym}_k)^T \overline{\Psi}(\widetilde{ym}_j) =$$

$$\frac{1}{P} \sum_{j=1}^{P} \beta_j \overline{\Psi}(\widetilde{ym}_k)^T \sum_{p=1}^{P} \overline{\Psi}(\widetilde{ym}_p)\overline{\Psi}(\widetilde{ym}_p)^T \overline{\Psi}(\widetilde{ym}_j), \tag{20}$$

with $K_{jp} = \overline{\Psi}(\widetilde{ym}_j)^T \overline{\Psi}(\widetilde{ym}_p)$.

We formulate the equation above as:

$$\overline{\Psi}(\widetilde{ym}_k) \cdot Cv = \frac{1}{P} \sum_j \beta_j \sum_p \overline{A}_{kp} \overline{A}_{pj}. \tag{21}$$

So we obtain

$$P\alpha K\beta = K^2\beta, \tag{22}$$

with K as the kernel matrix.

For dereverberation a test sample $\overline{\Psi}(\widetilde{ym})$, it is projected onto the Eigen-vectors V eigenvalues. The projection is presented in terms of kernel functions based on the Eq. (19):

$$\gamma_k = (V^k)^T \overline{\Psi}(\widetilde{ym}) = \sum_{j=1}^{P} \beta_j^k k(\widetilde{ym}, \widetilde{ym}_j). \tag{23}$$

Therefore, the projected sample in feature space is equal to:

$$O_n \overline{\Psi}(\widetilde{ym}) = \sum_{k=1}^{n} \beta_k V^k, \tag{24}$$

where O_n presents the projection operator.

For dereverberation, we try to find a sample y that satisfies $\overline{\Psi}(y) = O_n \overline{\Psi}(\widetilde{ym})$.

We consider the Gaussian kernel function:

$$K(\widetilde{ym}_j, \widetilde{ym}_p) = \exp\left(-\frac{||\widetilde{ym}_j - \widetilde{ym}_p||^2}{c}\right), \tag{25}$$

where c presents the variance.

The dereverberant speech is determined by an iterative update equation of y

$$y_{t+1} = \frac{\sum_{j=1}^{P} \eta_j k\left(y_t, \widetilde{ym}_j\right) \widetilde{ym}_j}{\sum_{j=1}^{P} \eta_j k\left(y_t, \widetilde{ym}_j\right)}, \tag{26}$$

where $\eta_j = \sum_{k=1}^{n} \gamma_k \beta_j^k$.

We indicate that the pre-image y is a linear combination of the input speech $\widetilde{ym}_j$.

The KPCA matrix was estimated only from the reverberant speech recordings. In order to perform the matrix, K, we choose $P = 1500$ frames from the training data, and we used the Gaussian kernel function. The KPCA uses 16 dimensions Mel-scale filter bank output.

3. Experiment Results

We evaluated the proposed dereverberation speech approach with the TIMIT database [29]. As we compare our approach to the state-of-the-art algorithms when the signals are sampled at 8 kHz [28] and [31] so we down-sample the speech signals from 16 kHz to 8 kHz. We use 32 English speakers of this database.

We apply the standard objective measures to determinate the quality of noisy speech enhanced by our proposed approach. The reverberant speech signal is generated by convolving the room impulse response function and the clean speech with $T_{60} = 0.3, 0.5$, and 0.7 s. In our approach, we use the mirror image model proposed in [30] to generate the RIR function. The RIR was created in a room with dimensions $6 \times 4 \times 3$ m, the microphone was installed at $4 \times 1 \times 2$ m, and the speaker was placed at $2 \times 3 \times 1.5$ m, see Fig. 4.

To evaluate the performance of our approach and to compare it with two existing methods named the Wu method [28], and MAP and variational deconvolution method (MA-VD) [31]. In [28], the authors proposed a two-step single-channel dereverberation method (Wu method) whose first step applied an adaptive inverse filtering scheme by kurtosis maximization. Then, they introduce the spectral subtraction technique in the second stage to improve the dereverberation performance for long reflections. In [31], the authors used a cost functions obtained from variational estimation and maximum posteriori to formulate a dereverberation in frequency bin.

We use the standard objective metrics namely respectively Global Signal Noise Ratio in dB (SNR),

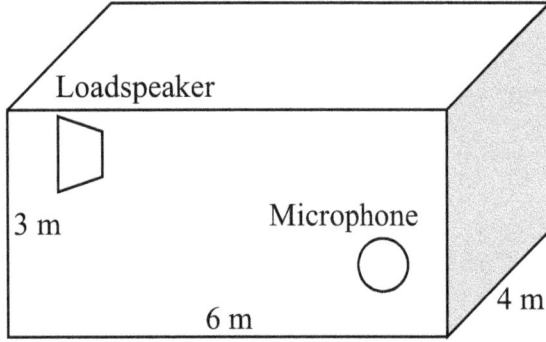

Fig. 4: Geometry of a varechoic room of the dimensions $6 \times 4 \times 3$ meters. Wall reflection coefficients are equal to 0.75. The microphone is at $4 \times 1 \times 2$ meters. The loudspeaker is at $2 \times 3 \times 1.5$ meters.

the Perceptual Evaluation of Speech Quality (PESQ) that measure the overall quality of the speech, and the Speech to Reverberation Ratio (SRR) with the three reverberation times.

Table 1 presents the results of global SNR in dB obtained by using the Wu, MA-VD, and proposed approach (Pro-App) for speech dereverberation using monaural speech recordings with the TIMIT database. We referred to the degraded reverberant speech by the word "Rev".

Tab. 1: Performance comparison of global SNR, in the presence of reverberant speech.

T_{60} (s)	Rev	Wu [28]	MA-VD [31]	Pro-App
	Global SNR (dB)			
0.3	−2.87	1.84	4.77	6.08
0.5	−3.65	1.02	3.69	5.34
0.7	−4.39	0.61	2.86	3.15

It can be seen that our approach gives better SNR results than the others methods at SNR levels varied from −2.87 at −4.39 dB.

Table 2 gives the results in the case of our approach and that of Wu and Wang [28] and MA-VD [31].

Tab. 2: Performance comparison of PESQ, in the presence of reverberant speech.

T_{60} (s)	Rev	Wu [28]	MA-VD [31]	Pro-App
	PESQ			
0.3	2.24	2.36	2.45	2.74
0.5	2.07	1.24	1.73	1.87
0.7	1.93	1.53	1.68	1.91

In Tab. 2, the PESQ scores show the intelligibility of our approach compared to that produced by the state-of-the-art methods. However, in the two reverberation times ($T_{60} = 0.5$ s and 0.7 s), the reverberant speech obtains the greater PESQ scores. This results show that the PESQ measure is not correlated with dereverberation.

Also, we compare the proposed approach with other methods in terms of SRR measures. The Speech to Reverberation Ratio (SRR) is a speech-based measure of reverberation that can be calculated even when the effect of the dereverberation method cannot be described in the impulse response. The SRR can be defined as:

$$SRR = 20 \log_{10} \left(\frac{||x||_2}{||y - x||_2} \right), \qquad (27)$$

where is the clean speech signal, and is the estimated output of our dereverberation approach.

Tab. 3: Performance comparison of SRR, in the presence of reverberant speech.

T_{60} (s)	Rev	Wu [28]	MA-VD [31]	Pro-App
	SSR (dB)			
0.3	−2.07	−1.46	0.68	1.42
0.5	−3.29	−2.95	−2.17	0.06
0.7	−5.08	−5.30	−5.13	−1.61

Results of the Tab. 3 indicate that our approach outperforms the two state of the art methods with SRR measures.

We observe that in the three reverberation times, the proposed approach demonstrate a greater SRR score compared to their corresponding methods. The SRR level of the reverberant speech is much lower than that of our approach. However, the two methods [28], and [31] fail to conserve its performance for $T_{60} = 0.7$ s. It can be seen that our approach outperforms the results of the compared methods when we added the KPCA technique to the first stage.

We can conclude that the proposed approach can suppress the long reflections in the first stage by applying a pre-processing procedure followed by a delayed long-term LP. The KPCA stage is used to enhance the obtained signal by the first stage. Furthermore, the proposed approach is more effective in suppressing the long reflections. However, the two state-of-the-art algorithms yield satisfactory results only when the T_{60} is short.

In order to show the enhanced speech obtained by using the proposed approach, spectrograms of the original speech, the reverberant speech, and the enhanced speech signals obtained by using the inverse filter, and the dereverberation by our approach after applying the kernel PCA are presented in Fig. 5.

The spectrogram of our proposed approach shows the inverse filter reconstructs the harmonic structure of the original speech. After the application of the kernel PCA algorithm, the spectrogram improves that the reverberation is greatly reduced in our proposed approach.

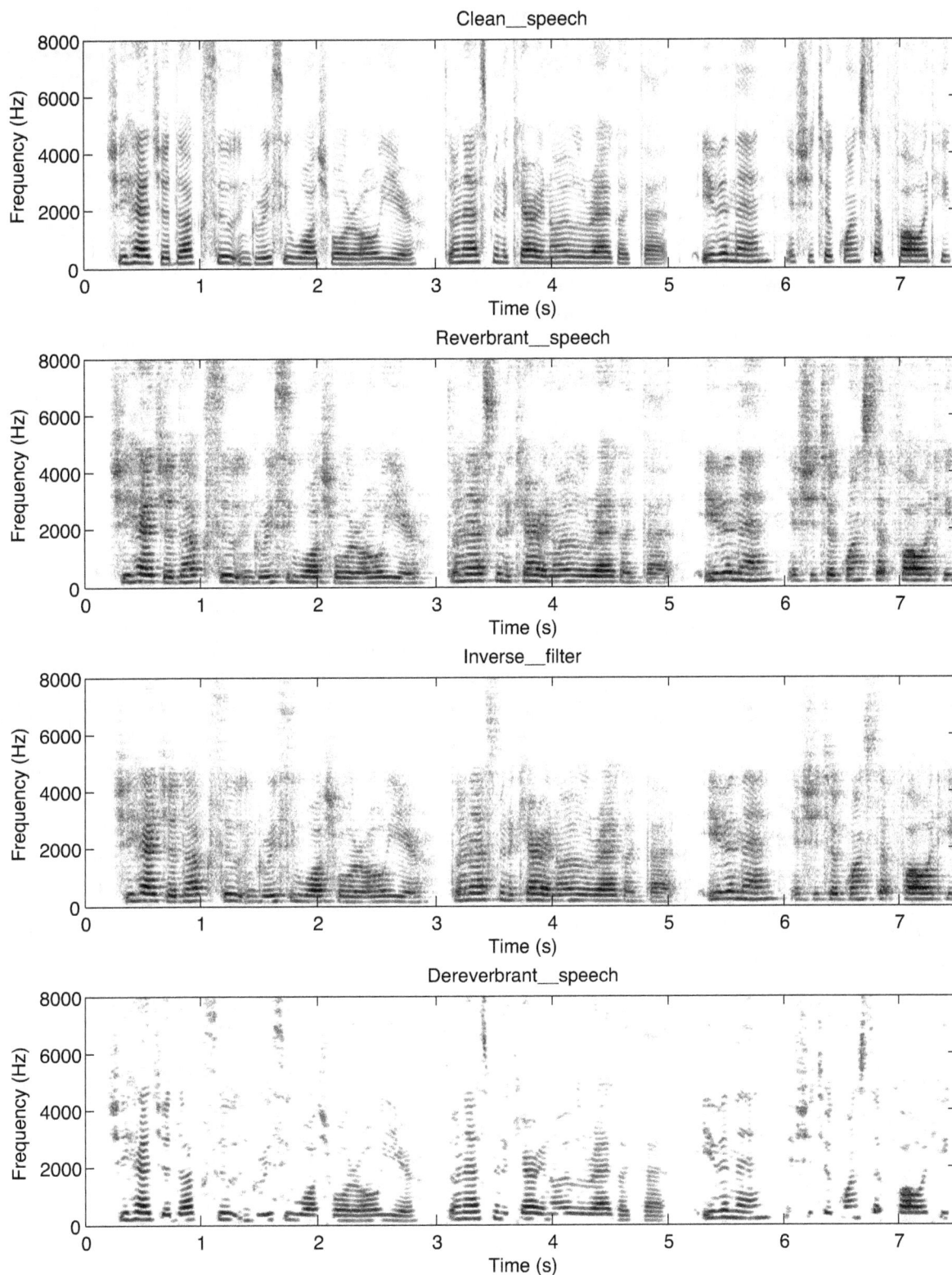

Fig. 5: Spectrograms of speech signal. (a) Spectrogram of clean speech signal. (b) Spectrogram of reverberant speech signal. (c) Spectrogram of dereverberant speech signal by inverse filter. (d) Spectrogram of enhanced dereverberant speech signal by our approach.

4. Conclusion

In this paper, we present a speech dereverberation approach of monaural recordings. The proposed approach can be summarized into two stages. First, we apply an inverse filter of reverberant speech with introducing a pre-processing procedure. Second, we propose to use the kernel principal component analysis algorithm to enhance the obtained speech by inverse filter.

The experimental results show that our approach achieves a significantly better speech quality than the state of the art algorithms and show a significant improvement in terms of these measures with minimal residual distortion. Also, the existing methods do not obtain better results for rooms with T_{60} of more than 0.5 s.

The future work consists in processing reverberant and noisy speech in real-time while keeping reasonable distortion levels.

References

[1] HUNT, M. J. and C. LEFEBVRE. Speaker dependent and independent speech recognition experiments with an auditory model. In: *International Conference on Acoustics, Speech, and Signal Processing.* New York: IEEE, 1988, pp. 215–218. ISSN 0736-7791. DOI: 10.1109/ICASSP.1988.196552.

[2] KLATT, D. Prediction of perceived phonetic distance from critical-band spectra: A first step. In: *IEEE International Conference on Acoustics, Speech, and Signal Processing.* Paris: IEEE, 1982, pp. 1278–1281. DOI: 10.1109/ICASSP.1982.1171512.

[3] BRANDSTEIN, M. S. and S. M. GRIEBEL. *Nonlinear, Model-Based Microphone Array Speech Enhancement.* New York: Springer, 2000. ISBN 978-1-4613-4656-2.

[4] ALLEN, J. B. *Synthesis of pure speech from a reverberant signal.* US patent. US3786188.

[5] YEGNANARAYANA, B. and P. S. MURTHY. Enhancement of reverberant speech using LP residual signal. *IEEE Transactions on Speech and Audio Processing.* 2000, vol. 8, iss. 3, pp. 267–281. ISSN 1063-6676. DOI: 10.1109/89.841209.

[6] GILLESPIE, B. W., H. S. MALVAR and D. A. F. FLORENCIO. Speech dereverberation via maximum-kurtosis subband adaptive filtering. In: *IEEE International Conference on Acoustics,* *Speech, and Signal Processing.* Salt Lake City: IEEE, 2001, pp. 3701–3704. ISBN 0-7803-7041-4. DOI: 10.1109/ICASSP.2001.940646.

[7] GAUBITCH, N. D. *Blind identification of acoustic systems and enhancement of reverberant speech.* London, 2006. Dissertation. Imperial College London.

[8] GAUBITCH, N. D. and P. A. NAYLOR. Spatiotemporal Averaging Method for Enhancement of Reverberant Speech. In: *15th International Conference on Digital Signal Processing.* Cardiff: IEEE, 2007, pp. 607–610. ISBN 1-4244-0881-4.

[9] NAKATANI, T., M. MIYOSHI and K. KINOSHITA. Single-Microphone Blind Dereverberation. *Speech Enhancement.* Berlin: Springer, 2005, pp. 247–270. ISBN 978-3-540-24039-6.

[10] HABETS, E. A. P., S. GANNOT, I. COHEN, P. C. W. SOMMEN. Joint Dereverberation and Residual Echo Suppression of Speech Signals in Noisy Environments. *IEEE Transactions on Audio, Speech, and Language Processing.* 2008, vol. 16, iss. 8, pp. 1433–1451. ISSN 1558-7916. DOI: 10.1109/TASL.2008.2002071.

[11] LEBART, K., J. M. BOUCHER and P. N. DENBIGH. A new method based on spectral subtraction for speech dereverberation. *Acta Acoust.* 2001, vol. 87, no. 3, pp. 359–366. ISSN 1610-1928.

[12] KINOSHITA K., T. NAKATANI and M. MIYOSHI. Spectral subtraction steered by multi-step forward linear prediction for single channel speech dereverberation. In: *IEEE International Conference on Acoustics Speech and Signal Processing Proceedings.* Toulouse: IEEE, 2006, pp. 817–820. ISBN 1-4244-0469-X. DOI: 10.1109/ICASSP.2006.1660146.

[13] HIKICHI, T., M. DELCROIX and M. MIYOSHI. Inverse filtering for speech dereverberation less sensitive to noise and room transfer function fluctuations. *EURASIP Journal on Advances in Signal Processing.* 2007, vol. 2007, iss. 1, pp. 1–4. ISSN 1110-8657. DOI: 10.1155/2007/34013.

[14] NAKATANI, M., B.-H. JUANG, T. YOSHIOKA, K. KINOSHITA, M. DELCROIX and M. MIYOSHI. Speech dereverberation based on maximum-likelihood estimation with time-varying Gaussian source model. *IEEE Transactions on Audio, Speech, and Language Processing.* 2008, vol. 16, iss. 8, pp. 1512–1527. ISSN 1558-7916. DOI: 10.1109/TASL.2008.2004306.

[15] GOMEZ, R., J. EVEN and H. SARUWATARI. Rapid unsupervised speaker adaptation robust in

reverberant environment conditions. In: *9th Annual Conference of the International Speech Communication Association*. Brisbane: ISCA, 2008, pp. 187–190. ISBN 978-1615673780.

[16] JOLIFFE, I. T. *Principal component analysis*. 2nd ed. New York: Springer Verlag, 2002. ISBN 978-0-387-95442-4.

[17] MISRA, M., H. H. YUE, S. J. QIN and C. LING. Multivariate process monitoring and fault diagnosis by multi-scale PCA. *Computer & Chemical Engineering*. 2002, vol. 26, iss. 9, pp. 1281–1293. ISSN 0098-1354. DOI: 10.1016/S0098-1354(02)00093-5.

[18] BENABDERRAHMANE, Y., S.-A. SELOUANI and D. D. OSHAUGHNESSY. Blind speech separation for convolutive mixtures using an oriented principal components analysis method. In: *18th European Signal Processing Conference*. Aalborg: EURASIP, 2010, pp. 1553–1557. ISBN 978-3-642-03882-2.

[19] ABOLHASSANI, A. H., S.-A. SELOUANI and D. OSHAUGHNESSY. Speech enhancement using PCA and variance of the reconstruction error model identification. In: *IEEE Workshop on Automatic Speech Recognition & Understanding*. Kyoto: IEEE, 2007, pp. 19–23. ISBN 978-1-4244-1746-9. DOI: 10.1109/ASRU.2007.4430077.

[20] TAKIGUCHI, T. and Y. ARIKI. PCA-Based Speech enhancement for distorted speech recognition. *Journal of multimedia*. 2007, vol. 2, no. 5, pp. 13–18. ISSN 1796-2048. DOI: 10.4304/jmm.2.5.13-18.

[21] SHINDE, V. D., C. G. PATIL and S. D. RUIKAR. Wavelet based multi-scale principal component analysis for speech enhancement. *International Journal of Engineering Trends and Technology*. 2012, vol. 3, iss. 3, pp. 397–400. ISSN 2231-5381.

[22] VERTELETSKAYA, E. and B. SIMAK. Spectral subtractive type speech enhancement methods. *Advances in Electrical and Electronic Engineering*. 2010, vol. 8, no. 3, pp. 66–72. ISSN 1804-3119.

[23] BEN MESSAOUD, M. A. and A. BOUZID. Speech Enhancement Based on Wavelet Transform and Improved Subspace Decomposition. *Journal Audio Engineering Society*. 2015, vol. 63, iss. 12, pp. 990–1000. ISSN 1549-4950.

[24] GERKMANN, T. and R. C. HENDRIKS. Unbiased MMSE-Based Noise Power Estimation with Low Complexity and Low Tracking Delay. *IEEE Transactions on Audio, Speech, and Language Processing*. 2012,

vol. 20, iss. 4, pp. 1383–1393. ISSN 1558-7916. DOI: 10.1109/TASL.2011.2180896.

[25] KODRASI, I., S. GOETZE and S. DOCLO. Regularization for Partial Multichannel Equalization for Speech Dereverberation. *IEEE Transactions on Audio, Speech, and Language Processing*. 2013, vol. 21, iss. 9, pp. 1879–1890. ISSN 1558-7916. DOI: 10.1109/TASL.2013.2260743.

[26] ZHENG, C., R. PENG, J. LI and X. LI. A Constrained MMSE LP Residual Estimator for Speech Dereverberation in Noisy Environments. *IEEE Signal Processing Letters*. 2014, vol. 21, iss. 12, pp. 1462–1466. ISSN 1070-9908. DOI: 10.1109/LSP.2014.2340396.

[27] KINOSHITA, K., M. DELCROIX, T. NAKATANI and M. MIYOSHI. Suppression of Late Reverberation Effect on Speech Signal Using Long-Term Multiple-step Linear Prediction. *IEEE Transactions on Audio, Speech, and Language Processing*. 2009, vol. 17, iss. 4, pp. 534–545. ISSN 1558-7916. DOI: 10.1109/TASL.2008.2009015.

[28] WU, M. and D. WANG. A one-microphone algorithm for reverberant speech enhancement. In: *IEEE International Conference on Acoustics, Speech, and Signal Processing*. Hong Kong: IEEE, 2003, pp. 892–895. ISBN 0-7803-7663-3. DOI: 10.1109/ICASSP.2003.1198925.

[29] GAROFOLO, J. *Getting started with the DARPA timit cd-rom: An acoustic phonetic continuous speech database*. Gaithersburgh: National Institute of Standards and Technology.

[30] ALLEN, J. B. and D. A. BERKLEY. Image method for efficiently simulating small-room acoustics. *Journal of the Acoustical Society of America*. 1979, vol. 65, iss. 4, pp. 943–950. ISSN 0001-4966. DOI: 10.1121/1.382599.

[31] JUKIC, A., T. VAN WATERSCHOOT, T. GERKMANN and S. DOCLO. Speech dereverberation with convolutive transfer function approximation using MAP and variational deconvolution approaches. In: *14th International Workshop on Acoustic Signal Enhancement*. Juan-les-Pins: IEEE, 2014, pp. 50–54. ISBN 978-1-4799-6808-4. DOI: 10.1109/IWAENC.2014.6953336.

About Authors

Mohamed Anouar BEN MESSAOUD received the Ph.D. degree in Electrical Engineering from the National School of Engineers of Tunis in 2011. His research interests include topics in speech analysis and applications to engineering and computer science, particularly to pitch and multi-pitch estimation, voicing decision, speech separation, and also speech enhancement.

Aicha BOUZID received her Ph.D. in Electrical Engineering from the National School of Engineers of Tunis (ENIT) Tunisia in 2004. Currently, she is working as Professor in ENIT. Her current research interests include Speech Processing and Image Processing.

Comparison of Fractional Order Modelling and Integer Order Modelling of Fractional Order Buck Converter in Continuous Condition Mode Operation

Ali MOSHAR MOVAHHED[1], Heydar TOOSIAN SHANDIZ[1], Syed Kamal HOSSEINI SANI[2]

[1]School of Electrical & Robotic Engineering, Shahrood University of Technology, Shahrood, Iran
[2]Department of Electrical Engineering, Faculty of Engineering, Ferdowsi University of Mashhad, Mashhad, Iran

a_mosharmovahhed@yahoo.com, htshandiz@shahroodut.ac.ir, k.hosseini@um.ac.ir

Abstract. *Producing a mathematical model with high accuracy is the first and important step in control of systems. Nowadays fractional calculus has been in the spotlight and it has a lot of application especially in control engineering. Fractional modelling on one of the conventional converters is done in this paper. Fractional state space model and related fractional transfer functions for a fractional DC/DC Buck converter is established and achieved results are compared to integer order models. At the end of this paper Oustaloup's recursive approximation is introduced and imposed for one of gathered fractional transfer function.*

Keywords

Approximation, averaged circuit model, buck converter, fractional calculus, fractional transfer function.

1. Introduction

The irregular pollution of fossil fuel energy and climate changing with greenhouse gases have put renewable energy sources in the spotlight [1]. Wind turbines in both offshore and onshore are one of the most impressive approach for producing energy with renewable sources [1]. Standalone wind turbine systems are recommended and useful in remote area and other locations with suitable potential of wind speed. In standalone systems, battery absorbs the produced electricity by wind turbine and charge control tasks to ensure stability and reach MPPT [2]. Full-power converters mostly with

PMSG are used in variable speed wind turbines [3]. In stand-alone systems DC/DC converters are one of the most important component which are playing and vital role.

The main purpose of DC/DC power electronic converters which are applied widely in switching power supplies and dc motor's drives is to control the amplitude of output voltage and current. In many cases the input power is unregulated and control scheme is essential to yield a regulated and appropriate output power [4]. Most of the engineers and researches have considered integer order models for all systems including power electronic systems and devices which are made up of fractional components in nature [5]. During the last few decades' fractional calculus has opened new horizons in all engineering branches. It seems necessary to replace integer order modeling analysis and modeling by fractional order. The conventional wisdom is that the electrical elements like inductors and capacitors are integer order in nature but the reality is different. Lots of researches have been done in order to prove the fractionality of electrical elements. In [6], [7], [8] and [9] Wesrelund et al. measured practically the fractional order of capacitors and inductors with different dielectrics and core coil. For instance, at 1 kHz frequency and room temperature. They found that the fractional order is 0.9776 for capacitor with polyvinylidenefluoride as dielectric or 0.99911 for capacitor with polysulfide as dielectric and some other capacitor with different dielectric was measured and also 0.97 is the amount of fractionality which gathered for an inductor with air core coil. The fractional property is not terminating to electrical elements and it's possible to expand it to all systems. Fractional order modeling is more accurate than integer order model in many real dynamical circuits [7]. The common integer order models for many electrical circuits are accurate enough because the inductors and capacitors which exist in markets are in fractional order of near 1but for circuits which are made up of fractional components it's essential to describe by fractional order models although in general fractional order modeling can cause more accuracy even for systems with integer order components [8]. It's proved that all systems have fractional characteristic in nature but they are different in amount of being fractional thus integer order models can describe features of many systems which have less fractionalities [9]. In this paper an assumptive fractional order Buck converter is modeled and frequency analysis upon achieved fractional order transfer function is done. Fractional model for Buck converter is compared with integer order model and the gathered results are shown in figures. The perturbation of inductor current and capacitor voltage is computed for fractional Buck converter and finally Oustaloup's recursive approximation is done for one of gathered transfer function.

2. Fractional Calculus

More than three centuries fractional calculus was considered as a theoretical field without any practical applications but during the last three decades this mathematical branch has become more common and useful in lots of sciences and engineering fields such as reaction-diffusion system, electrical circuits, rotor bearing system, finance system, biological system, thermoelectric system, and so on [11]. Perhaps in near future conventional calculus replace by fractional calculus because of its ability to expand usual controllers in order to achieve better performance [12]. Over the last few years the applications of fractional control spread out due to its robust performance [13]. Some fractional definitions which are used in this paper are presented in following equations.

Definition 1 *Rieman-Liouville fractional order integral is defined as bellow,*

$$I_c^\alpha F(t) \triangleq \frac{1}{\Gamma(\alpha)} \int_c^t (t-\tau)^{\alpha-1} f(\tau) d\tau,$$

$$t > c, \quad \alpha \in R^+,$$

(1)

$\Gamma(.)$ *is the Gamma function and α is the order of fractionality.*

Definition 2 *Most of the definition in fractional calculus relies on Gamma function which is well known [13]. The definition of Gamma function for real posi-*

tive number is presented in Eq. (2) [18].

$$\Gamma(n) = \int_0^\infty e^{-u} u^{n-1} du.$$

(2)

Remark 1 *If n belongs to natural numbers, Eq. (2) will change into factorial form [13].*

$$\Gamma(n) = (n-1)!, \quad n \in N.$$

(3)

Note that, substituting α by R^- in order to reach fractional order differential operator is not allowable.

Definition 3 *Rieman-Liouville definition for the fractional order derivative of order $\alpha \in R^+$ has the following form,*

$$_RD^\alpha f(t) \triangleq D^m I^{m-\alpha} f(t) =$$
$$= \frac{d^m}{dt^m} \left[\frac{1}{\Gamma(m-\alpha)} \int_0^t \frac{f(\tau)}{(t-\tau)^{\alpha-m-1}} d\tau \right],$$

(4)

where $m-1 < \alpha < m$, $m \in N$ and R is the representativeness of Rieman-Liouville definition.

Definition 4 *Caputo derivative is defined as Eq. (5).*

$$_cD^\alpha f(t) \triangleq I^{m-\alpha} D^m f(t) =$$
$$= \frac{1}{\Gamma(m-\alpha)} \int_0^t \frac{f^m(\tau)}{(t-\tau)^{\alpha-m+1}} d\tau,$$

(5)

where $m-1 < \alpha < m$, $m \in N$ and C shows the Caputo definition.

Definition 5 *Laplace transform is the next operator which is used in this paper. It is given for Caputo fractional-order derivative as defined in Eq. (6).*

$$\ell\left[_cD^\alpha f(t)\right] = s^\alpha F(s) - \sum_{k=0}^{m-1} s^{\alpha-k-1} f^{(k)}(0).$$

(6)

Definition 6 *Another significant definition is Mittag-Leffler function which is playing the role of an exponential function in integer order response [10]. One-parameter and two parameters Mittag-Leffler function is introduced in following Eq. (7).*

$$E_\alpha(t) = \sum_{k=0}^\infty \frac{t^k}{\Gamma(\alpha k + 1)},$$

$$E_{\alpha,\beta}(t) = \sum_{k=0}^\infty \frac{t^k}{\Gamma(\alpha k + \beta)},$$

(7)

$$\Re(\alpha) > 0, \quad \Re(\beta) > 0,$$

$E(.)$ *is the Mittag-Leffler function.*

3. State Space Model

Figure 1 shows the main scheme of a DC/DC Buck converter which is controlled by Pulse Width Modulation (PWM) unit.

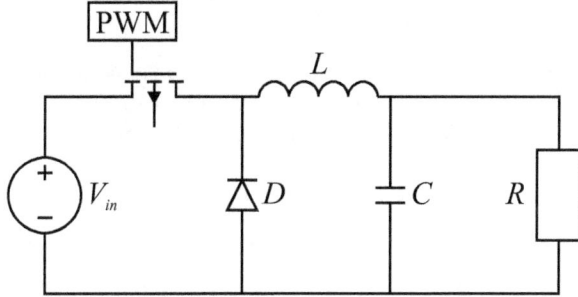

Fig. 1: Shows the circuit of DC/DC Buck converter.

In Eq. (8) and Eq. (9) the well-known equations which present the relationship between voltage and current of inductor and capacitor in fractional order condition is expressed respectively [8].

$$V_L(t) = L\frac{d^\alpha I_L(t)}{dt^\alpha}. \tag{8}$$

$$I_C(t) = C\frac{d^\beta V_C(t)}{dt^\beta}. \tag{9}$$

In which α, β indicates the fractional order of inductor and capacitor respectively. Capacitor voltage and inductor current consider as state space parameters and input voltage is allocated as input vector.

$$X = \begin{bmatrix} i_L(t) \\ v_C(t) \end{bmatrix}, \quad U = [V_{in}(t)]. \tag{10}$$

By using the following steps averaged model will be established. Two modes consider for DC/DC Buck converter, switch on and switch off mode.

3.1. Switch ON Mode

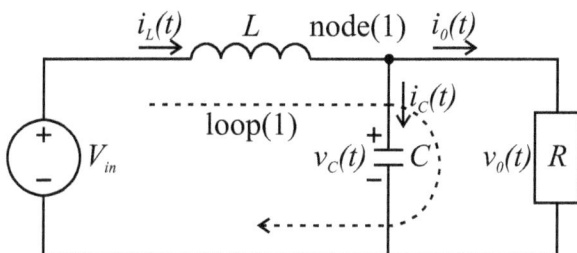

Fig. 2: Buck converter circuit in switch on mode.

Equation (11) is achieved by imposing voltage rule on circuit which is drawn in Fig. 2.

$$L\frac{d^\alpha i_L(t)}{dt^\alpha} = v_{in}(t) - v_0(t)$$
$$\rightarrow \frac{d^\alpha i_L(t)}{dt^\alpha} = \frac{v_{in}(t)}{L} - \frac{v_0(t)}{L}, \tag{11}$$

Equation (12) also is the result of imposing current rule on circuit which is drawn in Fig. 2.

$$C\frac{d^\beta v_C(t)}{dt^\beta} = i_L(t) - \frac{v_C(t)}{R}$$
$$\rightarrow \frac{d^\beta v_C(t)}{dt^\beta} = \frac{i_L(t)}{C} - \frac{v_C(t)}{RC}, \tag{12}$$

State space model can be written as bellow,

$$\begin{bmatrix} \dfrac{d^\alpha i_L(t)}{dt^\alpha} \\ \dfrac{d^\beta v_C(t)}{dt^\beta} \end{bmatrix} = \begin{bmatrix} 0 & -\dfrac{1}{L} \\ \dfrac{1}{C} & -\dfrac{1}{RC} \end{bmatrix} \begin{bmatrix} i_L(t) \\ v_C(t) \end{bmatrix} + \begin{bmatrix} \dfrac{1}{L} \\ 0 \end{bmatrix} v_{in}(t),$$

$$[v_0(t)] = [0 \quad 1] \begin{bmatrix} i_L(t) \\ v_C(t) \end{bmatrix}, \tag{13}$$

therefore

$$\mathbf{A}_1 = \begin{bmatrix} 0 & -\dfrac{1}{L} \\ \dfrac{1}{C} & -\dfrac{1}{RC} \end{bmatrix}, \quad \mathbf{B}_1 = \begin{bmatrix} \dfrac{1}{L} \\ 0 \end{bmatrix},$$
$$\mathbf{C}_1 = [0 \quad 1], \quad \mathbf{D}_1 = 0. \tag{14}$$

3.2. Switch OFF Mode

In this condition the same procedure as former condition is applied. Diode is passed the current as a short circuit path.

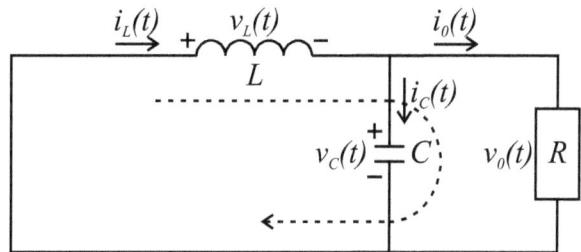

Fig. 3: Buck converter circuit in switch off mode.

Voltage rule:

$$v_C(t) = -L\frac{d^\alpha i_L(t)}{dt^\alpha} \rightarrow \frac{d^\alpha i_L(t)}{dt^\alpha} = -\frac{v_C(t)}{L}. \tag{15}$$

Current rule:

$$i_C(t) = i_L(t) - i_0(t) \rightarrow \frac{d^\beta v_C(t)}{dt^\beta} = \frac{i_L(t)}{C} - \frac{v_C(t)}{RC}. \quad (16)$$

Equation (15) and Eq. (16) is used to reach the state space model of DC/DC Buck converter in switch off mode condition.

$$\begin{bmatrix} \dfrac{d^\alpha i_L(t)}{dt^\alpha} \\ \dfrac{d^\beta v_C(t)}{dt^\beta} \end{bmatrix} = \begin{bmatrix} 0 & -\dfrac{1}{L} \\ \dfrac{1}{C} & -\dfrac{1}{RC} \end{bmatrix} \begin{bmatrix} i_L(t) \\ v_C(t) \end{bmatrix} + \begin{bmatrix} 0 \\ 0 \end{bmatrix} v_{in},$$

$$[v_0(t)] = \begin{bmatrix} 0 & 1 \end{bmatrix} \begin{bmatrix} i_L(t) \\ v_C(t) \end{bmatrix} + [0]v_{in}. \quad (17)$$

Thus,

$$\mathbf{A}_2 = \begin{bmatrix} 0 & -\dfrac{1}{L} \\ \dfrac{1}{C} & -\dfrac{1}{RC} \end{bmatrix}, \quad \mathbf{B}_2 = \begin{bmatrix} 0 \\ 0 \end{bmatrix},$$

$$\mathbf{C}_2 = \begin{bmatrix} 0 & 1 \end{bmatrix}, \quad \mathbf{D}_2 = 0. \quad (18)$$

As can be seen in fractional state space model which was obtained for DC/DC Buck converter, the additional parameters (α, β) indicates the fractional order of inductor and capacitor respectively and in comparison to integer order model it is more complicated but this complexity lead us to more accuracy.

4.   Averaged State Space Model of Fractional Order Buck Converter

According to averaging formula, it's possible to substitute averaged value of each parameter in state space model of DC/DC Buck converter.

$$\langle x(t) \rangle_T = \frac{1}{T} \int\limits_{t}^{t+T} x(\tau)d\tau, \quad (19)$$

where x is an arbitrary variable of the Buck converter.

By averaging a circuit variable over a period of switching, all the high frequency switching harmonics will be removed [7].

$$\frac{d^\alpha \langle x(t) \rangle_T}{dt^\alpha} = \left\langle \frac{d^\alpha x(t)}{dt^\alpha} \right\rangle_T. \quad (20)$$

Equation (20) can be proved easily [7].

Each variable of Buck converter is made up of AC and DC components which are presented in Eq. (21).

$$\langle i_L(t) \rangle = I_L + \widehat{i}_L(t), \quad \langle v_C(t) \rangle = V_C + \widehat{v}_C(t),$$

$$\langle v_{in}(t) \rangle = V_{in} + \widehat{v}_{in}(t), \quad \langle d(t) \rangle = D + \widehat{d}(t), \quad (21)$$

where I_L and $\widehat{i}_L(t)$ are DC and AC component of inductor current respectively.

It's time to build the averaged state space model of Buck converter by using Eq. (14) and Eq. (18) which depend on switch ON and switch OFF mode.

$$d(t) = \frac{T_{on}}{T}, \ d'(t) = \frac{T_{off}}{T}, \ d(t) + d'(t) = 1,$$

$$\mathbf{A} = d(t)\mathbf{A}_1 + d'(t)\mathbf{A}_2,$$

$$\mathbf{B} = d(t)\mathbf{B}_1 + d'(t)\mathbf{B}_2, \quad (22)$$

$$\mathbf{C} = d(t)\mathbf{C}_1 + d'(t)\mathbf{C}_2.$$

Thus,

$$\mathbf{A} = d(t)\begin{bmatrix} 0 & -\dfrac{1}{L} \\ \dfrac{1}{C} & \dfrac{1}{RC} \end{bmatrix} + d'(t)\begin{bmatrix} 0 & -\dfrac{1}{L} \\ \dfrac{1}{C} & \dfrac{1}{RC} \end{bmatrix} =$$

$$= \begin{bmatrix} 0 & -\dfrac{1}{L} \\ \dfrac{1}{C} & \dfrac{1}{RC} \end{bmatrix}. \quad (23)$$

$\mathbf{B}$ and $\mathbf{C}$ matrices are gathered as same as $\mathbf{A}$.

$$\mathbf{B} = d(t)\begin{bmatrix} \dfrac{1}{L} \\ 0 \end{bmatrix} + d'(t)\begin{bmatrix} 0 \\ 0 \end{bmatrix} = \begin{bmatrix} \dfrac{d(t)}{L} \\ 0 \end{bmatrix}, \quad (24)$$

$$\mathbf{C} = d(t)\begin{bmatrix} 0 & 1 \end{bmatrix} + d'(t)\begin{bmatrix} 0 & 1 \end{bmatrix} = \begin{bmatrix} 0 & 1 \end{bmatrix}.$$

Hence, the averaged state space model can be written as follow according to Eq. (23) and Eq. (24).

$$\begin{bmatrix} \dfrac{d^\alpha i_L(t)}{dt^\alpha} \\ \dfrac{d^\beta v_C(t)}{dt^\beta} \end{bmatrix} = \begin{bmatrix} 0 & -\dfrac{1}{L} \\ \dfrac{1}{C} & -\dfrac{1}{RC} \end{bmatrix} \begin{bmatrix} i_L(t) \\ v_C(t) \end{bmatrix} + \dots$$

$$\dots + \begin{bmatrix} \dfrac{d(t)}{L} \\ 0 \end{bmatrix} v_{in}(t). \quad (25)$$

Equations (20) and Eq. (21) are used in the averaged state space model of DC/DC Buck converter.

$$\begin{bmatrix} \dfrac{d^\alpha \langle i_L(t) \rangle}{dt^\alpha} \\ \dfrac{d^\beta \langle v_C(t) \rangle}{dt^\beta} \end{bmatrix} = \begin{bmatrix} 0 & -\dfrac{1}{L} \\ \dfrac{1}{C} & -\dfrac{1}{RC} \end{bmatrix} \begin{bmatrix} \langle i_L(t) \rangle \\ \langle v_C(t) \rangle \end{bmatrix} + \dots$$

$$\dots + \begin{bmatrix} \left\langle \dfrac{d(t)}{L} \right\rangle \\ 0 \end{bmatrix} \langle v_{in}(t) \rangle. \quad (26)$$

The averaged value of used variables in Eq. (21) is replaced by their AC and DC components then the Eq. (27) is reached.

$$\begin{bmatrix} \dfrac{d^\alpha (I_L + \hat{i}_L(t))}{dt^\alpha} \\ \dfrac{d^\beta (V_C + \hat{v}_C(t))}{dt^\beta} \end{bmatrix} = \begin{bmatrix} 0 & -\dfrac{1}{L} \\ \dfrac{1}{C} & \dfrac{1}{RC} \end{bmatrix} \cdots$$

$$\cdot \begin{bmatrix} I_L + \hat{i}_L(t) \\ V_C + \hat{v}_C(t) \end{bmatrix} + \begin{bmatrix} \dfrac{D + \hat{d}(t)}{L} \\ 0 \end{bmatrix} [V_{in} + \hat{v}_{in}(t)]. \quad (27)$$

5. DC Analysis

In DC analysis, AC value of variables will be omitted and each averaged value substitute by its DC component. Due to being zero in Caputo derivative of DC component, the following equation can be achieved.

$$0 = \mathbf{A} \begin{bmatrix} I_L \\ V_C \end{bmatrix} + \mathbf{B}U, \quad (28)$$

$$\begin{bmatrix} I_L \\ V_C \end{bmatrix} = -\mathbf{A}^{-1}\mathbf{B}U,$$

$$\begin{bmatrix} I_L \\ V_C \end{bmatrix} = -\begin{bmatrix} -\dfrac{L}{R} & C \\ -L & 0 \end{bmatrix} \begin{bmatrix} \dfrac{D}{L} \\ 0 \end{bmatrix} V_{in}. \quad (29)$$

Finally,

$$\rightarrow \begin{bmatrix} I_L \\ V_C \end{bmatrix} = \begin{bmatrix} \dfrac{DV_{in}}{R} \\ DV_{in} \end{bmatrix}. \quad (30)$$

6. Small Signal Analysis

By omitting the derivative of DC component of variables and multiplied small signal component in small signal analysis, AC part of equations is derived [15].

$$\frac{d^\alpha \hat{i}_L(t)}{dt^\alpha} = \frac{1}{L}\left(-\hat{v}_C(t) + D\hat{v}_{in}(t) + \hat{d}V_{in} \right), \quad (31)$$

$$\frac{d^\beta \hat{v}_C(t)}{dt^\beta} = \frac{1}{C}\left(\hat{i}_L(t) - \frac{1}{R}\hat{v}_C(t) \right). \quad (32)$$

Averaged state space model for small signal analysis rewrite as bellow.

$$\begin{bmatrix} \dfrac{d^\alpha \hat{i}_L(t)}{dt^\alpha} \\ \dfrac{d^\beta \hat{v}_C(t)}{dt^\beta} \end{bmatrix} = \begin{bmatrix} 0 & -\dfrac{1}{L} \\ \dfrac{1}{C} & -\dfrac{1}{RC} \end{bmatrix} \begin{bmatrix} \hat{i}_L(t) \\ \hat{v}_C(t) \end{bmatrix} + \cdots$$

$$\cdots + \begin{bmatrix} \dfrac{D}{L} \\ 0 \end{bmatrix} \hat{v}_{in}(t) + \begin{bmatrix} \dfrac{V_{in}}{L} \\ 0 \end{bmatrix} \hat{d}(t). \quad (33)$$

In order to analyze nonlinear circuits by linear methods, it's essential to find out the transfer function of nonlinear systems with linearization for linear analysis then achieved results are able to impose on exact and nonlinear models. By assuming zero initial condition and using the definition of Caputo derivative, calculation of transfer functions for DC/DC Buck converter is mentioned as follow.

By imposing Laplace transform to Eq. (31) and Eq. (32) following statements can be reached.

$$s^\alpha \hat{i}_L(s) = -\frac{1}{L}\hat{v}_C(s) + \frac{D}{L}\hat{v}_{in}(s) + \frac{v_{in}}{L}\hat{d}(s), \quad (34)$$

$$\hat{i}_L(s) = Cs^\beta \hat{v}_C(s) + \frac{1}{R}\hat{v}_C(s). \quad (35)$$

By substituting Eq. (35) into Eq. (34) and equalizing to zero, transfer function of output voltage to input voltage will be gathered.

$$s^\alpha \left(Cs^\beta \hat{v}_C(s) + \frac{1}{R}\hat{v}_C(s) \right) = -\frac{1}{L}\hat{v}_C(s) + \frac{D}{L}\hat{v}_{in}(s)$$

$$\rightarrow \hat{v}_C(s)\left(Cs^{\alpha+\beta} + \frac{1}{R}s^\alpha + \frac{1}{L} \right) = \frac{D}{L}\hat{v}_{in}(s). \quad (36)$$

Thus,

$$G_{\hat{v}_0 - \hat{v}_{in}} = \frac{\hat{v}_0(s)}{\hat{v}_{in}(s)}\Big|_{\hat{d}(s)=0} = \frac{D}{LCs^{\alpha+\beta} + \dfrac{L}{R}s^\alpha + 1}. \quad (37)$$

In order to explain the relationship between output voltage and duty cycle, Eq. (35) is substituted into Eq. (34) and in this part perturbation of input voltage is ignored.

$$Cs^{\alpha+\beta}\hat{v}_C(s) + \frac{1}{R}s^\alpha \hat{v}_C(s) = -\frac{1}{L}\hat{v}_C(s) + \frac{V_{in}}{L}\hat{d}(s)$$

$$\rightarrow \hat{v}_C(s)\left(Cs^{\alpha+\beta} + \frac{1}{R}s^\alpha + \frac{1}{L} = \frac{V_{in}}{L}\hat{d}(s) \right). \quad (38)$$

Therefore,

$$G_{\hat{v}_0 - \hat{d}}(s) = \frac{\hat{v}_0(s)}{\hat{d}(s)}\Big|_{\hat{v}_{in}(s)=0} =$$

$$= \frac{V_{in}}{LCs^{\alpha+\beta} + \dfrac{L}{R}s^\alpha + 1}. \quad (39)$$

From Eq. (35),

$$\hat{i}_L(s) = Cs^\beta \hat{v}_C(s) + \frac{1}{R}\hat{v}_C(s)$$

$$\rightarrow G_{\hat{v}_C - \hat{i}_L} = \frac{\hat{v}_C(s)}{\hat{i}_L(s)} + \frac{R}{RCs^\beta + 1}. \quad (40)$$

Equation (40) shows the transfer function of output voltage to inductor current.

7. Computation of Inductor Current

As mentioned before, there are two modes operation in DC/DC Buck converter. In this research Continues Condition Mode (CCM) is considered. In this condition the inductor current never touches zero value and always has an amount.

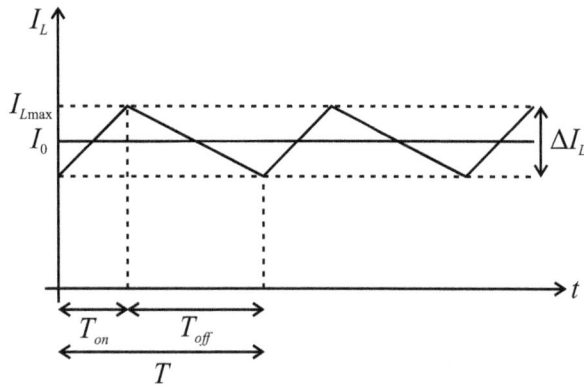

Fig. 4: Inductor current of DC/DC Buck converter.

T indicates the period of switching in ms.

Figure 4 shows the perturbation of inductor current in CCM mode operation. Computational method for finding inductor current is presented as follow. During the period which switch is on (T_{on}), variation of inductor current based on fractional calculus is as bellow:

$$v_L(t) = L\frac{d^\alpha i_L(t)}{dt^\alpha},$$
$$v_L(t) = v_{in}(t) - v_0(t). \tag{41}$$

By averaging the both side of Eq. (41) next statement is gathered.

$$\langle v_L(t)\rangle = L\frac{d^\alpha \langle i_L(t)\rangle}{dt^\alpha},$$
$$\rightarrow \frac{\langle v_{in}(t) - v_0(t)\rangle}{L} = \frac{d^\alpha \widehat{i}_L(t)}{dt^\alpha}. \tag{42}$$

By imposing fractional integrator of order α to both sides of Eq. (42), the next statement will be reached.

$$I_C^\alpha \frac{\langle v_{in}(t) - v_0(t)\rangle}{L} = I^\alpha\left(\frac{d^\alpha \widehat{i}_L(t)}{dt^\alpha}\right) = \Delta I_L(t). \tag{43}$$

By assuming that variation of output voltage and input voltage is almost constant and using Caputo definition for integrator, the left side of Eq. (43) can be written as Eq. (44).

$$I_C^\alpha \frac{\langle v_{in}(t) - v_0(t)\rangle}{L} =$$
$$= \frac{1}{\Gamma(\alpha)}\frac{\langle v_{in}(t) - v_0(t)\rangle}{L}\int_0^{DT}(t-\tau)^{\alpha-1}d\tau. \tag{44}$$

DT shows the time which switch is on in (ms) and D is the DC component of duty cycle. In order to get into closed relation for above equation, firstly it's necessary to expand the inner expression of integrator which is used in Eq. (44) by Taylor expansion around zero point then calculate fractional integrator of order α. Matlab is used in this section.

$$\Delta I_L = \frac{(v_{in} - v_0)(DT)^\alpha}{\Gamma(\alpha)L}. \tag{45}$$

Easily can be seen in Fig. 4 the maximum amount of inductor current.

$$I_{L\max} = I_L + \frac{1}{2}\Delta I_L,$$
$$I_{L\max} = \frac{Dv_{in}}{R} + \frac{1}{2}\frac{(v_{in} - v_0)}{\Gamma(\alpha)L}(DT)^\alpha. \tag{46}$$

Equation (45) and Eq. (46) indicates that the amount of inductor current and its variation depends on not only inductance of inductor but also the amount of fractional order of inductor.

8. Computation of Capacitor Voltage

One of the most important factors for designing Buck converters in order to find the value of inductance and capacitance of inductor and capacitor respectively is the amount of inductor current and capacitor voltage perturbation. Variation of inductor current mentioned in pervious section and capacitor voltage is explained in this section. In Fig. 1 following statement is obvious.

$$i_C(t) = -i_0(t) + i_L(t),$$
$$\rightarrow \frac{d^\beta v_C(t)}{dt^\beta} = -\frac{v_C(t)}{R} + \frac{i_L(t)}{C}. \tag{47}$$

By using Eq. (47) it's possible to gather the capacitor voltage. Adomian decomposition method is used in [7], [8] and [9] in order to solve the fractional differential equation. This method is a very strong approach for solving the nonlinear equations in case of linear analytically and also it's possible to use it for such Fractional Differential Equation (FDE) [16]. Due to have a solution for Eq. (47) Adomian decomposition method is used. The general form of fractional equation which is solved by Adomian decomposition method is available in Eq. (48).

$$D^\alpha x(t) = Ax(t) + f(t),$$
$$0 < \alpha < 1,\ 0 < t < T, \tag{48}$$

$f(.)$ is a function of time and A is a coefficient for x.

Mentioned method in case of Caputo fractional derivative for solving the Eq. (48) is as follow [17].

$$x(t) = x_h(t) + x_p(t),$$
$$x_h(t) = E_{\alpha,1}(At^\alpha)C, \tag{49}$$
$$x_p(t) = t^{\alpha-1}E_{\alpha,\alpha}(At^\alpha) \cdot f(t),$$

where C is the initial condition and E is the Mittag-Leffler function.

Note that the following condition must be satisfied.

$$\exists\ \epsilon > 0,\ M > 0,\ \beta > -\alpha. \tag{50}$$

Such that, $|f_i(t)| \le Mt^\beta$.

During the period which switch is on, Eq. (51) is solved. Note that the Eq. (51) for both periods which switch is on and off is the same because of circuit topology.

$$D^\beta V_C(t) = -\frac{V_C(t)}{RC} + \frac{i_L(t)}{C}. \tag{51}$$

Equation (52) shows the numerical solution of Eq. (51) according to Adomian decompodition method which mentioned in Eq. (49).

$$V(t) = E_{\beta,1}\left(-\frac{t^\beta}{RC}\right)V_0 + \dots$$
$$\dots + t^{\beta-1}E_{\beta,\beta}\left(\frac{t^\beta}{RC}\right) \cdot \frac{i_L(t)}{C}. \tag{52}$$

According to Fig. 5 it's possible to find out the output voltage variation [4]. Output voltage and capacitor voltage are the same.

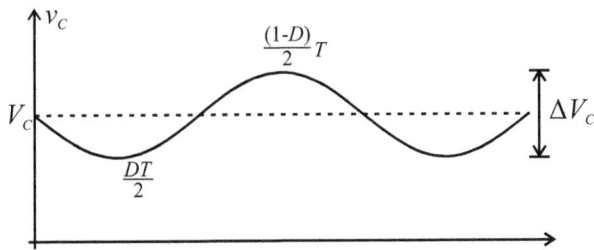

Fig. 5: Capacitor voltage of DC/DC Buck converter.

$$V\left(\frac{DT}{2}\right) = E_{\beta,1}\left(-\frac{\left(\frac{DT}{2}\right)^\beta}{RC}\right)V_0 + \left(\frac{DT}{2}\right)^{\beta-1}$$
$$\dots E_{\beta,\beta}\left(-\frac{\left(\frac{DT}{2}\right)^\beta}{RC}\right) \cdot \frac{i_L\frac{DT}{2}}{C}. \tag{53}$$

And,

$$V\left(\frac{1-D}{2}T\right) = E_{\beta,1}\left(-\frac{\left(\frac{1-D}{2}T\right)^\beta}{RC}\right)V_0 + \dots$$
$$+ \left(\frac{1-D}{2}T\right)^{\beta-1} + E_{\beta,\beta}\left(-\frac{\left(\frac{1-D}{2}T\right)^\beta}{RC}\right) \dots$$
$$\cdot \frac{i_L\left(\frac{1-D}{2}T\right)}{C}. \tag{54}$$

As it can be seen in Fig. 5 the amount of inductor current in $\frac{DT}{2}, \frac{(1-D)}{2}T$ is the same approximately. Therefore,

$$\Delta V = V\left(\frac{1-D}{2}T\right) - V\left(\frac{D}{2}T\right),$$
$$I\left(\frac{1-D}{2}T\right) \approx I\left(\frac{D}{2}T\right). \tag{55}$$

Then,

$$\Delta V = \left[\left[E_{\beta,1}\left(-\frac{\left(\frac{1-D}{2}T\right)^\beta}{RC}\right)\dots\right.\right.$$
$$\left.\left.-E_{\beta,1}\left(-\frac{\left(\frac{D}{2}T\right)^\beta}{RC}\right)\right]V_0 \cdot \frac{I_0}{C}\cdot\dots\right.$$
$$\left[\left(\frac{1-D}{2}\right)^{\beta-1}E_{\beta,\beta}\left(-\frac{\left(\frac{1-D}{2}T\right)^\beta}{RC}\right)\dots\right.$$
$$\left.\left.-\left(\frac{D}{2}T\right)^{\beta-1}E_{\beta,\beta}\left(-\frac{\left(\frac{D}{2}T\right)^\beta}{RC}\right)\right]\right]. \tag{56}$$

9.   Approximation

It's possible to describe the dynamical behavior of a fractional order transfer function by an integer order transfer function. Approximation is important because of the numerical solution methods for integer order differential equations are more known and available and also it can found in a lot of common software [13].

9.1. Oustaloup's Recursive Approximation

Mentioned filter is capable to approximate a fractional transfer function with high accuracy. For instance fractional element s^α will be described by $(2N+1)$ integer zeros and poles during the specified frequency interval $[\omega_l, \omega_h]$.

$$s^\alpha = K \prod_{k=-N}^{N} \frac{s + \omega_k'}{s + \omega_k}. \tag{57}$$

Such that,

$$\omega_k = \omega_l \left[\frac{\omega_h}{\omega_l} \right]^{\frac{K+N+\frac{1}{2}(1+\alpha)}{2N+1}},$$

$$\acute{\omega}_k = \omega_l \left[\frac{\omega_h}{\omega_l} \right]^{\frac{K+N+\frac{1}{2}(1-\alpha)}{2N+1}}, \tag{58}$$

$$K = (\omega_h)^\alpha.$$

The order of mentioned filter is $(2N+1)$ and by imposing each input signal to approximated integer transfer function, the output signal will be the response of fractional transfer function [14].

10. Simulation Results

A Buck Converter with following features is assumed:

- $L = 0.236$ mH,
- $C = 47$ mF,
- $R = 0.1\ \Omega$,
- $V_{in} = 28 - 70$ V,
- Output voltage 24 V,
- Switching frequency 30 kHz.

The main reason of current essay is to express the importance of fractional modeling for a fractional system in nature. It's assumed that is possible to have fractional elements like capacitor and inductor although it's not valid in real world up to now. In order to achieve this purpose, Buck converter with above features which used in a 1 kW vertical axis wind turbine assumed but the amount of fractionality for its elements (α, β) is hypothetical values duo to the mentioned Buck converter is made up of integer order components and integer order model has accurate enough for this system. In this section a comparison is done on fractional modeling and integer modeling for studied system. Bode diagram which is mentioned in Fig. 6 is related to transfer function of output voltage to input voltage.

$$G_{\widehat{V}_0 - \widehat{V}_{in}} = \frac{0.352}{1.1092e - 5s + 2.36e - 3s^{0.5} + 1}, \tag{59}$$

$$\alpha = \beta = 0.5.$$

Equation (59) is gathered by replacing the buck parameters into Eq. (37) and the averaged value for duty cycle is considered as $D = 0.352$.

If the fractional characteristic of mentioned Buck converter is omitted, the quantity of (α, β) will be equaled to 1 and Eq. (59) is converted to an integer order model of Buck converter which is well known equa-

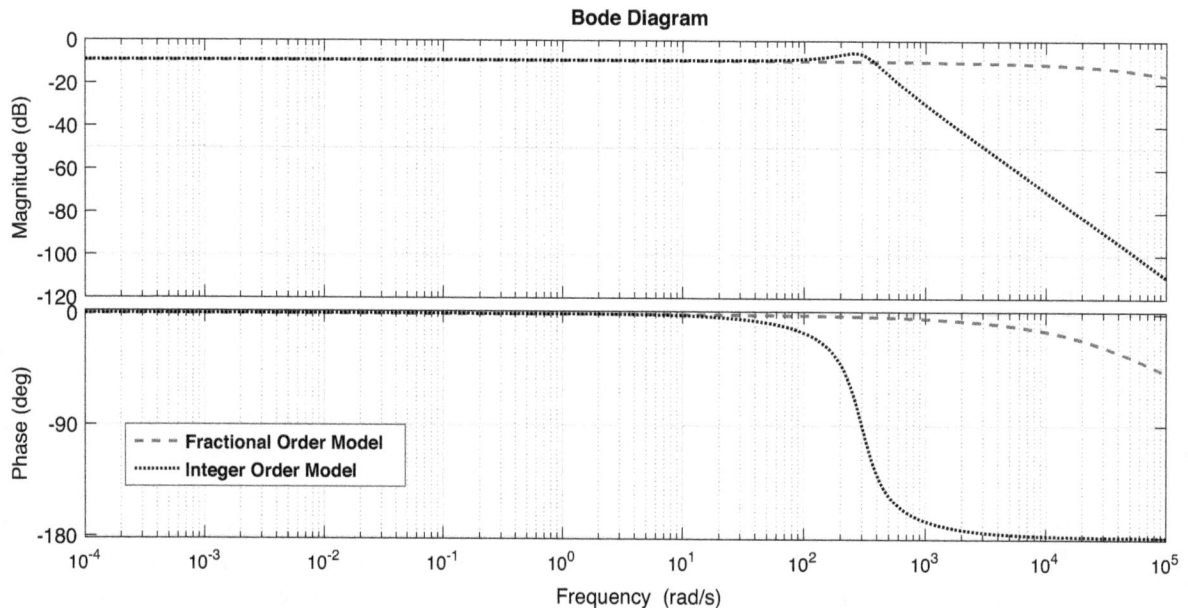

Fig. 6: Frequency response of fractional and integer order model.

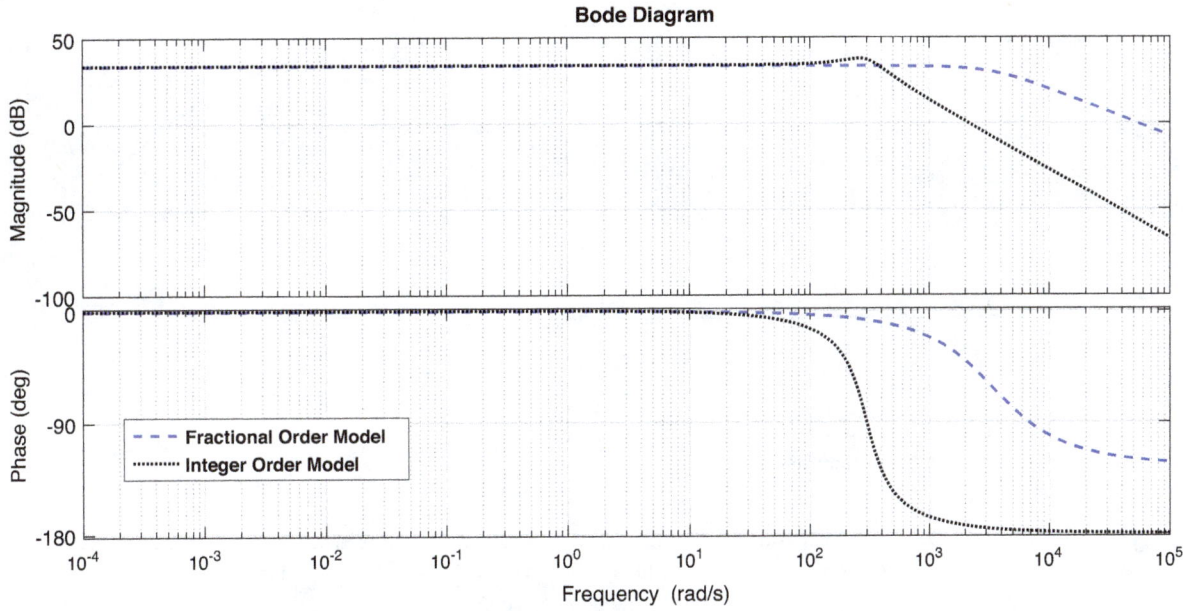

Fig. 7: Frequency response of fractional and integer model.

tion but in this part fractional characteristic is considered for Buck converter and this feature is represented by (α, β).

Fractional operation and simulation can be done easily in Matlab. Some important commands are presented: Fotf, Fomcon, Bode, Oustapp, Plot.

As can be seen in Fig. 6 integer order modeling for a system which is fractional order in nature (transfer function which mentioned in Eq. (59) causes remarkable difference between real system and integer modeled system.

Frequency analysis has been done on another mentioned transfer function. Assume that the DC value of input voltage is 50 V and fractional order of inductor and capacitor are 0.7 therefore,

$$G_{\widehat{V}_0 - \widehat{d}}(s) \;\; = \frac{50}{1.1092e - 5s^{1.4} + 2.36e - 3s^{0.7} + 1},$$
$$\alpha = \beta = 0.7. \tag{60}$$

Figure 7 indicates the frequency response of Eq. (60).

The value of (α, β) is hypothetical and just used for simulation and comparison. In order to establishing the fractional order controllers and modeling frac-

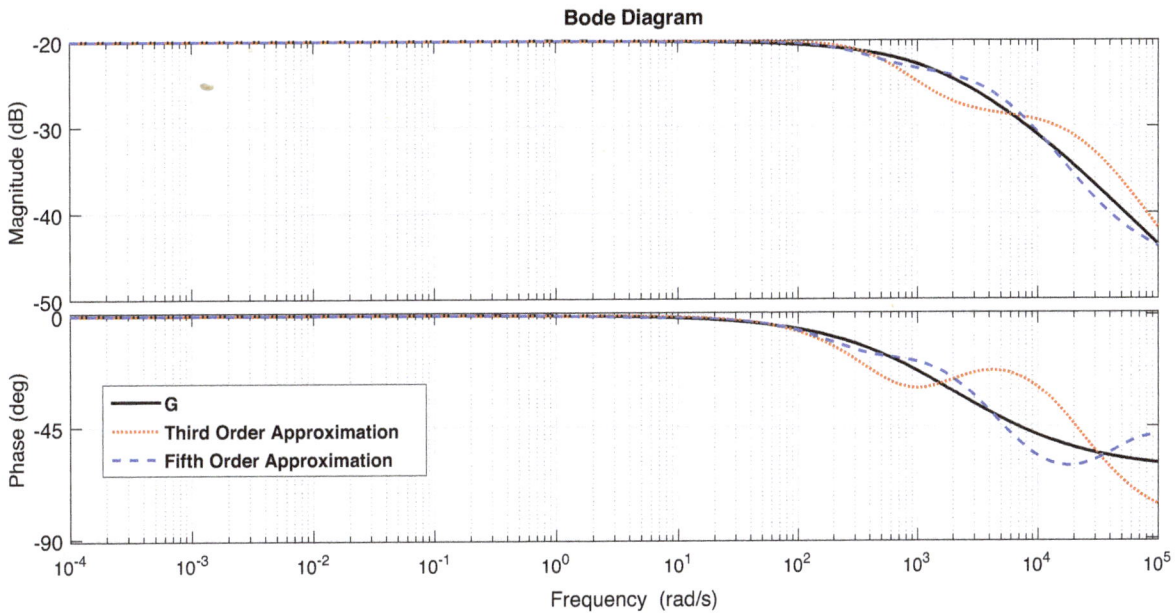

Fig. 8: Frequency response of approximated transfer functions.

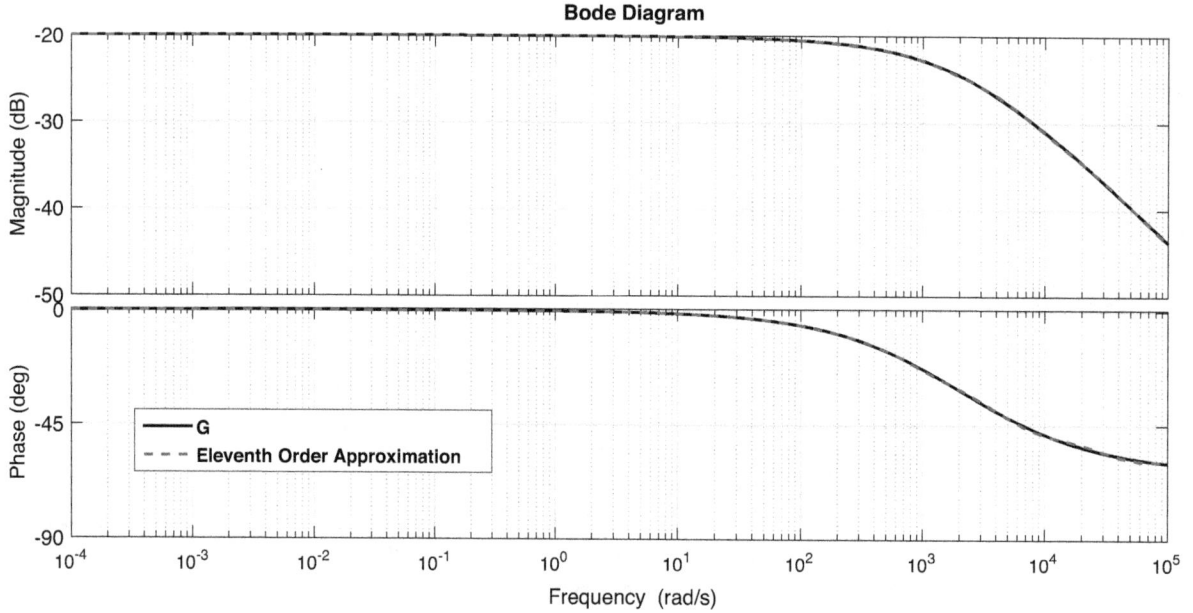

Fig. 9: Frequency response of the most accurate approximation.

tional systems in practical way, it's essential to convert them into integer order transfer function. As mentioned before Oustaloup's recursive approximation can be employed to approximate a fractional order transfer function during a specific frequency interval with $(2N + 1)$ zeros and poles. Here $G_{\widehat{V}_C-\widehat{i}_L}$ with $\beta = 0.7$ is considered and approximated by third and fifth integer order transfer function and then their frequency response are compared with the frequency response of fractional transfer function which gathered by using Matlab. As can be seen in Fig. 8 by increasing the order of approximated transfer function the amount of error is decreased.

$$G_{\widehat{V}_C-\widehat{i}_L} = \frac{0.1}{4.7e - 3s^{0.7} + 1}, \quad (61)$$

$$TF = \frac{2.67e - 4(s + 3.16e + 6)(s + 1468)(s + 0.6813)}{(s + 2.25e + 4)(s + 554)(s + 0.6764)}. \quad (62)$$

Equation (62) shows a third integer order model which approximated for $G_{\widehat{V}_C-\widehat{i}_L}$ in Eq. (61). In practical applications, it's not possible to use fractional equations or controllers and converting into integer order equations and controllers which are the same in characteristics is essential.

In order to have a model with high accuracy an eleventh order approximated transfer function is chosen and is compared by fractional transfer function which is achieved by Matlab.

Figure 9 shows the comparison between eleventh order transfer function which approximated for G and fractional order transfer function (G).

In order to emphasis the importance of fractional order modeling for a fractional order Buck converter, real time simulation was done and compare to integer order model with the same duty cycle. Closed loop transfer function for both Eq. (63) and Eq. (64) was gathered and their step response is ploted.

$$G_{\widehat{V}_0-\widehat{V}_{in}} = \frac{0.352}{1.1092e - 5s + 2.36e - 3s^{0.5} + 1},$$
$$\alpha = \beta = 0.5. \quad (63)$$

$$G_{\widehat{V}_0-\widehat{V}_{in}} = \frac{0.352}{1.1092e - 5s^2 + 2.36e - 3s + 1}. \quad (64)$$

Equation (63) and Eq. (64) show the fractional order model and integer order model of fractional order Buck converter with ($\alpha = \beta = 0.5$) respectively.

Input voltage is 70 V for simulation.

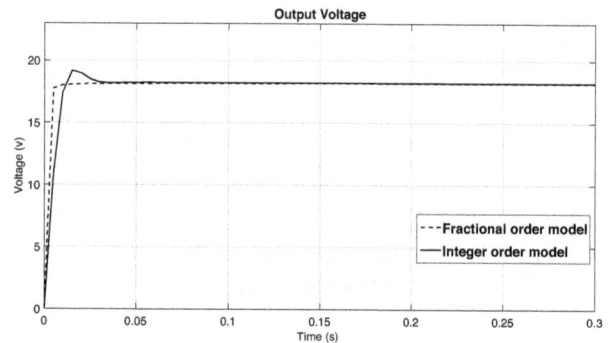

Fig. 10: Fractional order in comparison to integer order model.

Figure 10 shows the closed loop response to constant input voltage and the same duty cycle was imposed to both integer and fractional order model. As mentioned befor, assumed Buck converter was fractional in nature ($\alpha = \beta = 0.5$) then the fractional and integer order model has some differences in output response. Integer order model for fractional Buck converter caused modeling error and its not accurate enough.

11. Conclusion

Although fractional calculus related to a few centuries ago but its useful and important applications are in center of attention especially in engineering nowadays. As mentioned before all systems and components are fractional in nature, but they are different in amount of fractionality [9]. In this paper a DC/DC Buck converter with fractional order components was assumed and frequency analysis along the linear modeling is done. The remarkable difference between integer order model and fractional order model can be seen easily, but for lots of systems integer order modeling is satisfied our purposes and they have enough accuracy because fractionality order of their elements such as capacitors and inductors are near to one. It doesn't mean the fractional order modeling is useless because it can describe the characteristics of systems and circuits with more accuracy but for common and available systems, integer order has enough accuracy. If an extra finesse is required or fractional order elements with fractionality far from one is used, then fractional order modeling will be essential. At the end of research, real time comparison was done for closed loop transfer function of output voltage to input voltage of Buck converter although designing of controller was not the research purpose. Establishing the fractional order model for Buck converter and comparing it to integer order model was the main goal for this research but Fig. 10 shows the non-negligible error in real time simulation for closed loop system which is caused by modeling accuracy. It's obvious that integer order model for a fractional order Buck converter with ($\alpha = \beta = 0.5$) is not a suitable model and makes error.

Acknowledgment

The authors express their thanks to head of sun air research center of Ferdowsi University of Mashhad for supporting us in data acquisition from the real system in that center.

References

[1] SEIXAS, M., R. MELICIO and V. M. F. MENDES. Offshore wind turbine simulation: multibody drive train. Back-to-back NPC (Neutral Point Clamped) converters. Fractional order control. *Energy*. 2014, vol. 69, iss. 1, pp. 357–369. ISSN 0360-5442. DOI: 10.1016/j.energy.2014.03.025.

[2] KUO-YUAN, L., C. YAOW-MING and C. YUNG-RUEI. MPPT battery charger for stand-alone wind power system. *IEEE Transactions on power electronics*. 2011, vol. 26, iss. 6, pp. 1631–1638. ISSN 0885-8993. DOI: 10.1109/TPEL.2010.2088405.

[3] MELICIO, R., V. M. F. MENDES and J.P.S. CATALAO. Computer simulation of wind power systems: power electronics and transient stability analysis. In: *International Conference on Power Systems Transients*. Kyoto: IEN, 2009, pp. 1–7.

[4] MOHAN, N., M. UNDELAND and W. P. ROBBINS. *Power electronics*. New York: John Wiley & Sons, 1995. ISBN 0-471-58408-8.

[5] JONSCHER, A. K. Dielectric relaxation in solids. *Journal of Physics D: Applied Physics*. 1999, vol. 32, iss. 14, pp. R57–R70. ISSN 0022-3727. DOI: 10.1088/0022-3727/32/14/201.

[6] WESTERLUND, S. and L. EKSTAM. Capacitor Theory. *IEEE Transactions on Dielectrics and Electrical Insulation*. 1994, vol. 1, iss. 5, pp. 826–839. ISSN 1070-9878. DOI: 10.1109/94.326654.

[7] CHANDRAN, R., V. PRASAD and P. VINCENT. Modeling of Fractional order Buck-Boost Converter. *International Journal of Engineering and Advanced Technology (IJEAT)*. 2014, vol. 3, iss. 3, pp. 129–133. ISSN 2249-8958.

[8] WANG, F. and X. MA. Modeling and Analysis of the Fractional Order Buck Converter in DCM Operation by using Fractional Calculus and the Circuit-Averaging Technique. *Journal of Power Electronics*. 2013, vol. 13, iss. 6, pp. 1008–1015. ISSN 1598-2092. DOI: 10.6113/JPE.2013.13.6.1008.

[9] NING-NING, Y., L. CHONG-XIN and W. CHAO-JUN. Modeling and dynamic analysis of the fractional-order Buck-Boost converter in continues conduction mode. *Chinese Physical Society and IOP Publishing Ltd*. 2012, vol. 21, iss. 8, pp. 1–7. ISSN 1674-1056. DOI: 10.1088/1674-1056/21/8/080503.

[10] MONJE, C. A. *Fractional-order systems and controls: fundamentals and applications*. London: Springer, 2010. ISBN 978-1-84996-334-3.

[11] SHOJA MAJIDABAD, S., H. TOOSIAN SHAN-DIZ and A. HAJIZADEH. Decentralized sliding mode control of fractional-order large-scale nonlinear systems. *Nonlinear Dynamics.* 2014, vol. 77, iss. 1, pp. 119–134. ISSN 0924-090X. DOI: 10.1007/s11071-014-1278-4.

[12] KARANJKAR, D. S., S. CHATTERJI and P. R. VENKATESWARAN. Trends in fractional order controllers. *International journal of emerging technology and advanced engineering.* 2012, vol. 2, iss. 3, pp. 383–389. ISSN 2250-2459.

[13] VALERIO, D. and J. SA DA COSTA. Introduction to single-input, single-output fractional control. *IET Control Theory & Applications.* 2011, vol. 5, iss. 8, pp. 1033–1057. ISSN 1751-8644. DOI: 10.1049/iet-cta.2010.0332.

[14] BUCHADE, P. C., V. A. VYAWAHARE and V. V. BHOLE. Fractional-Order control of voltage source inverter (VSI) using Bode's ideal transfer function. In: *International Conference on Circuits, Systems, Communication and Information Technology Applications (CSCITA).* Mumbai: IEEE, 2014, pp. 403–407. ISBN 978-1479924950. DOI: 10.1109/CSCITA.2014.6839294.

[15] NAMJOO, N. *In-direct Adaptive control of non-inverting Buck-Boost converter.* Shahrood, 2013. Thesis. Shahrood University of Technology, Faculty of Electrical and Robotic Engineering. Supervisor: Amin Hajizadeh.

[16] DAFTARDAR-GEJJI, V. and H. JAFARI. Adomian decomposition: a tool for solving a system of fractional differential equations. *Journal of Mathematical Analysis and Applications.* 2005, vol. 301, iss. 2, pp. 508–518. ISSN 0022-247X. DOI: 10.1016/j.jmaa.2004.07.039.

[17] JUNSHENG, D., A. JIANYE and X. MINGYU. Solution of system of fractional differential equations by Adomian decomposition method. *Applied Mathematics-A Journal of Chinese Universities.* 2007, vol. 22, iss. 1, pp. 7–12. ISSN 1005-1031.

[18] MAHMOOD, A. K. and B. F. MOHAMMED. Design Optimal Fractional Order PID Controller Utilizing Particle Swarm Optimization Algorithm and Discretization Method. *International Journal of Emerging Science and Engineering (IJESE).* 2013, vol. 1, iss. 10, pp. 87–92. ISSN 2319-6378.

About Authors

Ali MOSHAR MOVAHHED was born in Iran in 1989. He received his B.Sc. degree in control engineering from Sadjad University of Technology, Mashhad, Iran, in 2012. He is pursuing M.Sc. in control engineering in Shahrood university of technology, Shahrood, Iran since 2013. His area of interest includes fractional control, renewable energy especially wind turbines and power electronics.

Heydar TOOSIAN SHANDIZ has received Bsc and M.Sc. degree in electrical engineering from Ferdowsi Mashad University in Iran. He has graduated Ph.D. in Instrumentation from UMIST, Manchester UK in 2000. He has been associate professor in Shahrood University of Technology, Iran. His fields of research are Fractional Control System, identification systems, adaptive control, Image and signal processing, neural networks and Fuzzy systems.

Syed Kamal HOSSEINI SANI received the B.Sc. degree in electrical engineering from Ferdowsi University of Mashhad, Mashhad, Iran, in 1995, the M.Sc. degree in control engineering from KNT University in 1998 and the Ph.D. degree in control engineering from Tarbiat Modares University of Iran in 2006. He is currently an Assistant Professor in the Department of Electrical Engineering, Ferdowsi University of Mashhad. His research interests include wind turbines, digital and adaptive control, applied control, and power converter systems.

3

Fire Risk in MTBF Evaluation for UPS System

Stefano ELIA, Alessio SANTANTONIO

Department DIAEE – Electrical Engineering Section, Sapienza University,
Piazzale Aldo Moro 5, 00185 Rome, Italy

stefano.elia@uniroma1.it, alessio.santantonio@outlook.it

Abstract. *The reliability improvement of no-break redundant electrical systems is the first aim of the proposed strategy. The failure of some UPS (Uninterruptible Power Supply) system may lead to the fire occurrence. The most used electrical configurations are presented and discussed in the paper. The innovation of the proposed method consists of taking into account the fire risk to improve the accuracy of wiring configuration and components' failure rate. Thorough research on MTBF (Mean Time Between Failure) data has been performed for each wiring component and UPS. The fire risk is taken into account introducing an equivalent fire block in the Reliability Block Diagram scheme; it has an MTBF value calculated form yearly statistics of UPS fire events. The reliability of the most used UPS electrical configurations is evaluated by means of the RBD method. Different electrical systems have been investigated and compared based on MTBF. The importance of fire compartmentation between two or more UPS' connected in parallel is proved here.*

Keywords

Electrical installation, failure rate, fire risk, MTBF, no-break power system, RBD, redundant electrical system, reliability, UPS.

1. Introduction

Ten years of maintenance activity in no-break electrical installations reveals a lot of design errors and a lack of reliability. Currently, it is common to evaluate the reliability only for the equipment and for electrical system' components. Today, it can be noticed that there is a defect in reliability evaluation of all project choices. Many real fire case occurrences on UPS were caused by power electronic or battery fail. The producers do not consider fire hazard significant for the UPS device

and perform tests only in standard and good maintenance conditions. The UPS machines are very vulnerable and subjected to inadequate maintenance, overvoltages, high temperature, working conditions, and other electrical system malfunctions. Nowadays redundant UPS' are not isolated by a fire compartment. No-break systems are often designed following a wrong guideline. The standard procedure considered for reliability improvement consists of a simple installation of two UPS', which are connected in parallel. UPS power electronic, Control Unit, manual and static bypass, batteries and other system components are usually installed in the same room. All devices are, therefore, exposed to the same fire risk. Working UPS unit may be involved in fire event caused by the failure of another UPS unit in the same room, thus rendering the system redundancy ineffective. Moreover, an emergency manual bypass wired out of the UPS room is never installed. In those conditions a fire event entails certainly the complete failure of the no-break system. A short and contained fire is also sufficient to generate smoke and risk of toxic air; in this case nobody can access the UPS room and technicians are obliged to communicate immediately with the director to inform him of the imminent failure and recommend to stop all current operations. A lot of fire case studies have been investigated; the complete failure of the no break electrical system was often due to the lack of fire compartmentations between two UPS.

2. Reliability Model

The reliability evaluation for each electrical configuration is based on the Reliability Block Diagram model (RBD) [1], [2]. By means of this method, the *MTBF* of UPS, Control Unit, batteries, switches, and other components are represented. Data on UPS fire occurrence frequency were obtained during ten years of consultant activity in the hospitals. In about a hundred of case studies the existence of two UPS' room fire events is proved per year (2 % per year). Two hypotheses are necessary to evaluate the failure rate [3] by statistical data on fire risk. Firstly, the failure rate is constant in time. Secondly, break components are not repairable but quickly replaceable ($MTTR = 0$). The average fire failure rate is defined as the ratio between number of fires and number of studied events per time. It is shown in Eq. (1):

$$\lambda_{AVG} = \frac{N_F}{N_{TOT} \cdot T} \ , \tag{1}$$

where N_F is the quantity of fire events, N_{TOT} is the number of observed systems and T is the observation time. Subsequently the $MTBF$ value is calculated [4] in Eq. (2):

$$MTBF_F = \frac{1}{\lambda_{AVG}} = \frac{N_{TOT} \cdot T}{N_F} = \frac{100 \cdot 8760}{2} = \tag{2}$$

$$= 438000 \ \text{h}.$$

Using the Eq. (2), the reliability calculation can be based on different fire statistics. Moreover, calculation can be developed implementing a parametric analysis varying the value of fire statistic. The reliability evaluation model is used based on these hypotheses. Firstly, UPS is only considered as a no break system that avoids voltage dips. Secondly, the continuous energy source is based only on power supplier's grid or emergency diesel generator.

3. MTBF Data

$MTBF$ data on studied components have been deduced by an accurate statistical survey. Used data were obtained from: Gold Book [5], some papers [1], [6], [7], [8], [9] and many datasheets. The UPS' $MTBF$ is evaluated including the presence of the on-board automatic static bypass and batteries. $MTBF$ value for the Fire Risk Factor block is pointed out by the maintenance activity experience, Eq. (2). The average values of $MTBF$ for all components are pointed out in Tab. 1.

Tab. 1: Average $MTBF$ of no-break system's components.

Components	Symbol	Failure rate [failure·h^{-1}]	MTBF [h]
Battery (lead acid)	BAT	$8.52086 \cdot 10^{-7}$	1173590
Circuit breaker	CB	$4.348 \cdot 10^{-6}$	229991
Complete UPS module (internal STS and batteries included)	UPS	$1.3779 \cdot 10^{-5}$	72574
Control Unit	CU	$1.33333 \cdot 10^{-6}$	750000
Fire Risk Factor	FRF	$2.28311 \cdot 10^{-6}$	438000
Inverter	INV	0.00002	50000
Rectifier	REC	0.00002	50000
Static Transfer Switch	STS	$9.79499 \cdot 10^{-6}$	102093
Switchgear Bus Bar	SBB	$1.08334 \cdot 10^{-6}$	923068

4. MTBF Evaluation of Various No-Break Electrical Systems

4.1. One UPS

The reliability of a base configuration with only one UPS is shown in this subsection.

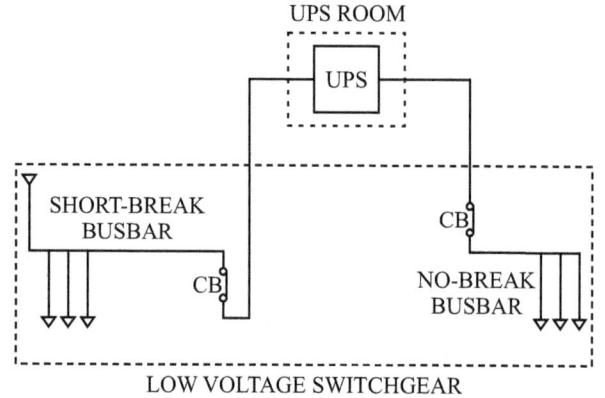

Fig. 1: No-break system with one UPS.

The Reliability Block Diagram of the one UPS configuration is shown in Fig. 2.

Fig. 2: RBD scheme of no-break system with one UPS.

The calculation shown in Eq. (3) reveals the total $MTBF$ for the system configuration with only one UPS.

$$MTBF_T =$$

$$\frac{1}{\cfrac{\cfrac{1}{MTBF_{SBB}} + \cfrac{1}{MTBF_{CB}} + \cfrac{1}{MTBF_{FRF}} + \cdots}{1}} = \tag{3}$$

$$\cdots \cfrac{1}{MTBF_{UPS}} + \cfrac{1}{MTBF_{CB}} + \cfrac{1}{MTBF_{SBB}}$$

$$= 37140 \ \text{h} = 4.2 \ \text{years}.$$

4.2. Two UPS Without Fire Compartmentations

The configuration with two UPS' connected in parallel is considered here. Each UPS has a rated power greater than the load demand. Machines are installed in the same room together with the batteries and without any fire compartmentations. The respective configuration scheme is shown in Fig. 3. The RBD scheme of that

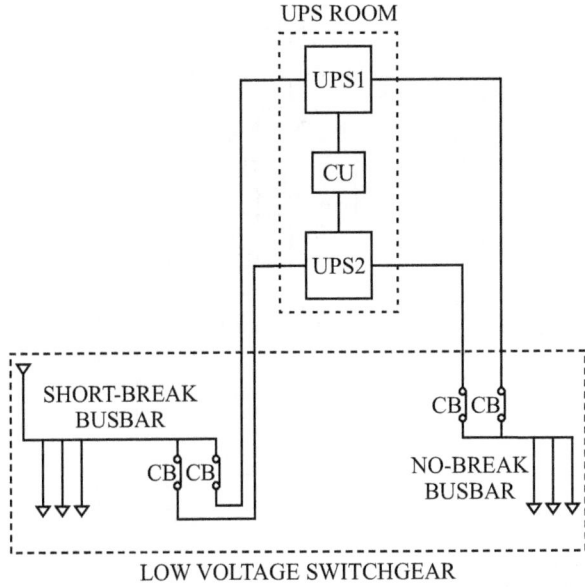

Fig. 3: No-break system with two UPS without fire compartmentations.

system is shown in Fig. 4, where it must be highlighted that the Fire Risk Factor of both UPS' influences the entire system.

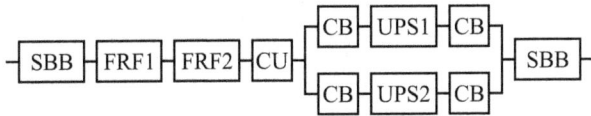

Fig. 4: RBD scheme of no-break system with two UPS without fire compartmentations.

The *MTBF* for that configurations is can be computed with the use of Eq. (4), Eq. (5), Eq. (6) and Eq. (7):

$$MTBF_{UPS} =$$

$$= \frac{1}{\frac{1}{MTBF_{CB}} + \frac{1}{MTBF_{UPS}} + \frac{1}{MTBF_{CB}}} = \quad (4)$$

$$= 44494 \text{ h}.$$

$$R_{UPS1UPS2} = R_{UPS1} + R_{UPS2} - R_{UPS1} \cdot R_{UPS2}, \quad (5)$$

where R [8] is the reliability and it is defined as

$$R = e^{-\frac{1}{MTBF} \cdot t}.$$

$$MTBF_{UPS1UPS2} = \int_0^\infty R_{UPS1UPS2} \cdot dt = \quad (6)$$

$$= 66741 \text{ h}.$$

$$MTBF_T =$$

$$= \frac{1}{\frac{1}{MTBF_{SBB}} + \frac{1}{MTBF_{FRF1}} + \frac{1}{MTBF_{FRF2}} + \cdots}$$

$$\cdots \frac{1}{\frac{1}{MTBF_{CU}} + \frac{1}{MTBF_{UPS1UPS2}} + \frac{1}{MTBF_{SBB}}} =$$

$$= 43385 \text{ h} = 5 \text{ years}.$$

(7)

Only one Control Unit is usually installed for an emergency load switching between two UPS', that choice makes the *MTBF* worse.

4.3. Two UPS in Different Fire Compartments

The reliability of two fire compartmented UPS' is studied here. In Fig. 5 and Fig. 6, there are shown the system configuration and the RBD scheme respectively.

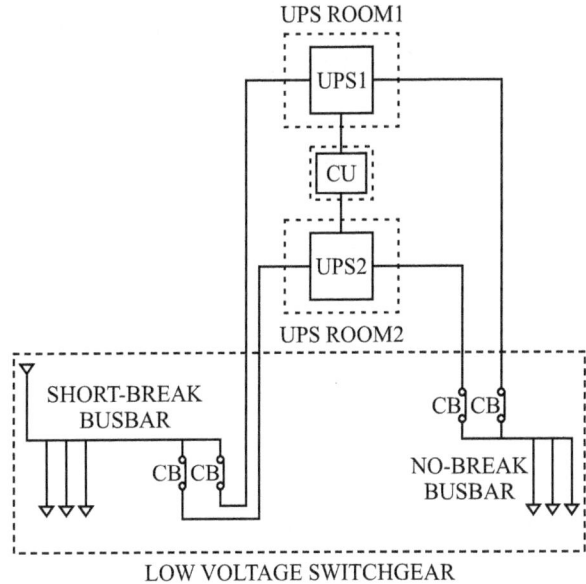

Fig. 5: No-break system composed of two UPS with fire compartmentations.

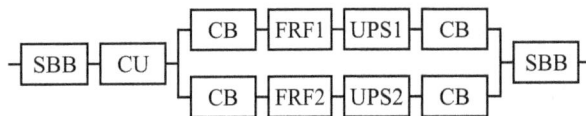

Fig. 6: RBD scheme of no-break system with two fire compartmented UPS.

In this case, each Fire Risk Factor is related exclusively to the respective UPS. Calculations to evaluate the total $MTBF$ are represented in Eq. (8), Eq. (9), Eq. (10) and Eq. (11):

$$MTBF_{UPS} = \cfrac{1}{\cfrac{1}{MTBF_{CB}} + \cfrac{1}{MTBF_{FRF}} + \cdots}$$

$$\cfrac{1}{\cdots \cfrac{1}{MTBF_{UPS}} + \cfrac{1}{MTBF_{CB}}} = 40391 \text{ h.} \qquad (8)$$

$$R_{UPS1UPS2} = R_{UPS1} + R_{UPS2} - R_{UPS1} \cdot R_{UPS2}. \qquad (9)$$

$$MTBF_{UPS1UPS2} = \int_0^\infty R_{12} \cdot dt = 60587 \text{ h.} \qquad (10)$$

$$MTBF_T = \cfrac{1}{\cfrac{1}{MTBF_{SBB}} + \cfrac{1}{MTBF_{CU}} + \cdots}$$

$$\cfrac{1}{\cdots \cfrac{1}{MTBF_{UPS1UPS2}} + \cfrac{1}{MTBF_{SBB}}} = \qquad (11)$$

$$= 49987 \text{ h} = 5.7 \text{ years.}$$

4.4. Two UPS' and One STS with Fire Compartmentations

Another system improvement consists of installing a safety external bypass over two UPS'. All these devices must be installed in different fire compartmented rooms. In the Fig. 7 and Fig. 8 the configuration and the RBD scheme are shown respectively.

According to the proposed system scheme, calculations to evaluate total $MTBF$ are shown in Eq. (12), Eq. (13), Eq. (14), Eq. (15), Eq. (16), Eq. (17), Eq. (18) and Eq. (19).

$$MTBF_{UPS} = \cfrac{1}{\cfrac{1}{MTBF_{CB}} + \cfrac{1}{MTBF_{FRF}} + \cdots}$$

$$\cfrac{1}{\cdots \cfrac{1}{MTBF_{UPS}} + \cfrac{1}{MTBF_{CB}}} = 40391 \text{ h.} \qquad (12)$$

$$R_{UPS1UPS2} = R_{UPS1} + R_{UPS2} - R_{UPS1} \cdot R_{UPS2}. \qquad (13)$$

$$MTBF_{UPS1UPS2} = \int_0^\infty R_{12} \cdot dt = 60587 \text{ h.} \qquad (14)$$

$$MTBF_{UPS1UPS2CU} =$$

$$= \cfrac{1}{\cfrac{1}{MTBF_{UPS1UPS2}} + \cfrac{1}{MTBF_{CU}}} = 56058 \text{ h.} \qquad (15)$$

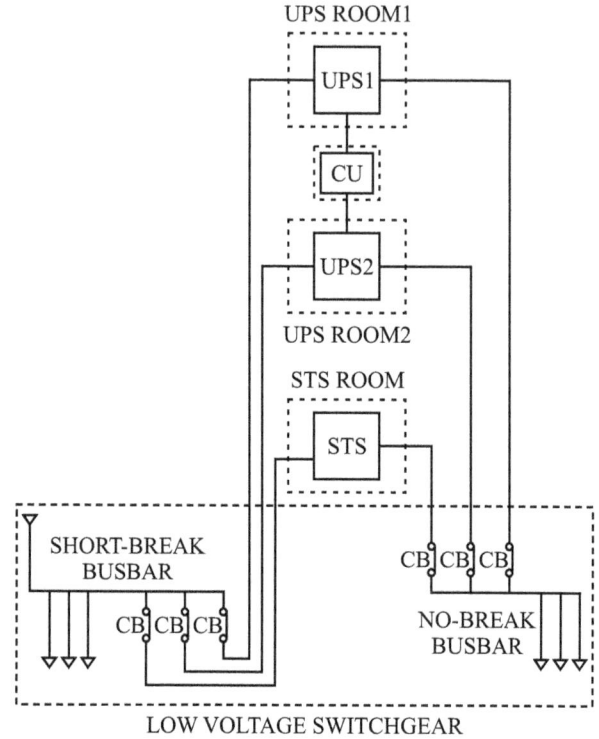

Fig. 7: No-break system made up of two UPS and one STS with fire compartmentations.

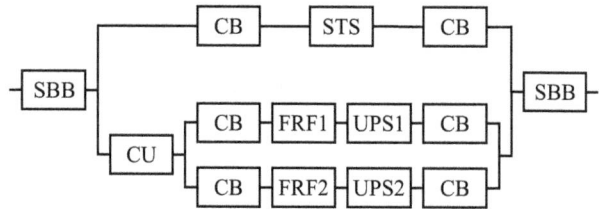

Fig. 8: RBD scheme for no-break system with fire compartmented UPS and STS.

$$MTBF_{STS} =$$

$$= \cfrac{1}{\cfrac{1}{MTBF_{CB}} + \cfrac{1}{MTBF_{STS}} + \cfrac{1}{MTBF_{CB}}} = \qquad (16)$$

$$= 54080 \text{ h.}$$

$$R_{UPS1UPS2CUSTS} = R_{UPS1UPS2CU} +$$

$$+ R_{STS} - R_{UPS1UPS1CU} \cdot R_{STS}. \qquad (17)$$

$$MTBF_{UPS1UPS2CUSTS} =$$

$$= \int_0^\infty R_{UPS1UPS2CUSTS} \cdot dt = 82616 \text{ h.} \qquad (18)$$

$$MTBF_T =$$
$$= \frac{1}{\dfrac{1}{MTBF_{SBB}} + \dfrac{1}{MTBF_{UPS1UPS2CUSTS}} + \cdots} \quad (19)$$
$$\cdots \frac{1}{\dfrac{1}{MTBF_{SBB}}} = 70070 \text{ h} = 8 \text{ years.}$$

The total $MTBF$ for compartmented system is 8 years. Calculations have been also performed for the case of absence of fire compartmentations. Installing two UPS' and one STS in the same room results in $MTBF$ of 6.4 years.

4.5.   Three UPS' with Fire Compartmentations

The system configuration made of three UPS' installed in different rooms is studied here. The configuration scheme is shown in Fig. 9.

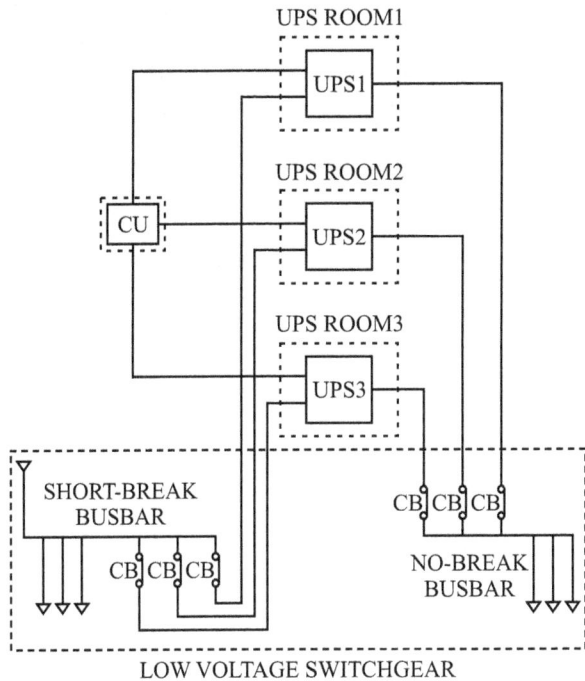

Fig. 9: No-break system made of three fire compartmented UPS.

The respective RBD scheme is shown in Fig. 10.

The total $MTBF$ of the three UPS' configuration is computed using the Eq. (20), Eq. (21), Eq. (22),

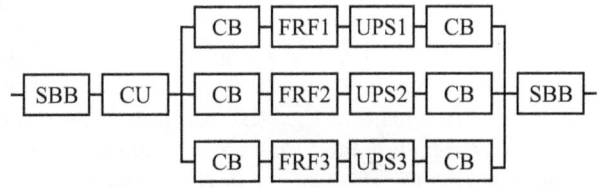

Fig. 10: RBD scheme for fire compartmented three UPS configuration.

Eq. (23), and Eq. (24).

$$MTBF_{UPS} =$$
$$= \frac{1}{\dfrac{1}{MTBF_{CB}} + \dfrac{1}{MTBF_{FRF}} + \cdots} \quad (20)$$
$$\cdots \frac{1}{\dfrac{1}{MTBF_{UPS}} + \dfrac{1}{MTBF_{CB}}} = 40391 \text{ h.}$$

$$R_{UPS1UPS2} = R_{UPS1} + R_{UPS2} - R_{UPS1} \cdot R_{UPS2}. \quad (21)$$

$$R_{UPS1UPS2UPS3} = R_{UPS1UPS2} + R_{UPS3} - \\ - R_{UPS1UPS2} \cdot R_{UPS3}. \quad (22)$$

$$MTBF_{UPS1UPS2UPS3} =$$
$$= \int_0^\infty R_{UPS1UPS2UPS3} \cdot dt = 76743 \text{ h.} \quad (23)$$

$$MTBF_T = \frac{1}{\dfrac{1}{MTBF_{SBB}} + \dfrac{1}{MTBF_{CU}} + \cdots}$$
$$\cdots \frac{1}{\dfrac{1}{MTBF_{UPS1UPS2UPS3}} + \dfrac{1}{MTBF_{SBB}}} = \quad (24)$$
$$= 60494 \text{ h} = 6.9 \text{ years.}$$

The total $MTBF$ for three compartmented UPS' is 6.9 years. In case of absence of fire compartments for the same configuration the reliability is studied; and $MTBF$ of 5.1 years is obtained for three UPS installed in the same room. With respect to the configuration with two UPS' and one STS, a little decrease of reliability is to be noticed. This is due to the better $MTBF$ of the STS compared to the UPS. On the opposite, this three system UPS permits all maintenance operations during working activities.

5.   Conclusions

Revised statistical data on $MTBF$ components used in no-break systems have been summarized here. A method to take into account the fire risk in a Reliability Block Diagram model has been performed. The reliability of seven different UPS configurations has been studied by means of the RBD method. The total *MTBF* has been computed for each configuration taking into account the effect of the devices' fire compartmentation. The results of this comparative analysis are shown in Fig. 11.

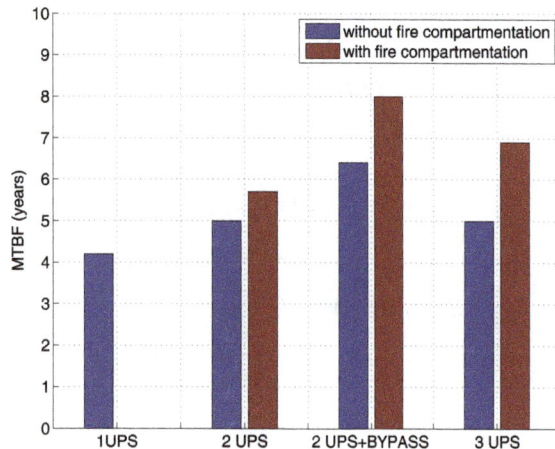

Fig. 11: *MTBF* for different no-break power systems by fire compartmentation.

These *MTBF* results can be also converted to yearly failure probability values by means of Eq. (25):

$$FP_\% = \frac{1}{MTBF} \cdot 100. \qquad (25)$$

These results demonstrate the importance of fire compartmentation for reliability improving in redundant UPS systems, especially for hospitals and safety systems. The more suitable configuration consists of two UPS and one STS fire compartmented, which achieves the best *MTBF*. Moreover, the three compartmented UPS' configuration and permits a complete maintenance during system's operation.

Acknowledgment

In this part of the article the authors can express their gratitude to projects from the results of which the article has been written, or to people who have contributed to the results published in the article. It is not allowed to insert any figures or logos here.

References

[1] RAHMAT, M. K., S. JOVANOVIC and K. L. LO. Uninterruptible Power Supply (UPS) system configurations: Reliability comparison. In: *2010 IEEE International Conference on Power and Energy.* Kuala Lumpur: IEEE, 2010, pp. 835–840. ISBN 978-1-4244-8947-3. DOI: 10.1109/PECON.2010.5697695

[2] CHIESA S. *Affidabilita, sicurezza e manutenzione del progetto dei sistemi.* 1st ed. Turin: CLUT, 2008. ISBN 978-8879922647.

[3] *Military Handbook: Electronic Reliability Design Handbook, MIL-HDBK-338B.* Washington: Department of Defense, 1998.

[4] NIL, M., M. NIL and B. CAKIR. MTBF analysis in OEM Company: Applications to ZCZVT PWM soft-transition inverters. In: *9th International Conference on Environment and Electrical Engineering.* Prague: IEEE, 2010, pp. 353–356. ISBN 978-1-4244-5370-2. DOI: 10.1109/EEEIC.2010.5489940.

[5] IEEE Std 493-1990. *Recommended practice for the design of reliable industrial and commercial power systems.* New York: IEEE, 2007. ISBN 978-0-7381-7083-1.

[6] RAHMAT, M. K., S. JOVANOVIC and K. L. LO. Reliability Comparison of Uninterruptible Power Supply (UPS) System Configurations. In: *Telecommunications Energy Conference 'Smart Power and Efficiency' (INTELEC).* Hamburg: IEEE, 2013, pp. 1–6, ISBN 978-3-8007-3500-6.

[7] RAHMAT, M. K., S. JOVANOVIC and K. L. LO. Reliability and Availability modelling of Uninterruptible Power Supply (UPS) systems using Monte-Carlo simulation. In: *2011 5th International Power Engineering and Optimization Conference.* Shah Alam: IEEE, 2011, pp. 267–272. ISBN 978-1-4577-0355-3. DOI: 10.1109/PEOCO.2011.5970403.

[8] BODI, F. "DC-grade" reliability for UPS in telecommunications data centers. In: *INTELEC 07 - 29th International Telecommunications Energy Conference.* Rome: IEEE, 2007, pp. 595–602. ISBN 978-1-4244-1627-1. DOI: 10.1109/INTLEC.2007.4448849.

[9] RAHMAT, M. K., S. JOVANOVIC and K. L. LO. Reliability Modelling of Uninterruptible Power Supply Using Probability Tree Method. In: *Proceedings of the 41st International Universities Power Engineering Conference.* Newcastle: IEEE, 2006, pp. 603–607. ISBN 978-186135-342-9. DOI: 10.1109/UPEC.2006.367549.

About Authors

Stefano ELIA was born in Rome, Italy. He received his Degree in 1999 and the Ph.D. in 2003 in Electrical Engineering from Sapienza University of Rome. Now is employed as a Professor at the Electrical Engineering Institute of Sapienza University. His research interests include power electronics, electrical systems, maintenance, reliability, safety, measures, energy saving and renewable energy.

Alessio SANTANTONIO was born in Caserta, Italy. He is an Electrical Engineer graduated at Sapienza University in 2014. Now is employed as laboratory assistant at the Electrical Engineering Institute of Sapienza University. He collaborates with some electricians' companies on maintenance procedure for electrical power systems. His research interests include power systems, power quality, reliability, maintenance.

4

ON-LINE EFFICIENCY IMPROVEMENT OF INDUCTION MOTOR VECTOR CONTROLLED

Djamel BENOUDJIT, Mohamed Said NAIT-SAID, Said DRID, Nasreddine NAIT-SAID

LSP-IE Laboratory, Electrical Engineering Department, Faculty of Technology, University of Batna 2, Rue Chahid Mohamed El-HadiBoukhlouf, 05000 Batna, Algeria

d_benoudjit@yahoo.fr, medsnaitsaid@yahoo.fr, s_drid@yahoo.fr, n_naitsaid@yahoo.com

Abstract. *Efficiency improvement is an important challenge for electric motor driven systems. For an induction motor, operation under rated conditions (at rated load with rated flux) is very efficient. However, in many situations, operation with rated flux causes low efficiency especially at light load ranges. In these applications, induction motor should operate at reduced flux which causes a balance between iron losses and copper losses leading to an improved efficiency. This paper concerns energy optimization, i.e. efficiency improvement is carried out via a controller designed on the basis of imposing the rated power factor, by finding a relationship between rotor flux and torque current component which can optimize the compromise between torque and efficiency in steady state as well as in transient state. Experimental results are presented to prove the effectiveness and validity of the proposed controller.*

Keywords

Induction motor, power factor, Robust Optimizer Efficiency Factor (ROEF), vector control.

1. Introduction

Electrical motors consume more than 50 % of the electric energy used by the industry sector [1] and [2]. Applications utilizing induction motors include electric vehicles, water pumps, etc. These applications require different power levels and energy-efficient motor. Therefore, motor energy saving solutions using control techniques that maximize the motor efficiency are highly required [2] and [3].

One of the most important advantages of energy saving work is the ability to decrease power losses and electrical energy consumption.

Motor efficiency is defined as the ratio of mechanical output power to the electrical input power. It can be thus improved by reducing the electrical input power by means of minimizing the total losses.

To provide a high efficiency for electric drives it is necessary to find a system which may be optimal from the viewpoint of both engineering and economics.

Energy conservation and sufficient operation of a drive can be obtained, starting from the high quality of motor, inverter (improvement of its waveforms) and design, or by selecting an appropriate control method respecting the applied motor load. In the vector controlled method of induction motor, to get high dynamic performances, the flux is generally maintained at nominal value in order to give the maximum torque abilities of the machine especially at lower speed regime. But this fact decreases the motor efficiency especially at light load. To solve this problem, diverse methods and approaches have been developed in order to maximise the machine efficiency [2], [4], [5], [6], [7], [8], [9] and [10]. In these methods, the efficiency of induction motor drives under variable operating conditions can easily be improved by varying the flux level that guarantees loss minimization. At light load induction motor should operate at adapted and reduced flux, and for operation under rated conditions with rated load and rated speed the flux should be increased. However, it should be noted that the loss amount that can be reduced by adjusting the flux level significantly to the light load case more than at high load condition.

The efficiency optimization, for an induction motor lightly loaded, is realized by an optimum balance of the iron and copper losses. The algorithms are then developed by controlling different variables, such as rotor flux, power factor, etc. [10], [12] and [13].

Based on simple state control approach, this paper describes the experimental validation of an efficiency optimization controller for the vector controlled induction motor drives.

Efficiency improvement is carried out via an optimizer efficiency factor controller (ROEF) designed on the basis of setting the power factor equal to its rated value. In order to prove the validity of the proposed controller, the vector control scheme of the induction motor drive with and without the optimization algorithm is experimentally implemented using a digital signal processor board DS1103 for a laboratory 1.1 kW squirrel-cage Induction Motor (IM) vector controlled.

The paper is organized as follows. In Section 2., we briefly review the indirect rotor-flux-oriented control of induction motor drives. The proposed efficiency optimization approach is explained in Section 3. In Section 4., the experimental platform will be first presented. Some experimental results and their detailed discussions are then given. Conclusion is done in Section 5.

2. Vector Controlled Induction Motor

Field orientation is a technique that provides independent control of torque and flux, which is similar to a separately excited DC motor. The Park model of an induction motor can be represented according to the usual d-axis and q-axis components in synchronous rotating frame with:

$$\begin{cases} \bar{v}_s = \left(R_s + \sigma L_s \frac{d}{dt}\right)\bar{i}_s + \frac{d}{dt}\bar{\phi}_r + j\omega_s\sigma L_s\bar{i}_s \\ \qquad + j\omega_s\bar{\phi}_r, \\ (1-\sigma)L_s\bar{i}_s = \left(\frac{d}{dt}T_r + 1\right)\bar{\phi}_r + j\omega_{sl}\bar{\phi}_r, \end{cases} \quad (1)$$

where $\omega_{sl} = \omega_s - p\Omega$ is the slip frequency, ω_s is the synchronous angular speed, p is the number of pole pairs, Ω is mechanical rotor speed, $\bar{v}_s$ is stator voltage vector $(v_{sd} + jv_{sq})$, $\bar{i}_s$ is stator current vector $(i_{sd} + ji_{sq})$, $\bar{\phi}_r$ is rotor flux vector, R_s is the stator resistance, σ is the redefined leakage inductance, L_s is the stator inductance, T_r is the rotor constant time and j is imaginary unit, satisfying $j^2 = -1$). The developed torque is then:

$$T_e = p\frac{M}{L_r}\Im_m\left[\bar{i}_s\right]\cdot\bar{\phi}_r^*, \quad (2)$$

where M - is the mutual inductance, L_r - the rotor inductance. Orientation flux process is given by [14]:

$$\phi_{rq} = 0 \text{ and } \phi_{rd} = \phi_r. \quad (3)$$

Hence, the rotor flux can be controlled directly from the stator direct current component i_{sd}, while

the torque can be linearly controlled from the stator quadrate current component i_{sq} when the rotor flux is maintained constant. The basic formulations of filed oriented control are:

$$\frac{\phi_r}{i_{sd}} = \frac{(1-\sigma)L_s}{1+T_r \cdot s}, \quad (4)$$

where $s = \dfrac{d}{dt}$: Laplace operator.

The slip frequency is given by:

$$\omega_{sl} = \frac{(1-\sigma)L_s}{T_r}\cdot\frac{i_{sq}}{\phi_r}. \quad (5)$$

The rotor flux orientation angle can be derived from Eq. (5) as:

$$\theta_s = \int \left(\omega_{sl} + p\Omega\right)dt. \quad (6)$$

In many applications such as electric vehicle, the loads can be largely varied. From this point of view, the induction motor does not operate normally in filed weakening region, thus the flux must be maintained constant to its rated value.

Consequently, once the flux ϕ_r is established to its rated value $\phi_r^* = M \cdot i_{sd}^*$ from i_{sd}^* - command in open loop control characterized by Eq. (3), the torque can be controlled linearly from i_{sq}^* - command such that:

$$T_e = p\frac{M}{L_r}\phi_r^* i_{sq}^*. \quad (7)$$

Considering the previous assumptions, the stator voltage equation can be written as:

$$\bar{v}_s = \left(R_s + \sigma L_s s\right)\bar{i}_s + \bar{e}_s, \quad (8)$$

where $\bar{e}_s$ denotes the vector nonlinear coupling term given as:

$$\bar{e}_s = j\omega_s\left(\sigma L_s\bar{i}_s + \bar{\phi}_r^*\right). \quad (9)$$

The real time $\bar{e}_s$ is compensation, from the feedback measured stator currents and the computed stator frequency, conducts to define a linear control between the component of stator current and its voltage, respectively. So, $\bar{u}_s$ denotes the new stator voltage input from which we can define the following first order transfer function:

$$T(s) = \left(\frac{\bar{i}_s}{\bar{u}_s}\right) = \frac{T_o}{1+\tau_\sigma s}, \quad (10)$$

where $\bar{u}_s = \bar{v}_s - \bar{e}_s$, $T_o = \dfrac{1}{R_s}$ and $\tau_\sigma = \dfrac{\sigma L_s}{R_s}$.

If real time $\bar{e}_s$ - compensation is well realized, the bloc diagram of the linear and decoupled control of induction motor based in Indirect Field Oriented Control (IFOC) strategy can be stated as shown in Fig. 1.

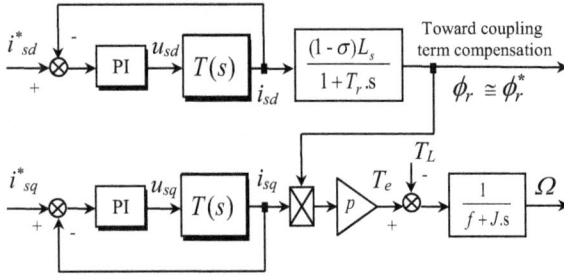

Fig. 1: Block diagram of the equivalent control of IFOC-IM.

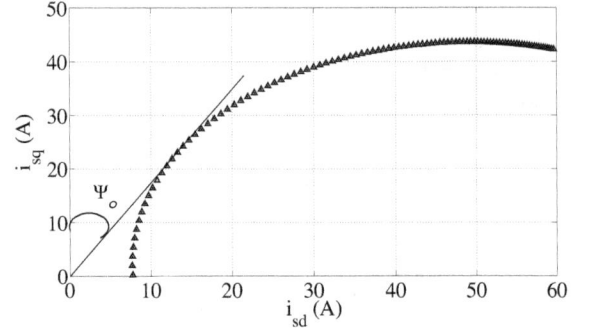

Fig. 2: Simplified circle diagram.

The PI Proportional Integral controllers are useful to control the stator current components and are determined from the pole placement method.

Assume that the induction machine is magnetized before through the input command of the rated direct stator current i_{sd}^* as shown in Fig. 1.

Therefore, the rotor flux can reach its rated value ϕ_r^* when the direct stator current is rapidly controlled from the PI - controller, hence we can simply write $i_{sd} = i_{sd}^*$. In such a way, the torque control is completely and linearly referred to the quadrate stator current component. On the other hand, if the inner quadrate stator current closed loop is faster than the external speed one, we can write also $i_{sq} = i_{sq}^*$. By this manner, an open loop transfer function between i_{sq} - command and motor speed can be defined as:

$$G\left(s\right) = \left.\frac{\Omega}{i_{sq}^*}\right|_{T_L=0} = \frac{p\phi_r^*}{Js+f}. \tag{11}$$

3. Efficiency Optimization Approach

The best exploitation of an induction machine is when we can maintain a certain balance between active power generating electromagnetic torque and reactive power that products the flux. The simultaneous existence of these two powers is vital for an induction motor.

Figure 2 shows a simplified induction motor circle diagram, which is a graphical approach depicting the operation characteristics of an induction motor. Using this diagram, all the performance characteristics of an induction motor like power factor, efficiency etc. can be predicted.

As can be seen on Fig. 2, Ψ_o is the external load angle giving optimal power factor. This is obtained when the stator current vector becomes tangent to the circle, which carries out the optimality active-reactive.

The reactive power Q can be written with stator voltage and stator current components by:

$$Q = v_{sd}i_{sq} - v_{sq}i_{sd}, \tag{12}$$

and the active power P as:

$$P = v_{sd}i_{sq} + v_{sq}i_{sd}. \tag{13}$$

The angle Ψ_o which constitutes the external load angle giving the optimal power factor is defined by the following relationship:

$$\tan\left(\Psi_o\right) = k_o = \frac{Q}{P} = \frac{\sin\Psi_o}{\cos\Psi_o} = \sqrt{\frac{1}{\left(\cos\Psi_o\right)^2} - 1}. \tag{14}$$

These obvious relations Eq. (11) and Eq. (12) lead to:

$$i_{sd} = \frac{1 - k_\alpha k_o}{k_\alpha + k_o} i_{sq}, \tag{15}$$

where $k_\alpha = \frac{v_{sq}}{v_{sd}} = \tan(\alpha)$.

The active-reactive optimization operation becomes simply:

$$i_{sd}^* = K_{ROEF} \, i_{sq}^*, \tag{16}$$

where * indicates input command variable.

The index (ROEF) means Robust Optimizer Efficiency Factor. The angle α is only given by the stator voltage components measurement.

One main advantage of the efficiency optimizer controller is that it is independent of machine parameter variations.

This efficiency controller in conjunction with the indirect vector control technique for induction motor drives has been employed in our configuration.

Figure 3 illustrates the particular characteristics of induction motor vector control describing by formulations Eq. (16), Eq. (17) and Eq. (18). Based on the

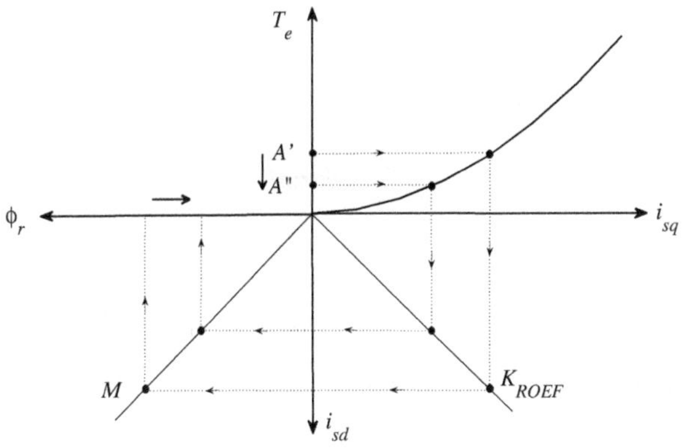

Fig. 3: Vector control characteristics.

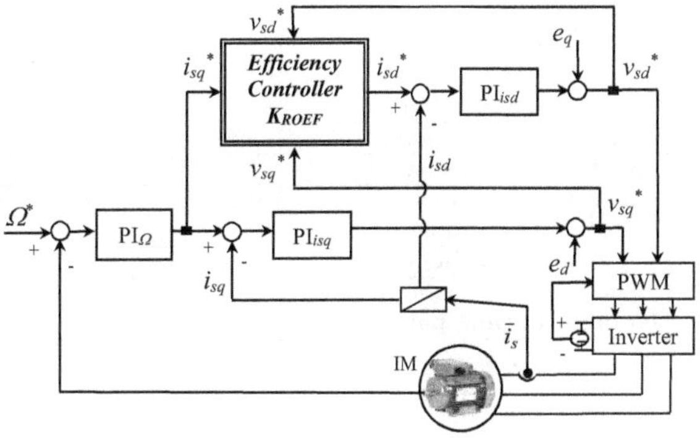

Fig. 4: ROEF controller implementation.

analysis of vector control principle, these characteristics are described by the following expressions.

When the rotor flux modulus ϕ_r is maintained constant (or established to its rated value).

$$\phi_r^* = M \cdot i_{sd}^*. \tag{17}$$

The electromagnetic torque can be controlled linearly from i_{sq}^* - command such that:

$$T_e = K_t \cdot i_{sq}^{*2}, \tag{18}$$

where:

$$K_t = p\frac{M}{L_r}\phi_r^* = p\frac{M^2}{L_r}K_{ROEF} = \text{Torque constant.} \tag{19}$$

In the vector controlled technique of induction motor, to get high performance, the flux is generally maintained constant equal to its nominal value. In this situation the induction motor run efficiently around the nominal point. However, the motor has lower efficiency especially at light load where rated flux is fixed.

This problem particularly at lower load needs special attention and the level flux may be adjusted to the load torque, as explained in Fig. 3.

So, as it can be observed on Fig. 3, if we want to reduce the load torque by moving the point A' to the point A", the stator current is reduced and the rotor flux also reduced, which includes the reduction of the total losses and consequently, the machine efficiency is more improved.

From Fig. 4, it is easy to notice that the proposed controller is implemented in a same way as in a conventional structure of a field oriented control for an induction motor.

Between the stator quadrate component of torque control (i_{sq}^* - command) and the stator direct current component controlling the flux (i_{sd}^* - command) e_d, e_q denote the compensated vector nonlinear coupling term components.

4. Experimental Validation and Discussion

In order to validate the efficiency improvements obtained by the proposed algorithm, the experimental setup shown in Fig. 5 has been used. It consists of a three-phase network supply, three-phase bridge PWM inverter (1 kVA, 2 kHz) based on six IGBT power components with filter capacitor.

The phase currents are measured with LEM-I sensor attached to the bridge, a dspace DS1103 Digital Signal Processor (DSP) based real-time Data Acquisition Control (DAC) system, and MATLAB/Simulink environment (PC), three-phase squirrel-cage induction motor 1.1 kW. The parameters of the induction motor are given next in Tab. 1.

Various tests have been carried out to confirm the validity of the proposed controller in comparison to the induction motor Vector Control (VC) with and without the optimization algorithm (ROEF controller). In the following, the results of one significant test will be presented.

Let us take a given speed profile defined as illustrated in Fig. 6 by reference trajectory Ω^*. Before 4 second, the motor is magnetized to its rated value; after there the motor starts with a constant acceleration until attains the reference speed of 100 rad·s^{-1}, with a light load torque corresponding to 0.73 Nm applied to the motor. At around 22 s, we activated the efficiency optimization algorithm.

Figure 7 shows an experimental result of motor efficiency. It is easy to notice that the efficiency under the action of the proposed optimization controller (with ROEF) is higher than the vector control without the controller at light loads. This is due to the optimization controller action allowing the reduction of the flux component of the stator current in d-q synchronously rotating frame (i_{sd}).

Tab. 1: Induction motor parameters.

Power	P_N	1.1	(kW)
Speed	n_N	1400	(rev·min^{-1})
Stator phase resistance	R_s	7.5	(Ω)
Rotor phase resistance	R_r	3.8	(Ω)
Mutual inductance	M	0.303	(H)
Stator inductance	L_s	0.3594	(H)
Rotor inductance	L_r	0.3032	(H)
Rotor inertia	J	0.0052	(kg·m^2)
Friction coefficient	f	0.0005	(Nm·s·rd^{-1})
Number of pole pairs	p	2	(-)

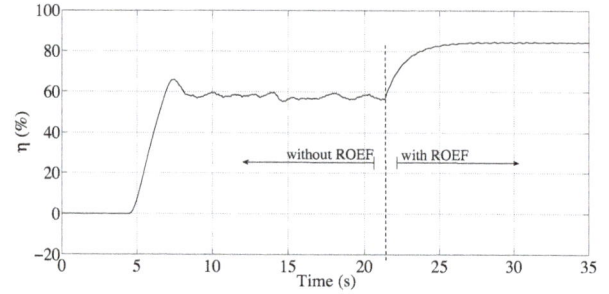

Fig. 7: Efficiency versus time.

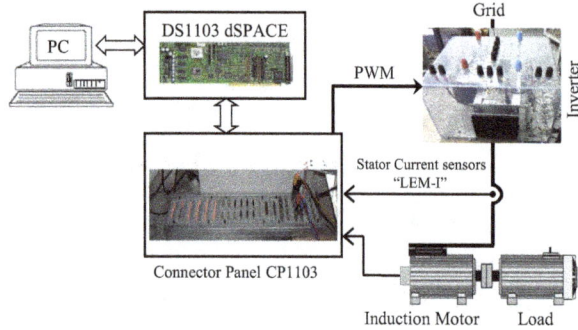

Fig. 5: Experimental setup.

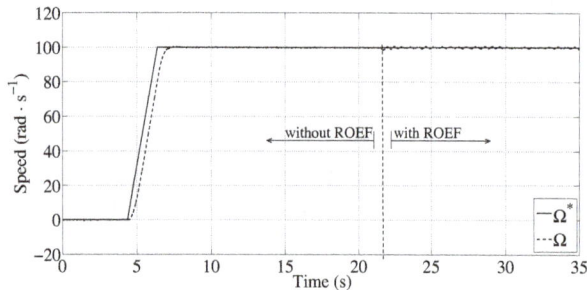

Fig. 8: Stator current direct component versus time.

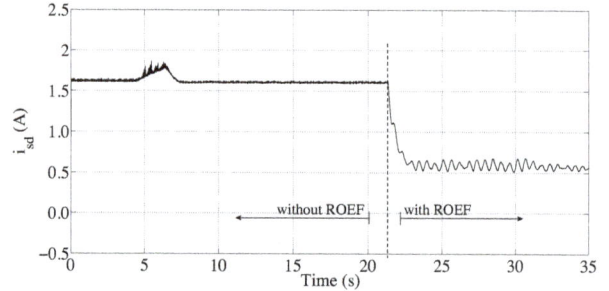

Fig. 9: Stator current versus time.

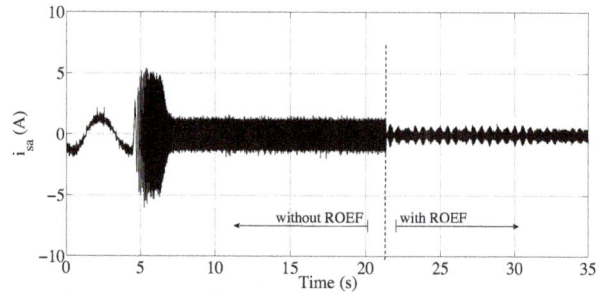

Fig. 6: Speed versus time.

As previously mentioned in Section 1. , it is known that efficiency improvement of induction motor drives can be realized via motor flux level.

This characteristic is demonstrated in Fig. 8 with and without the action of the efficiency optimization controller (ROEF). In these applications, induction motor should operate at reduced flux causes a balance between iron losses and copper losses, thereby improving the efficiency motor.

Consequently, as can be seen in Fig. 9, the terminal stator current is reduced and that the copper losses are decreased respectively to the flux component decreasing, which is represented by the stator current i_{sd} in Fig. 10.

It is also observed in this situation, the motor is operating with reduced flux level at light loads, which will certainly adapt and agree the torque output abilities.

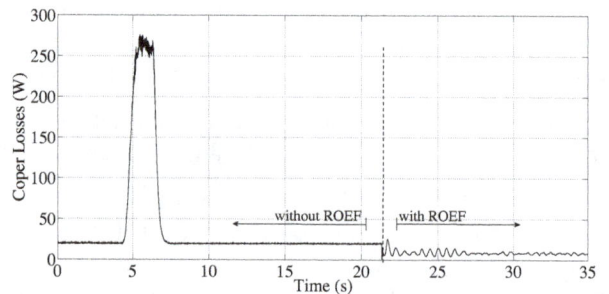

Fig. 10: Copper losses versus time.

This characteristic is well illustrated in Fig. 11 by the torque component current i_{sq}.

Finally, after various experimental tests, Fig. 12 shows comparison of motor efficiency evolution versus load torque T_L.

The results of this experiment demonstrate that when drive operates at light loads, the proposed algorithm (ROEF controller) enables important efficiency improvement (optimal efficiency) compared to the vec-

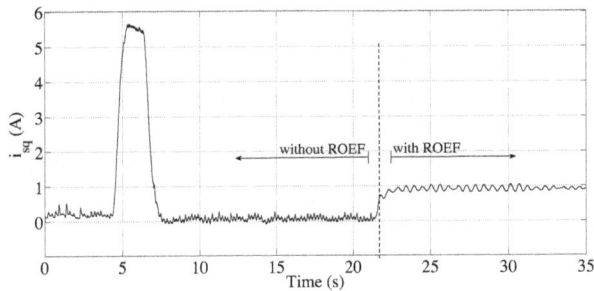

Fig. 11: Stator current quadrate component versus time.

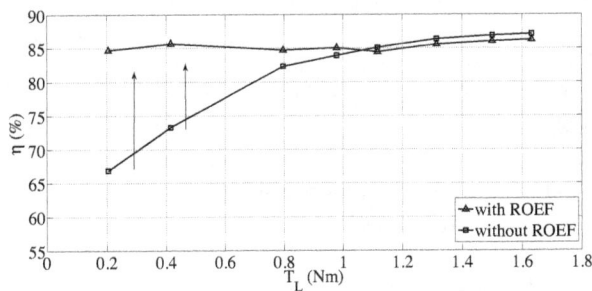

Fig. 12: Efficiency versus load torque.

tor control without ROEF. As an example, for a load toque of 0.2 Nm an increase in efficiency from 67.5 % to 84.5 % has been noticed.

The efficiency becomes smaller as the load torque increases. Indeed, for larger loads there is no efficiency improvement under the action of the proposed controller since the latter is only efficient at light loads.

This shows that the proposed controller (ROEF) is very suitable for the efficiency optimization of the induction motor drive, particularly at light load values. This will keep a certain balance between copper losses and iron losses, thereby a substantial power saving is achieved and the motor efficiency is consequently improved.

5.   Conclusion

An efficiency optimization controller for the induction motor drives has been presented in this paper. The approach is simple and practical implementation is also easy. In addition, the main advantage of the proposed controller is that it is not related to the machine parameters. Experimental results confirm the validity and usefulness of the proposed controller, especially a light load values. In this sense, under the action of the proposed controller, any energy gain of some percent becomes significant and a substantial power saving is achieved and the efficiency of the drive is consequently improved.

References

[1] CHAKRABORTY, C. and Y. HORI. Fast efficiency optimization techniques for the indirect vector-controlled induction motor drives. *IEEE Transactions on Industry Applications.* 2003, vol. 39, iss. 4, pp. 1070–1076. ISSN 0093-9994. DOI: 10.1109/TIA.2003.814550.

[2] TA, C.-M. and Y. HORI. Convergence improvement of efficiency-optimization control of induction motor drives. *IEEE Transactions on Industry Applications.* 2001, vol. 37, iss. 6, pp. 1746–1753. ISSN 0093-9994. DOI: 10.1109/28.968187.

[3] SAIDUR, R. A review on electrical motors energy use and energy savings. *IEEE Transactions on Industry Applications.* 2010, vol. 14, iss. 3, pp. 877–898. ISSN 1364-0321. DOI: 10.1016/j.rser.2009.10.018.

[4] ABRAHAMSEN, F., F. BlAABJERG, K. PEDERSEN and P. B. THOEGERSEN. Efficiency-optimized control of medium-size induction motor drives. *IEEE Transactions on Industry Applications.* 2001, vol. 37, iss. 6, pp. 1761–1767. ISSN 0093-9994. DOI: 10.1109/28.968189.

[5] LIM, S. and K. NAM. Loss-minimizing control scheme for induction motors. *IEEE Proceedings - Electric Power Applications.* 2004, vol. 151, iss. 4, pp. 385–397. ISSN 1350-2352. DOI: 10.1049/ip-epa:20040384.

[6] VUKOSAVIC, S. N. and E. LEVI. Robust DSP-Based Efficiency Optimization of a Variable Speed Induction Motor Drive. *IEEE Transactions on Industrial Electronics.* 2003, vol. 50, iss. 3, pp. 560–570. ISSN 0278-0046. DOI: 10.1109/TIE.2003.812468.

[7] SHUMA, C., L. CHEN and S. LIWEI. Study on efficiency calculation model of induction motors for electric vehicles. In: *Vehicle Power and Propulsion Conference.* Harbin: IEEE, 2008, pp. 1–5. ISBN 978-1-4244-1849-7. DOI: 10.1109/VPPC.2008.4677791.

[8] LI, J., L. XU and Z. ZHANG. A New Efficiency Optimization Method on Vector Control of Induction Motors. In: *IEEE International Conference on Electric Machines and Drives.* San Antonio: IEEE, 2005, pp. 1995–2001. ISBN 0-7803-8987-5. DOI: 10.1109/IEMDC.2005.195993.

[9] NAXIN, C., Z. CHENGHUI and Z. MIN. Optimal efficiency control of field-oriented induction

motor drive and rotor resistance adaptive identifying. In: *The 4th International Power Electronics and Motion Control Conference*. Xi'an: IEEE, 2004, pp. 414–419. ISBN 7-5605-1869-9.

[10] VAEZ-ZADE, S. and F. HENDI. A continuous efficiency optimization controller for induction motor drives. *Energy Conversion and Management*. 2005, vol. 46, iss. 5, pp. 701–713. ISSN 0196-8904. DOI: 10.1016/j.enconman.2004.05.006.

[11] SREEJETH, M., M. SINGH and P. KUMAR. Efficiency optimization of vector controlled induction motor drive. In: *38th Annual Conference on IEEE Industrial Electronics Society*. Montreal: IEEE, 2012, pp. 1758–1763. ISBN 978-1-4673-2420-5. DOI: 10.1109/IECON.2012.6388935.

[12] QU, Z., M. RANTA, M. HINKKANEN and J. LUOMI. Structure property relationships in polyethylene/montmorillonite nanodielectrics. *IEEE Transactions on Industry Application*. 2012, vol. 48, iss. 3, pp. 952–961. ISSN 0093-9994. DOI: 10.1109/TIA.2012.2190818.

[13] ABRAHAMSEN, F. *Energy Optimal Control of Induction Motor Drives*. 2nd ed. Aalborg: Institute of Energy Technology, Aalborg University, 2000. ISBN 87-89179-26-9.

[14] XIAO-FENG. X., L. GUO-FENG and H. RONG-TAI. A rotor field oriented vector control system for electric traction application. In: *Proceedings of the 2000 IEEE International Symposium Industrial Electronics*. Cholula: IEEE, 2000, pp. 294–299. ISBN 0-7803-6606-9. DOI: 10.1109/ISIE.2000.930529.

About Authors

Djamel BENOUDJIT was born in Batna, Algeria. He received his M.Sc. and Ph.D. degrees in electrical engineering from the University of Batna, Algeria, in 2005 and 2010 respectively. His research interests include modelling and control of electrical drives, control system applications, and electric vehicles.

Mohamed-Said NAIT-SAID was born in Batna, Algeria. He received his M.Sc. in Electrical and Computer Engineering from the Electrical Engineering Institute of Constantine University, Algeria, in 1992. He received his Ph.D. degree in Electrical and Computer Engineering from the University of Batna in 1999. Currently, he is a Full Professor at the Electrical Engineering Department, University of Batna 2. His research interests include electric machines, drives control, and diagnosis.

Said DRID was born in Batna, Algeria. He received his M.Sc. and Ph.D. degrees in Electrical Engineering from the University of Batna, Algeria, in 2000 and 2005, respectively. Currently, he is a Full Professor at the Electrical Engineering Department, University of Batna 2, Algeria. His research interests include electric machines and drives, renewable energy, field theory and computational electromagnetism.

Nasreddine NAIT-SAID was born in Batna, Algeria. He received his M.Sc. degree in Industrial Electricity in 1993 and Ph.D. degree in electrical engineering from the University of Batna, Algeria in 2003. Currently, he is a Full Professor at the Electrical Engineering Department, University of Batna 2. His research interests include application of AI techniques and control in the field of electrical machines.

5

Advanced Model of Squirrel Cage Induction Machine for Broken Rotor Bars Fault Using Multi Indicators

Ilias OUACHTOUK[1], Soumia EL HANI[1], Said GUEDIRA[2], Khalid DAHI[1], Lahbib SADIKI[1]

[1]Electrical Laboratory Researche, Ecole Normale Superieure de Enseignement Technique, Mohammed V University, Avenue des Nations Unies, Agdal, Rabat, Morocco
[2]Higher National School of Mines, Ecole Nationale Superieure des Mines de Rabat, Avenue Hadj Ahmed Cherkaoui, Agdal, Rabat, Morocco

ilias_ouachtouk@um5.ac.ma, s.elhani@um5s.net.ma, guedira@enim.ma, khalid.dahi@um5s.net.ma, Lahbib.sadiki@um5s.net.ma

Abstract. *Squirrel cage induction machine are the most commonly used electrical drives, but like any other machine, they are vulnerable to faults. Among the widespread failures of the induction machine there are rotor faults. This paper focuses on the detection of broken rotor bars fault using multi-indicator. However, diagnostics of asynchronous machine rotor faults can be accomplished by analysing the anomalies of machine local variable such as torque, magnetic flux, stator current and neutral voltage signature analysis. The aim of this research is to summarize the existing models and to develop new models of squirrel cage induction motors with consideration of the neutral voltage and to study the effect of broken rotor bars on the different electrical quantities such as the park currents, torque, stator currents and neutral voltage. The performance of the model was assessed by comparing the simulation and experimental results. The obtained results show the effectiveness of the model, and allow detection and diagnosis of these defects.*

Keywords

Diagnostic, neutral voltage signature, park currents, squirrel cage induction machine, stator current signature.

1. Introduction

The squirrel-cage induction machines are widely used and are the most common type of electrical rotating machine used in industry. However, due to the combination of poor working environment and installation, internal faults frequently occur on rotor, such as broken rotor bars, end ring connectors and eccentricities [1]. Detection of these faults is an absolute must in any real-life engineering system. Detection of broken rotor bar, particularly at an early stage, is rather difficult than stator faults [2] and [3]. This research is important because that even if though broken bars do not cause motor failures initially, they can significantly lower the efficiency and shorten the durability of induction machines.

In order to deal with the problems connected with these failures, a numerical simulation model is usually implemented to improve traditional techniques. In fact, some companies use simulation technique for designing their new product [4]. To detect the mechanical or electrical faults in induction machine, multiple methods have been utilized in the literature such as: Fast Fourier Transforms (FFT), Motor Current Signature Analysis (MCSA), Park Vectors, Stator voltages monitoring and recently Neutral Voltage (NV) [5] and [6]. Generally, MCSA is the most commonly used technique because it is simple and effective in appropriate conditions. However, this technique has significant limitations due to the increasing complexity of electrical machines and drives [10]. In order to reduce these limitations, the neutral voltage signature analysis has been used. This technique focuses on the use of voltage between the stator neutral voltage and an artificial supply neutral voltage [11] and [12].

To demonstrate the performance of the model, a comparison between simulation and experimental results have been verified, the obtained results show the effectiveness of the model, and allow detection and diagnosis of broken rotor bars defects.

2. Detection of Broken Rotor Bars

Among the widespread failures of the induction machine are the rotor faults, precisely broken rotor bars, end ring connectors and eccentricities. Rotor faults lead to speed fluctuation, torque pulsation, changes of the frequency component in the supplying neutral voltage and current of the motor, temperature increase, arcing in the rotor, and vibration of the machine. These side effects have been utilized in recent years for detecting and diagnosing this type of fault [13] and [14].

2.1. Line Current Spectrum Analysis

Motor Current Signature Analysis (MCSA), based on spectrum amplitude, have been widely used to detect broken rotor bars and end ring faults. This technique analyses the anomaly, which corresponds to broken bar faults in motor stator current spectrum, and then predicts the existence of the faults. Considering the speed ripple effect, it was reported that other frequency components of stator current due to rotor asymmetry could be observed around the fundamental at the following frequencies [8]:

$$f_b = (1 \pm 2 \cdot k \cdot s) f_s, \qquad (1)$$

where s is slip, f_s is supply frequency and $k = 1, 2, 3, \ldots$

Other higher harmonic components can be also induced nearby to the rotor slot harmonics in the stator current spectrum:

$$f_{hk} = f_s \left[\lambda \frac{N_r}{p} \left(1 - s\right) \pm 1 \pm 2 \cdot k \cdot s \right], \qquad (2)$$

where s is slip, f_s is supply frequency, λ is positive integer, N_r is number of rotor bars, p is number of pole pairs and $k = 1, 2, 3, \ldots$

2.2. Line Neutral Voltage Spectrum Analysis

The proposed approach based on spectral analysis of line-neutral voltage focuses on the use of voltage be-

tween the supply and the stator neutrals for broken rotor bars detection. Broken rotor bars causes asymmetries in the mutual inductance of the machine, which gives rise to reveal of additional components in the spectrum of the neutral voltage at frequencies given by the relation:

$$f_h = f_s \left[3h - (3h \pm 1) s \right], \qquad (3)$$

where s is slip, f_s is supply frequency and $h = 1, 3, 5, \ldots$

The speed ripple induced additional harmonic components around the previous frequency given by Eq. (3), and frequencies of all components can be expressed as follows:

$$f_h = f_s \left[3h \cdot (1 - s) \pm s (1 + 2k) \right], \qquad (4)$$

where s is slip, f_s is supply frequency, $k = 1, 2, 3, \ldots$ and $h = 1, 3, 5, \ldots$

The following Tab. 1 presents a summary of frequency components of motor current signature analysis and neutral voltage signature analysis.

3. Squirrel Cage Induction Motor Model

The model is built considering that both stator and rotor consist of multiple inductive circuits coupled together, and the current in each circuit is considered as an independent variable. We include in this model the most important supply voltage harmonics but also a large number of space harmonics. These harmonic spaces allow obtaining a machine model closer to the real one. However, the main information for the detection of broken bars is at the level of harmonic 3 in this voltage. The model of the induction motor takes into account the following assumptions [13], [14], [16] and [18]:

- saturation is neglected,

- uniform air gap,

- neglecting inter-bar currents,

- evenly distributed rotor bars,

- neglecting flux coupling between different winding without air gap crossing.

3.1. System of Equations

The Stator comprises conventional three phase windings, thus three circuits are required to represent the stator. The rotor consists of N_r identical and equally

Tab. 1: Frequency components of broken rotor bars motor faults.

Signature analysis	Components frequency	Harmonic components frequency
Motor Current	$f_b = (1 \pm 2 \cdot k \cdot s) f_s$	$f_{hk} = f_s \left\lvert \lambda \frac{N_r}{p}(1-s) \pm 1 \pm 2 \cdot k \cdot s \right\rvert$
Neutral voltage	$f_h = f_s [3h - (3h \pm 1) s]$	$f_h = f_s [3h \cdot (1-s) \pm s(1+2k)]$

spaced bars shorted together by two identical end rings [13]. Voltage equations for the motor can be written in vector-matrix form as follows:

$$\begin{cases} \mathbf{V}_s = \mathbf{R}_s \cdot \mathbf{I}_s + \dfrac{d\mathbf{\Phi}_s}{dt}, \\[4pt] \mathbf{V}_r = \mathbf{R}_r \cdot \mathbf{I}_r + \dfrac{d\mathbf{\Phi}_r}{dt}, \\[4pt] \mathbf{\Phi}_s = \mathbf{L}_s \cdot \mathbf{I}_s + \mathbf{M}_{sr} \cdot \mathbf{I}_r, \\[4pt] \mathbf{\Phi}_r = \mathbf{L}_r \cdot \mathbf{I}_r + \mathbf{M}_{rs} \cdot \mathbf{I}_s, \\[4pt] C_{em} = \dfrac{1}{2} \begin{bmatrix} \mathbf{I}_s \\ \mathbf{I}_r \end{bmatrix}^T \dfrac{d}{dq} \begin{bmatrix} \mathbf{L}_s & \mathbf{M}_{sr} \\ \mathbf{M}_{sr} & \mathbf{L}_r \end{bmatrix} \begin{bmatrix} \mathbf{I}_s \\ \mathbf{I}_r \end{bmatrix}, \\[4pt] W = \dfrac{dq}{dt}, \end{cases} \quad (5)$$

where $\mathbf{M}_{sr}$ is the Mutual matrix inductances between the stator and rotor, $\mathbf{R}_s$ is the Stator resistances matrix, $\mathbf{R}_r$ is the Rotor resistances matrix, $\mathbf{V}_r$ is the Rotor voltages vector, $\mathbf{V}_s$ is the Stator voltages vector, $\mathbf{\Phi}_s$ is the Stator flux vector and $\mathbf{\Phi}_r$ is the Rotor flux vector. The matrix $\mathbf{M}_{sr}$ depends on time, which necessitates the inversion of the inductance matrix $\mathbf{L}_s$ of dimension $\mathbf{N}_r + 4$ in each calculation. To make this matrix constant, we apply the Park transformation. The use of the Park transformation bypasses allows obtaining a system of equations with constant coefficients which facilitates their resolution.

3.2. Inductances Calculation

It is obvious that the calculation of all inductances is the key to successful simulation of an induction motor. These inductances are conveniently calculated using the Winding Function Approach. According to winding function theory, the inductance between any two windings i and j in any electric machine can be computed by the following equation [7] and [17]:

$$L_{ij}(\varphi) = \mu_0 \cdot L_r \int_0^{2\pi} \frac{n_i(\varphi, \theta) N_j(\varphi, \theta)}{e(\varphi, \theta)} d\theta, \quad (6)$$

where $\mu_0 = 4\pi \cdot 10^{-7}$ H·m^{-1}, e is the air gap length, θ is the particular rotor angular position, r is the average radius of the air gap, L the active stack length of the motor, φ angular position along the stator inner surface, and $n_i(\theta, \varphi), N_j(\theta, \varphi)$ is called the winding function of circuit, i and j represent the magnetomotive force distribution along the air gap for the unit current in winding.

3.3. Quadrature-Phase Model

The Park transformation is a well-known three-phase to two-phase transformation in machine analysis, consisting of the application of current, voltage and flux, a change of variable by involving the angle between the axis of the windings and the d and q axes. We transform the three-phase windings a, b and c in three orthogonal d, q and o windings, referred to as [13] and [17]:

- direct axis d,

- transverse axis q,

- homopolar axis o.

The mathematical model machine equations in the axis system d, q can be written in vector matrix form as follows:

$$\begin{bmatrix} V_{0s} \\ V_{ds} \\ V_{qs} \\ 0 \\ 0 \\ 0 \\ 0 \\ \vdots \\ \vdots \\ 0 \\ \cdots \\ 0 \\ 0 \end{bmatrix} = \mathbf{L}_{tr} \frac{d}{dt} \begin{bmatrix} I_{0s} \\ I_{ds} \\ I_{qs} \\ I_{r0} \\ I_{r1} \\ I_{r2} \\ I_{r3} \\ \vdots \\ \vdots \\ I_{rk} \\ \cdots \\ I_{r(N_r-1)} \\ I_e \end{bmatrix} + \mathbf{R}_{tr} \begin{bmatrix} I_{0s} \\ I_{ds} \\ I_{qs} \\ I_{r0} \\ I_{r1} \\ I_{r2} \\ I_{r3} \\ \vdots \\ \vdots \\ I_{rk} \\ \cdots \\ I_{r(N_r-1)} \\ I_e \end{bmatrix}, \quad (7)$$

and the Electromagnetic torque equation of the machine is defined as follows:

$$C_e = \sqrt{\frac{3}{2}} p L_{sr} \left(I_{qs} \sum_{k=0}^{N_r-1} I_{rk} \cos(Ka) \right. \\ \left. - I_{ds} \sum_{k=0}^{N_r-1} I_{rk} \sin(Ka) \right), \quad (8)$$

where V_{sd} is the Component according to the d axis voltage, V_{sq} is the Component according to the q axis voltage, I_{sq} is the Component of the rotor current along the axis d, I_{sd} is the Component of the stator current on the axis d, I_{rq} is the Rotor current component along the axis q, I_{rd} is the Component of the stator current on the axis q, C_{em} is the Electromagnetic torque, $\mathbf{L}_{tr}$ and $\mathbf{R}_{tr}$ are global matrices inductors and resistors obtained after the transformation of Park [13].

$$L_{sr} = \frac{4\mu_0 N_s Rl}{e\pi p^2} \sin\left(\frac{a}{2}\right),$$
$$a = \frac{2\pi}{N_r}, \tag{9}$$
$$K = 0, \ldots, N_r - 1.$$

3.4.   Line-Neutral Voltage Analysis

The approach of Neutral Voltage Signature Analysis focuses on the use of voltage between the supply and the stator neutrals [9], [11], [12] and [19]. This voltage is given by the following mathematical relationship:

$$\mathbf{V}_{3s} = \mathbf{R}_s \mathbf{i}_{3s} + \frac{d}{dt}\mathbf{\Psi}_{3s} + \mathbf{V}_{nn}, \tag{10}$$

$$\mathbf{V}_{nn} = \mathbf{R}_s \mathbf{i}_{3s} + L_a \frac{dI_{3s}}{dt} + \frac{dL_a}{d\theta}\Omega I_{3s} - \mathbf{V}_{supply}, \tag{11}$$

where $\mathbf{R}_s$ represents the stator-phase resistance, L_a his inductance, I_{sa} is the current passing through it, Ω is the rotation speed, θ is the angular position of the rotor and $\mathbf{V}_{supply}$ is simple voltage generated by network supply.

In order to depict the harmonic components related to broken rotor bars defects in the line neutral voltage, it is necessary to explore its theoretical formula.

As the stator windings are star-connected then:

$$\begin{cases} v_{sa} + v_{sb} + v_{sc} = 3v_{so}, \\ i_{sa} + i_{sb} + i_{sc} = 0, \end{cases} \tag{12}$$

and

$$v_{so} = \sum_{h=1}^{\infty} V_{soh} \cos\left(h\omega_s t + \varphi_h\right), \tag{13}$$

where v_{os} is the zero-sequence component of the supply voltage.

By the summation of the Eq. (10) and Eq. (13) we get:

$$v_{nn} = -\frac{1}{3}\left(\frac{d\Psi_{sa}}{dt} + \frac{d\Psi_{sb}}{dt} + \frac{d\Psi_{sc}}{dt}\right). \tag{14}$$

The mutual inductance as described by Eq. (15) presents harmonics with respect to the electrical angle θ, where $a = p(2\pi \cdot N_r^{-1})$ is the electrical angle of a rotor loop.

Therefore, the line voltage between neutrals can be written as Eq. (16), where V_{nn} is the potential difference between the neutral of star-connected stator and the neutral network in the case of a direct feed or artificial neutral in the case of a supply voltage by inverter results in healthy condition of induction motor.

4.   Simulation Results Analysis

For a squirrel cage induction motor with N_r bars, and one end ring current, Eq. (7) and Eq. (8) can be resolved using the fourth-order Runge-Kutta method. To validate the proposed model, the machine was first simulated under healthy condition. Then, the rotor faults under different broken rotor bars were simulated. The rotor has been presented by all the meshes allowing the representation of various faults, to simulate a broken rotor bar, the resistance of a bar of the cage is increased 40 times its healthy value ($R_{bb} = 40 \cdot R_b$) in the $\mathbf{R}_{tr}$ until the current in the bar is closest to zero.

4.1.   Simulation and Analysis of Healthy State

The simulation study was at: no load for 1 sec and then the motor is loaded with 15 Nm load using a machine of 3 phases, 50 Hz, 48 stator slots, 28 rotor bars and 2 poles machine. The Fig. 1(a), Fig. 1(b), Fig. 1(c),

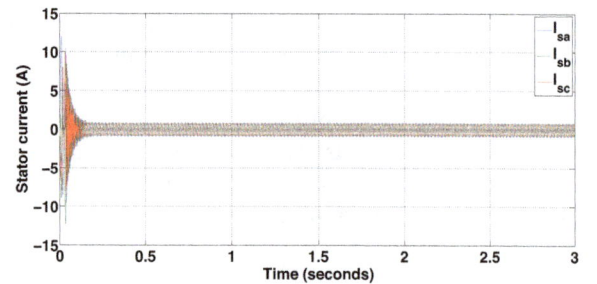

(a) Stator current at no load.

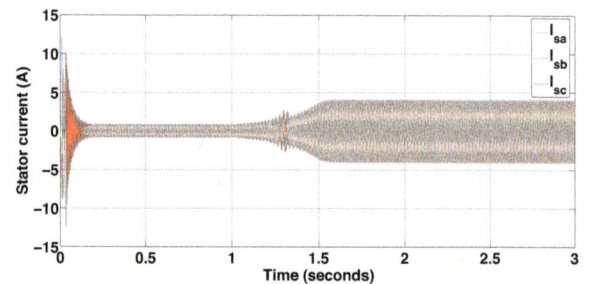

(b) Stator current when the motor is loaded.

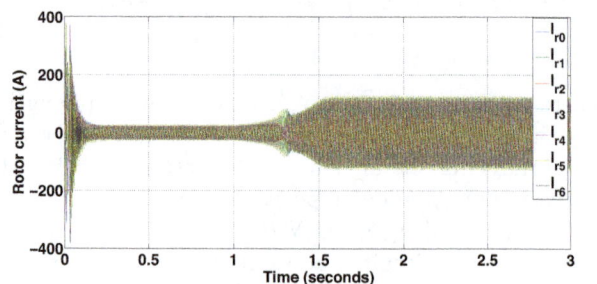

(c) Rotor currents.

$$\mathbf{M}_{sr} = \sum_{h=1}^{\infty} M_h^{sr} \cdot \begin{bmatrix} \cos(h \cdot (\theta + \phi_h)) & \dots & \cos(h \cdot (\theta + \phi_h + k \cdot a)) & \dots \\ \cos(h \cdot (\theta + \phi_h - \frac{2\pi}{3})) & \dots & \cos(h \cdot (\theta + \phi_h + k \cdot a - \frac{2\pi}{3})) & \dots \\ \cos(h \cdot (\theta + \phi_h + \frac{2\pi}{3})) & \dots & \cos(h \cdot (\theta + \phi_h + k \cdot a - \frac{2\pi}{3})) & \dots \end{bmatrix}. \tag{15}$$

$$\mathbf{V}_{nn} = -\sum_{h=1}^{\infty} M_{3h}^{sr} \begin{bmatrix} \cos(3h(\theta + \varphi_{3h})) & \dots & \cos(3h(\theta + \varphi_{3h} + ka)) & \dots \end{bmatrix} \frac{d}{dt}\mathbf{i}_{rk} + \frac{d\theta}{dt} \sum_{h=1}^{\infty} 3hs\omega_s M_{3h}^{sr}$$
$$\cdot \begin{bmatrix} \cos(3h(\theta + \varphi_{3h})) & \dots & \cos(3h(\theta + \varphi_{3h} + ka)) & \dots \end{bmatrix} \cdot \mathbf{i}_{rk} + \sum_{h=1}^{\infty} V_{soh} \cos(h\omega_s t + \varphi_h). \tag{16}$$

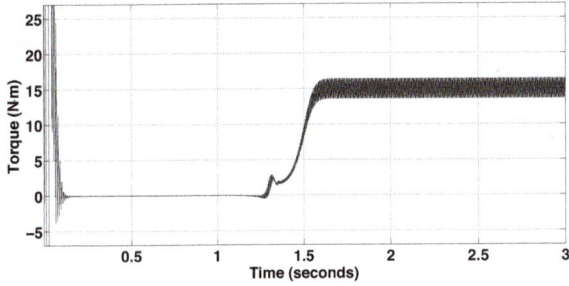

(d) Electromagnetic torque.

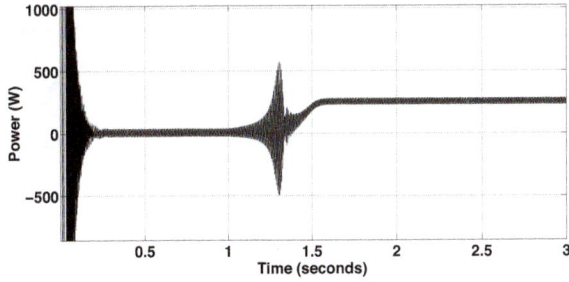

(e) Power.

Fig. 1: Simulation results of healthy state.

Fig. 1(d) and Fig. 1(e) show the simulation results in healthy condition of induction motor.

4.2. Simulation and Analysis of Broken Bars Fault in Motor

The motor is simulated at no load for 1 sec and then the motor is loaded with 15 Nm load, at 3 sec a completely broken rotor bars occur in a rotor induction motor. The Fig. 2(a), Fig. 2(b), Fig. 2(c), Fig. 2(d) and Fig. 2(e) show the simulation results for failures of the induction motor.

The impact of broken rotor bars fault on machine current can be examined through Park vector transformation approach. The current Park's vector for a healthy motor corresponds to a circle, whereas for a faulty one, the shape distorts depending upon the amount of fault level. To decide on the nature of the

fault occurred in induction machine, an analysis with the well-known FFT is done to decide on the fault and also its severity. In this case, broken bars related harmonic components are clearly located around the fundamental Fig. 4.

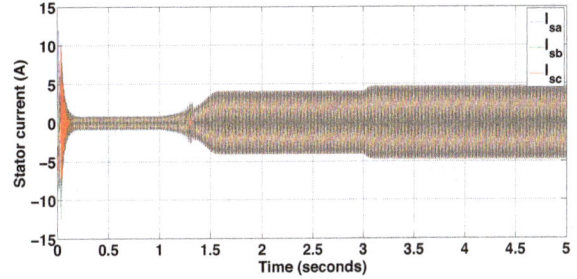

(a) Stator current.

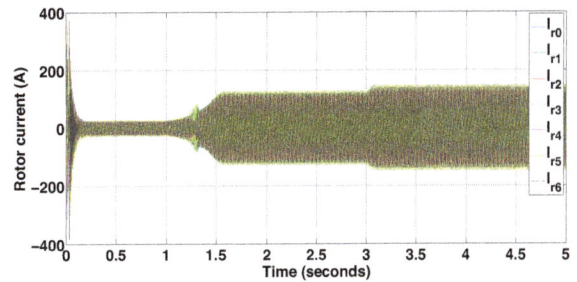

(b) Rotor currents.

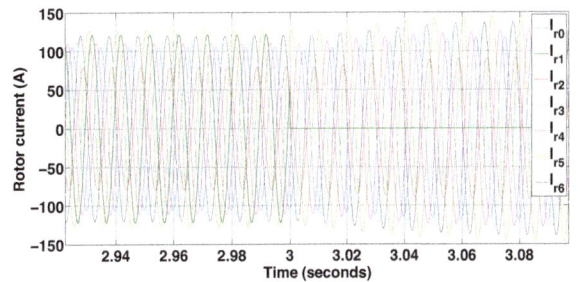

(c) Zoom of rotor currents.

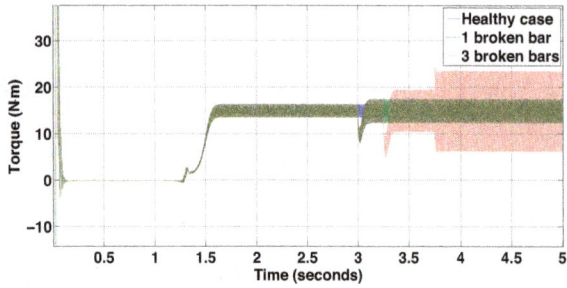

(d) Electromagnetic torque.

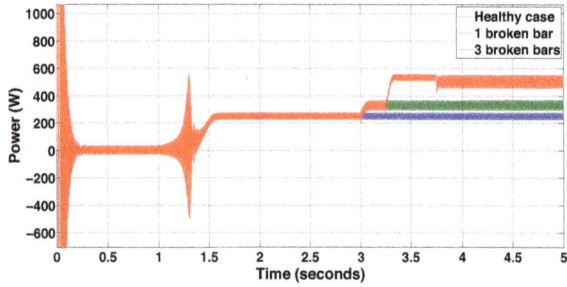

(e) Power.

Fig. 2: Simulation results of broken bars fault in motor.

5.   Experimental Results Analysis

In order to test the proposed model, an experimental system is configured as shown in Fig. 6. The experimental tests were developed on a 3 kW, 50 Hz, 220 V = 380 V, 4-poles Induction Machine. The motor was directly coupled to a direct current machine

acting as a load. Two voltage sensors are used to monitor the induction machine operation. The IM voltages are measured by means of the two sensors, which are used as inputs of the signal conditioning and the data acquisition board integrated into a personal computer.

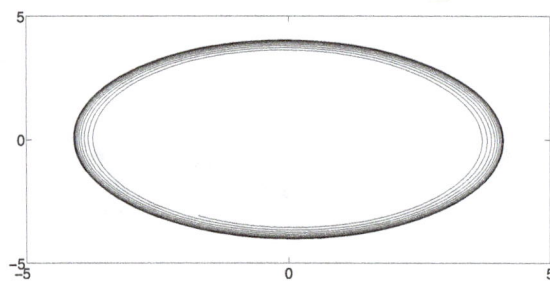

(a)

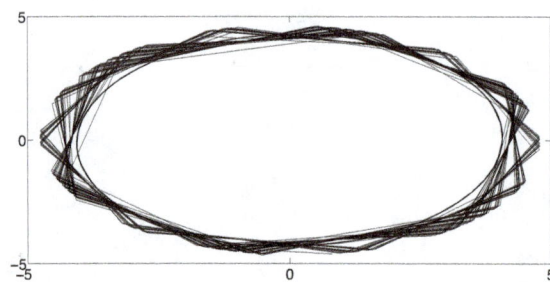

(b)

Fig. 4: Simulation: Normalized FFT spectrum of machine line current of IM with 3 broken rotor bars.

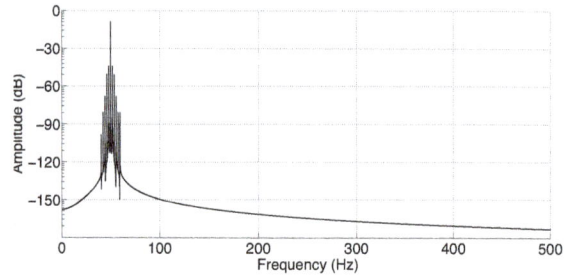

(a) Healthy case.

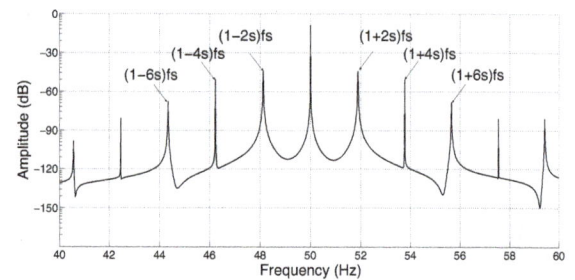

(b) Faulty case.

Fig. 3: Motor current's Park's vector representation.

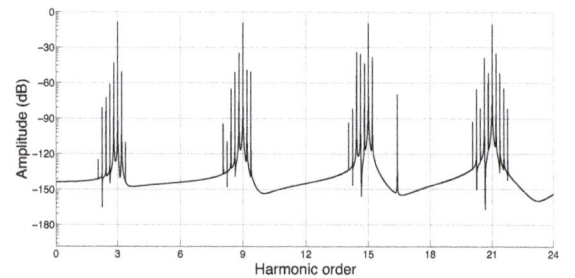

(a)

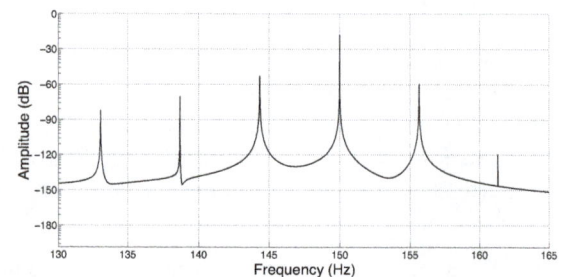

(b)

Fig. 5: (a) Simulation: Normalized FFT spectrum of machine line neutral voltage of IM with 3 broken rotor bars, (b) Zoom of line neutral voltage around the 3rd harmonic.

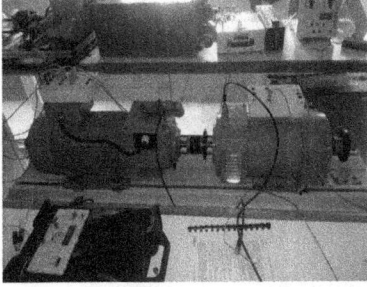

Fig. 6: Experimental setup.

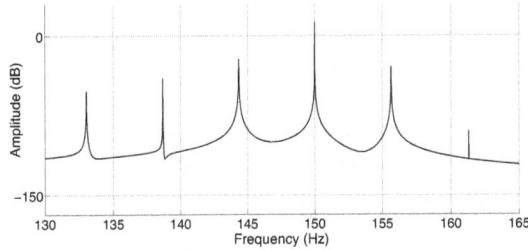

(a) Simulations results.

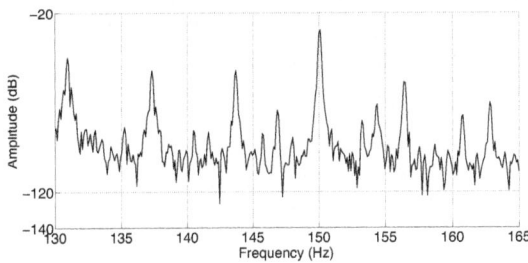

(b) Experimental results.

Fig. 7: Line neutral voltage comparison.

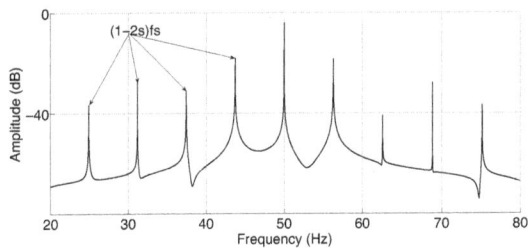

(a) Simulations results.

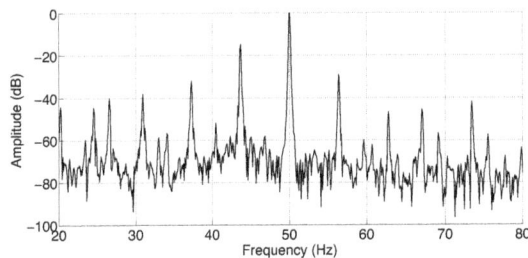

(b) Experimental results.

Fig. 8: Line current comparison.

6. Discussion

Broken rotor bars fault in induction machine lead to changes of the frequency component in the supplying neutral voltage and current of the motor, electromagnetic torque and power pulsation. The simulation results are shown in Fig. 1, Fig. 2, Fig. 3, Fig. 4, Fig. 5, Fig. 6, Fig. 7 and Fig. 8.

Figure 1 shows the simulation results of induction machine in healthy case. Figure 1(a): Stator current for no load normal condition, Fig. 1(b): Stator current when the motor is loaded, Fig. 1(c): Rotor currents, Fig. 1(d): Electromagnetic torque and Fig. 1(e): Power.

Figure 2 shows the simulation results in defect case, under one and three adjacent broken bars: Fig. 2(a): Stator current, Fig. 2(b): Rotor currents, Fig. 2(c): Zoom of rotor currents, Fig.1(d): Electromagnetic torque and Fig. 2(e): Power.

Figure 3 shows the motor current's Park's vector representation, Fig. 3(a): Healthy case, Fig. 3(b): Faulty case.

Figure 4(a) Simulation: Normalized FFT spectrum of machine line current of IM with 3 broken rotor bars and Fig. 4(b) show a zoom around the fundamental frequency.

Figure 5(a) shows the spectrum content of the line between neutrals ($\mathbf{V}_{nn}$), when a rotor dissymmetry is considered (constructional dissymmetry or broken rotor bar) under a balanced sinusoidal voltage supply. Equation (4) clarifies the frequencies of the additional harmonics, $f_h(3h, 0, \pm k)$, where $3h = 3, 9, 15, 21, 27, \ldots, \eta = 0$ and $k = 0, \pm 1, \pm 2, \pm 3, \pm 4 \ldots$. Figure 5(b) shows a zoom around the first-order harmonic $3h$, which gives $f_h(3, 0, \pm k) = 3 - (3h \pm 1)s + 2k_s$.

Figure 7 shows a Line neutral voltage comparison between the simulations results (Fig. 7(a) and experiment ones (Fig. 7(b)), we note a match between these results.

Figure 8 shows a line current comparison between the simulations results Fig. 8(a) and experiment ones Fig. 8(b), we note a match between these results. This model is helpful in quantifying the rotor slot harmonics under healthy as well as faulty condition.

7. Conclusion

In this paper, we presented a mathematical model and simulation of squirrel cage induction motor in healthy case and under defects of one and three adjacent broken bars. The particularity of the model is that it takes into account the line neutral voltage in the system equation, and it is based on a detailed modeling of the induction motors, represented by m stator phases and N_r rotor bars, which allows detecting and localization of the completely and partially broken rotor bars with multi indicators, without the need to change the model structure. The simulation results show the effectiveness of the model as it adapts with the presented problem and it shows a good match with the theoretical predictions. The accuracy of the simulation results is verified by the experimental results.

The results presented in this paper are promising and clear and thus the future research work should focus on the use of information collected from multiple sensors (indicators), such as current, voltage, torque, vibration, and temperature, to detect and identify motor faults.

References

[1] CHEN, S. and R. ZIVANOVIC. A novel high-resolution technique for induction machine broken bar detection. In: *Australasian Universities Power Engineering Conference*. Perth: IEEE, 2007, pp. 1–5. ISBN 978-0-646-49488-3. DOI: 10.1109/AUPEC.2007.4548040.

[2] CALIS, H. and A. CAKIR. Experimental study for sensorless broken bar detection in induction motors. *Energy Conversion and Management*. 2008, vol. 49, iss. 2, pp. 869–875. ISSN 0196-8904. DOI: 10.1016/j.enconman.2007.06.030.

[3] GODOY, W. F., I. N. DA SILVA, A. GOEDTEL, R. H. C. PALACIOS and T. D. LOPES. Application of intelligent tools to detect and classify broken rotor bars in three-phase induction motors fed by an inverter. *IET Electric Power Applications*. 2016, vol. 10, iss. 5, pp. 430–439. ISSN 1751-8679. DOI: 10.1049/iet-epa.2015.0469.

[4] KAZAKOV, J. and I. PALILOV. Research Related Electromechanical Processes in an Asynchronous Traction Motor - Asynchronous Generator with Common Shaft Based on Field Model. *Advances in Electrical and Electronic Engineering*. 2010, vol. 13, no. 5, pp. 442–446. ISSN 1804-3119. DOI: 10.15598/aeee.v13i5.1388.

[5] DAHI, K., S. ELHANI and S. GUEDIRA. Wound-rotor IM diagnosis method based on neutral voltage signal analysis. In: *40th Annual Conference of the IEEE Industrial Electronics Society*. Dallas: IEEE, 2014, pp. 965–971. ISBN 978-1-4799-4032-5. DOI: 10.1109/IECON.2014.7048618.

[6] MENACER, A., M.-S. NAIT-SAID, A. BENAKCHA and S. DRID. Stator current analysis of incipient fault into asynchronous motor rotor bars using Fourier fast transform. *Journal of Electrical Engineering*. 2004, vol. 55, no. 5–6, pp. 122–130. ISSN 1335-3632.

[7] JUNG, J.-H., J.-J. LEE and B.-H. KWON. Online Diagnosis of Induction Motors Using MCSA. *IEEE Transactions on Industrial Electronics*. 2006, vol. 53, iss. 6, pp. 1842–1852. ISSN 1557-9948. DOI: 10.1109/TIE.2006.885131.

[8] GU, F., T. WANG, A. ALWODAI, X. TIAN, Y. SHAO and A. D. BALL. A new method of accurate broken rotor bar diagnosis based on modulation signal bispectrum analysis of motor current signals. *Mechanical Systems and Signal Processing*. 2015, vol. 50, iss. 1, pp. 400–413. ISSN 0888-3270. DOI: 10.1016/j.ymssp.2014.05.017.

[9] KHEZZAR, A., M. E. K. OUMAAMAR and M. HADJAMI. Induction Motor Diagnosis Using Line Neutral Voltage Signatures. *IEEE Transactions on Industrial Electronics*. 2008, vol. 56, iss. 11, pp. 4581–4591. ISSN 1557-9948. DOI: 10.1109/TIE.2008.2010209.

[10] FILIPPETTI, F., A. BELLINI and G.-A. CAPOLINO. Condition monitoring and diagnosis of rotor faults in induction machines: State of art and future perspectives. In: *1st IEEE Workshop on Electrical Machines Design, Control and Diagnosis*. Paris: IEEE, 2013, pp. 196–209. ISBN 978-1-4673-5657-2. DOI: 10.1109/WEMDCD.2013.6525180.

[11] DAHI, K., S. ELHANI, S. GUEDIRA, L. SADIKI and I. OUACHTOUK. High-resolution spectral analysis method to identify rotor faults in WRIM using Neutral Voltage. In: *1st International Conference on Electrical and Information Technologies*. Marrakech: IEEE, 2015, pp. 82–87. ISBN 978-1-4799-7479-5. DOI: 10.1109/EITech.2015.7162988.

[12] OUMAAMAR, M. E. K., A. KHEZZAR, M. BOUCHERMA, H. RAZIK, R. N. ANDRIA-MALALA and L. BAGHLI. Neutral Voltage Analysis for Broken Rotor Bars Detection in Induction Motors Using Hilbert Transform Phase. In: *42nd Annual Meeting of the IEEE-Industry-Applications-Society*. New Orleans: IEEE, 2007, pp. 1940–1947. ISBN 978-1-4244-1259-4. DOI: 10.1109/07IAS.2007.295.

[13] OUACHTOUK, I., S. E. HANI, S. GUEDIRA, L. SADIKI and K. DAHI. Modeling of squirrel cage induction motor a view to detecting broken rotor bars faults. In: *1st International Conference on Electrical and Information Technologies*. Marrakech: IEEE, 2015, pp. 347–352. ISBN 978-1-4799-7479-5. DOI: 10.1109/EITech.2015.7163001.

[14] MUSTAFA, M. O., G. NIKOLAKOPOULOS and T. GUSTAFSSON. Broken bars fault diagnosis based on uncertainty bounds violation for three-phase induction motors. *International Transactions on Electrical Energy Systems*. 2013, vol. 25, iss. 2, pp. 402–409. ISSN 2050-7038. DOI: 10.1002/etep.1843.

[15] WALLACE, A. K. and A. WRIGHT. Novel Simulation of Cage Windings Based on Mesh Circuit Model. *IEEE Transactions on Power Apparatus and Systems*. 1974, vol. PAS-93, iss. 1, pp. 377–382. ISSN 0018-9510. DOI: 10.1109/TPAS.1974.293957.

[16] DELFORGE, C. and B. LEMAIRE-SEMAIL. Measuring the efficiency of decision making units. *IEEE Transactions on Magnetics*. 1995, vol. 31, iss. 3, pp. 2092–2095. ISSN 0018-9464. DOI: 10.1109/20.376457.

[17] FILIPPETTI, F., A. BELLINI and G.-A. CAPOLINO. Condition monitoring and diagnosis of rotor faults in induction machines: State of art and future perspectives. In: *1st IEEE Workshop on Electrical Machines Design, Control and Diagnosis*. Paris: IEEE, 2013, pp. 196–209. ISBN 978-1-4673-5657-2. DOI: 10.1109/WEMDCD.2013.6525180.

[18] SINGH, A., B. GRANT, R. DEFOUR, C. SHARMA and S. BAHADOORSINGH. A review of induction motor fault modeling. *Electric Power Systems Research*. 2016, vol. 133, iss. 1, pp. 191–197. ISSN 0378-7796. DOI: 10.1016/j.epsr.2015.12.017.

[19] DAHI, K., S. E. HANI, S. GUEDIRA and I. OUACHTOUK. A New Indicator for Rotor Asymmetries in Induction Machines Based on Line Neutral Voltage. *Research Journal of Applied Sciences, Engineering and Technology*. 2016, vol. 12, iss. 11, pp. 1136–1145. ISSN 2040-7467. DOI: 10.19026/rjaset.12.2855.

About Authors

Ilias OUACHTOUK was born in Foum Zguid-Tata, Morocco, in 1991. He received his M.Sc. degree in electrical engineering from the Mohammed V University in Rabat, Morocco, in 2014. Where he is currently working toward the Ph.D. degree, in the department of electrical engineering. Since 2015. His research interests include modeling and diagnosis of electrical drives, in particular, synchronous and asynchronous motors.

Soumia EL HANI has been Professor at the ENSET (Ecole Normale Superieure de l'Enseignement Technqique - Rabat, Morroco) since October 1992. She is a Research Engineer at the Mohammed V University in Rabat Morocco, in charge of the research team electromechanical, control and diagnosis, IEEE member, member of the research laboratory in electrical engineering at ENSET - Rabat. Author of several publications in the field of electrical engineering, including robust control systems, diagnosis and control systems of wind electric conversion. She has been general co-Chair the two editions of 'the International Conference on Electrical and Information Technologies', held in Marrakech, March 2015 and Tangier, May 2016 respectively.

Said GUEDIRA Professor at ENSMR (Ecole Nationale des Mines de Rabat Morroco) since 1983. Research Engineer and Director of the research laboratory CPS2I (Control, Potection et Surveillance des Installations Industrielles). Author of several publications in the field of Electrical Engineering, including diagnosis and control of mechatronic systems. His research interests are in the area of electromechanical engineering, Monitoring and Diagnosis of Mechatronic Systems.

Khalid DAHI was born in Errachidia, Morocco, in 1988. He received the M.Sc. degree in electrical engineering in 2012 from the Mohammed V University in Rabat- Morocco. Where he is currently working toward the Ph.D. degree in the department of electrical engineering. Since 2012, His research interests are related to electrical machines and drives, diagnostics of induction motors. His current activities include monitoring and diagnosis of induction machines in wind motor.

Lahbib SADIKI was born in Errachidia, Morocco, in 1987. He received the M.Sc. degree in electrical engineering in 2013 from the Mohammed V University in Rabat- Morocco. Where he is currently working toward the Ph.D. degree in the department of electrical engineering. Since 2014, His research interests are related to electrical machines and drives, diagnostics of induction motors. His current activities include monitoring and diagnosis of induction machines.

Appendix A

Tab. 2: Units for magnetic properties.

Symbol	Quantity	Numerical application
P	Power	5 kW
U	Power supply voltage	380 V
f	Current stator frequency	50 Hz
p	Number of pole pairs	2
N_r	Number of rotor bars	28
S_n	Number of stator slots	48
N_s	Number of turns in series per phase	80
e_a	Ring thickness	$0.28 \cdot 10^{-3}$ m
R_s	Resistance of a stator Phase	$1.5\ \Omega$
R_b	Resistance of a rotor bar	$96.9 \cdot 10^{-6}\ \Omega$
R_e	Resistance of a short circuit ring	$5 \cdot 10^{-6}\ \Omega$
L_b	Leakage inductance of a rotor bar	$0.28 \cdot 10^{-6}$ H
L_e	Leakage inductance of a ring of short circuit	$0.036 \cdot 10^{-6}$ H
j	Moment of inertia	0.0131 Kg·m^2
μ	Permeability	$4\pi \cdot 10^{-7} \dfrac{W_b}{\text{A} \cdot \text{m}}$

6

Speed Estimators Using Stator Resistance Adaption for Sensorless Induction Motor Drive

Hau Huu VO[1], Pavel BRANDSTETTER[2], Chau Si Thien DONG[1], Thinh Cong TRAN[1]

[1]Faculty of Electrical and Electronics Engineering, Ton Duc Thang University,
19 Nguyen Huu Tho, Dist. 7, Ho Chi Minh City, Vietnam
[2]Department of Electronics, Faculty of Electrical Engineering and Computer Science,
VSB–Technical University of Ostrava, 17. listopadu 15, 708 33 Ostrava, Czech Republic

vohuuhau@tdt.edu.vn, pavel.brandstetter@vsb.cz, dongsithienchau@tdt.edu.vn, trancongthinh@tdt.edu.vn

Abstract. The paper describes speed estimators for a speed sensorless induction motor drive with the direct torque and flux control. However, the accuracy of the direct torque control depends on the correct information of the stator resistance, because its value varies with working conditions of the induction motor. Hence, a stator resistance adaptation is necessary. Two techniques were developed for solving this problem: model reference adaptive system based scheme and artificial neural network based scheme. At first, the sensorless control structures of the induction motor drive were implemented in Matlab-Simulink environment. Then, a comparison is done by evaluating the rotor speed difference. The simulation results confirm that speed estimators and adaptation techniques are simple to simulate and experiment. By comparison of both speed estimators and both adaptation techniques, the current based model reference adaptive system estimator with the artificial neural network based adaptation technique gives higher accuracy of the speed estimation.

Keywords

Artificial neural network, induction motor drive, model reference adaptive system, sensorless control.

1. Introduction

The control and estimation of induction motor drives is almost an unbounded subject, and the technology has been developing very strong in last few decades. The induction motor drive with a cage type of machine has many applications in industry such as pumps and fans, paper and textile mills, subway and locomotive propulsions, electric and hybrid vehicles, machine tools and robotics, home appliances, heat pumps and air conditioners, rolling mills, wind generation systems. These applications often require adjustable speed and wide power range [1] and [2]. The control methods without speed encoder can be classified as follows:

- Methods without machine model: estimators with injection methods and estimators using artificial intelligence such as neural network [3].

- Methods with machine model [4], [5], [6] and [7]: open loop estimators, Model Reference Adaptive System (MRAS) and observers (such as extended Kalman filter, Luenberger observer, sliding mode observer). These methods are simple, but sensitive to variations of parameters of induction motor such as the vector control is very sensitive to variations in the rotor time constant [8].

2. Speed Estimators with Stator Resistance Adaptation

The Direct Torque and Flux control (DTC) has comparable performance with the vector control. In this control scheme, the torque and the stator flux are controlled by selecting voltage space vector of the inverter through a look-up table. The errors between the command torque and stator flux with estimated values will be processed by two hysteresis-band controllers: a flux controller and a torque controller.

The DTC technique has many advantages such as no feedback current control, no traditional Pulse Width

Modulation algorithm, and no vector transformation. However, the DTC is very sensitive to variation of stator resistance.

In general, methods of stator resistance estimation are similar to those utilized for rotor time constant (rotor resistance) estimation and include application of observers, extended Kalman filters, model reference adaptive systems, and artificial intelligence [9]. In this paper, two MRAS-based adaptation mechanisms with two MRAS speed estimators are implemented on MAT-LAB software: Proportional-Integral (PI) controller and Artificial Neural Network (ANN).

2.1. Reference Frame MRAS

The structure of a reference frame model reference adaptive system (RF-MRAS) with stator resistance adaptation for the rotor speed estimation is shown in Fig. 1. In the reference model, the stator voltages and stator currents are used for obtaining rotor flux vector components. In the adaptive model, these components of the estimated rotor flux vector can be gotten from stator currents together with the exact value of rotor speed.

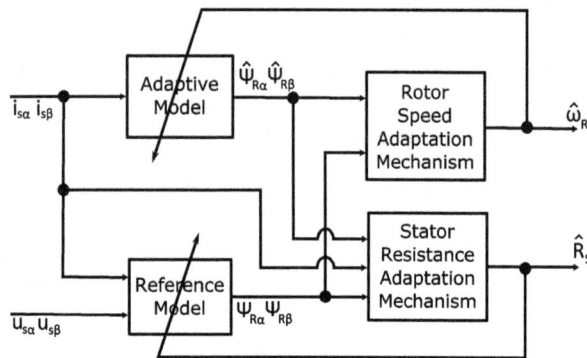

Fig. 1: Structure of RF-MRAS.

The correlative rotor flux vector components from the reference model and the adaptive model will be equal in ideal condition (such as the exact knowledge of stator resistance).

However, the accuracy of this method is decreased because of parameter variations (such as uncertainty of stator resistance or rotor time constant). Thus, for improving the performance of the RF-MRAS scheme, a stator resistance adaptation mechanism (SRAM) is added to update stator resistance.

The outputs of reference model and adaptive model are calculated according to the following equations:

$$\psi_{R\alpha} = \frac{L_r}{L_m} \int \left(u_{s\alpha} - \hat{R}_s i_{s\alpha} \right) dt - \sigma L_s i_{s\alpha},$$
$$\psi_{R\beta} = \frac{L_r}{L_m} \int \left(u_{s\beta} - \hat{R}_s i_{s\beta} \right) dt - \sigma L_s i_{s\beta}, \qquad (1)$$

$$\hat{\psi}_{R\alpha} = \int \left(\frac{L_m}{T_r} i_{s\alpha} - \frac{1}{T_r} \hat{\psi}_{R\alpha} - \hat{\omega}_r \hat{\psi}_{R\beta} \right) dt,$$
$$\hat{\psi}_{R\beta} = \int \left(\frac{L_m}{T_r} i_{s\beta} - \frac{1}{T_r} \hat{\psi}_{R\beta} + \hat{\omega}_r \hat{\psi}_{R\alpha} \right) dt. \qquad (2)$$

Using the Popov's criterion for hyperstability for a globally asymptotically stable system, the rotor speed adaptation mechanism uses the adaptive signal ξ to tune the rotor speed:

$$\xi = \hat{\psi}_{R\alpha} \psi_{R\beta} - \hat{\psi}_{R\beta} \psi_{R\alpha},$$
$$\hat{\omega}_r = K_P \xi + K_I \int_0^t \xi dt, \qquad (3)$$

where $K_P > 0$, $K_I > 0$.

2.2. Current Based MRAS

Figure 2 shows the structure of a current-based model reference adaptive system (CB-MRAS) with the stator resistance adaptation for the rotor speed estimation. This MRAS estimator uses stator currents as output quantities of the reference model [7].

The CB-MRAS scheme has one reference model (stator currents of induction motor) and two adaptive models (current model and current estimator).

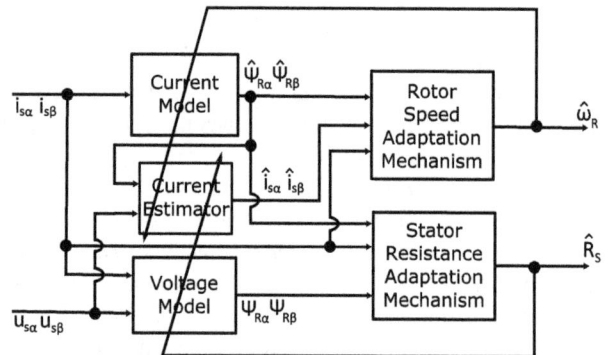

Fig. 2: Structure of CB-MRAS.

The outputs of current model are also described with Eq. (2); the outputs of the current estimator are calculated:

$$\hat{i}_{s\alpha} = \frac{1}{T_i} \int \left(K_1 u_{s\alpha} + K_2 \hat{\psi}_{R\alpha} + K_3 \hat{\omega}_R \hat{\psi}_{R\beta} - \hat{i}_{s\alpha} \right) dt,$$
$$\hat{i}_{s\beta} = \frac{1}{T_i} \int \left(K_1 u_{s\beta} + K_2 \hat{\psi}_{R\beta} - K_3 \hat{\omega}_R \hat{\psi}_{R\alpha} - \hat{i}_{s\beta} \right) dt, \qquad (4)$$

where:

$$K_1 = \frac{L_r}{C_1 L_m}, \quad K_2 = \frac{L_m}{C_1 (L_r \hat{R}_s T_r + L_m^2)},$$
$$K_3 = \frac{1}{C_1}, \quad T_i = \frac{L_s L_r - L_m^2}{C_1 L_m}, \quad C_1 = \frac{L_r \hat{R}_s}{L_m} + \frac{L_m}{T_r}. \qquad (5)$$

As in RF-MRAS scheme, stator resistance is also updated by a similar SRAM. Besides that, the voltage model is supplemented to obtain components of the rotor flux vector according to Eq. (1). The following equation is used by rotor speed adaptation mechanism for obtaining the rotor speed:

$$\xi = \left(i_{s\alpha} - \hat{i}_{s\alpha}\right)\hat{\psi}_{R\beta} - \left(i_{s\beta} - \hat{i}_{s\beta}\right)\hat{\psi}_{R\alpha},$$
$$\hat{\omega}_r = K_P\xi + K_I \int_0^t \xi dt, \tag{6}$$

where $K_{PRs} > 0$, $K_{IRs} > 0$.

2.3. MRAS-Based SRAMs

The inputs of the stator resistance adaptation mechanisms are outputs of reference and adaptive models for RF-MRAS (see Fig. 1), or current and voltage models for CB-MRAS (see Fig. 2). The adaptive signal, derived from the error between two models, is used to estimate stator resistance by PI adaptation mechanism:

$$\xi_{Rs} = \left(\psi_{R\alpha} - \hat{\psi}_{R\alpha}\right)i_{s\alpha} + \left(\psi_{R\beta} - \hat{\psi}_{R\beta}\right)i_{s\beta},$$
$$\hat{R}_s = K_{PRs}\xi_{Rs} + K_{IRs}\int_0^t \xi_{Rs}dt, \tag{7}$$

where $K_{PRs} > 0$, $K_{IRs} > 0$.

Equation (7) is discretized into the following equation with sampling period T_s:

$$\hat{R}_s(k) = \hat{R}_s(k-1) + K_{PRs}\xi_{Rs}(k)$$
$$+K_{IRs}T_s\frac{\xi_{Rs}(k) + \xi_{Rs}(k-1)}{2}. \tag{8}$$

Equation (9) is produced by generalizing Eq. (8):

$$\hat{R}_s(k) = f\left(\xi_{Rs}(k), \xi_{Rs}(k-1), \hat{R}_s(k-1)\right). \tag{9}$$

Based on Eq. (9), an ANN adaptation mechanism is proposed as in Fig. 3.

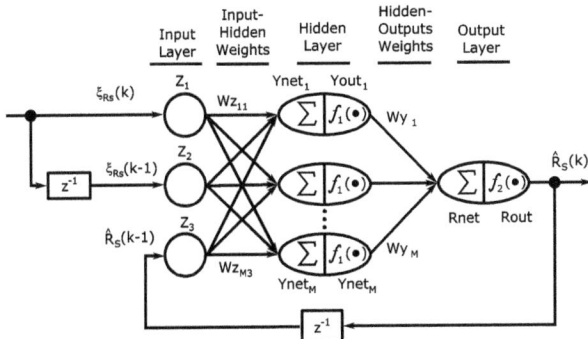

Fig. 3: Structure of ANN adaptation mechanism.

This feed-forward neural network has three layers: one input layer with three neurons (three parameters

of f), one hidden layer with M neuron(s) ($3M$ input-hidden weights and M hidden-output weights), and one output layer with one neuron (estimated value of stator resistance). The signal propagation process from the neurons of input layer to the neuron of output layer is done:

$$Ynet_j(k) = \sum_{i=1}^3 Wz_{ji}(k)Z_i(k),$$
$$Yout_j(k) = f_1(Ynet_j(k)),$$
$$Rnet(k) = \sum_{j=1}^M Wy_j(k)Yout_j(k), \tag{10}$$
$$\hat{R}_s(k) = Rout(k) = f_2(Rnet(k)),$$

where:

$$Z_1(k) = \xi_{Rs}(k),$$
$$Z_2(k) = \xi_{Rs}(k-1),$$
$$Z_3(k) = \hat{R}_s(k-1), \tag{11}$$
$$f_1(x) = \frac{2}{1+e^{-x}} - 1,$$
$$f_2(x) = \frac{R_s}{1+e^{-x}} + 0.5R_s.$$

Function f_2 is chosen with assumption that $\hat{R}_s \in (0.5R_s; 1.5R_s)$. The cost function E is minimized by updating weights of ANN according to Eq. (12):

$$E(k) = \frac{1}{2}\left[\bar{R}_s - \hat{R}_s(k)\right]^2,$$
$$Wy_j(k+1) = Wy_j(k) - \eta(k)\cdot\frac{\partial E(k)}{\partial Wy_j(k)}, \tag{12}$$
$$Wz_{ji}(k+1) = Wz_{ji}(k) - \eta(k)\cdot\frac{\partial E(k)}{\partial Wz_{ji}(k)}$$

where $\eta(k) > 0$ is the learning rate, and:

$$\frac{\partial E(k)}{\partial Wy_j(k)} = C_2C_3Yout_j(k),$$
$$\frac{\partial E(k)}{\partial Wz_{ji}(k)} = \frac{C_2C_3Wy_j(k)\left(1-Yout_j^2(k)\right)\cdot Z_i(k)}{2},$$
$$C_2 = -(\bar{R}_s - \hat{R}_s(k)), \tag{13}$$
$$C_3 = \frac{(Rout(k)-0.5R_s)(1.5R_s-Rout(k))}{R_s}.$$

The term C_2 is unknown. Assume that $\zeta_{Rs}(k) \approx \zeta_{Rs}(k-1)$, based on Eq. (8), this term can be replaced with the term $\zeta_{Rs}(k)$, and for faster convergence $C_2 = |\zeta_{Rs}(k)|\zeta_{Rs}(k)$. In this network, learning rate is also adapted by the following rule:

$$\Delta E(k) = Z_1^2(k) - Z_2^2(k),$$
$$\eta(k+1) = \begin{cases} \eta(k) + 0.005, & \text{if } \Delta E(k) < 0, \\ \eta(k) - 0.005\eta(k), & \text{if } \Delta E(k) > 0, \\ \eta(k), & \text{otherwise,} \end{cases} \tag{14}$$

where:

$$\Delta E(k) = Z_1^2(k) - Z_2^2(k). \tag{15}$$

3. Simulation Results

The designed control algorithms were simulated in sensorless control structure of the induction motor drive using Matlab-Simulink. Time courses of important quantities were obtained from the control structure with two MRAS speed estimators and two SRAMs at the jump of the load torque $T_L = 2$ Nm (see Fig. 4). The results in Fig. 5, Fig. 6, Fig. 7, Fig. 8, Fig. 9, Fig. 10 are done with $\bar{R}_S = 1.2R_S$, $M = 1$ and initial values of weights in ANN are random real number in range $(-10^{-3}, +10^{-3})$. Table 1 and Tab. 2 shows a comparison in maximum value of absolute of speed difference (MSD), steady-state error of estimated stator resistance (ESSR), maximum value of estimated stator resistance (MESR), root mean square error between reference torque and motor torque (RMSET).

Tab. 1: Comparison with RF-MRAS speed estimator.

Quantity	PI	ANN with M-neuron hidden layer				
		1	**2**	**3**	**4**	**5**
MSD [rpm]	3.76	3.41	3.41	3.4	3.4	3.4
ESSR [$\times 10^{-4} R_s$]	75	64.1	41.3	28.4	24.6	18.2
MESR [$\times R_s$]	1.28	1.48	1.49	1.49	1.49	1.49
RMSET [N_m]	0.27	0.27	0.27	0.27	0.26	0.27

Tab. 2: Comparison with CB-MRAS speed estimator.

Quantity	PI	ANN with M-neuron hidden layer				
		1	**2**	**3**	**4**	**5**
MSD [rpm]	1	0.54	0.97	0.77	0.74	0.85
ESSR [$\times 10^{-4} R_s$]	2	0.33	1.2	0.14	0.12	0.1
MESR [$\times R_s$]	1.28	1.48	1.49	1.49	1.49	1.5
RMSET [N_m]	0.26	0.26	0.26	0.26	0.26	0.26

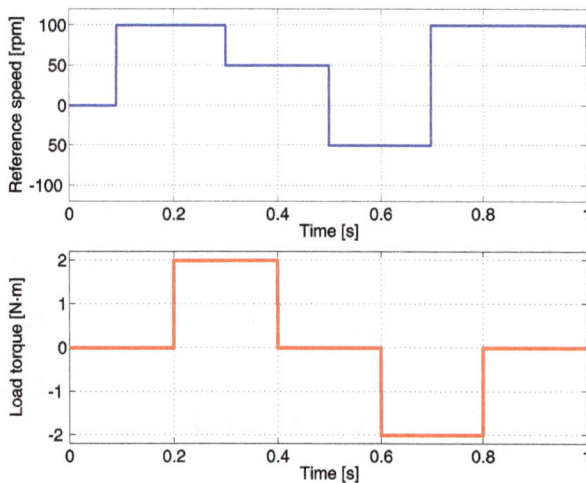

Fig. 4: Reference speed and load torque.

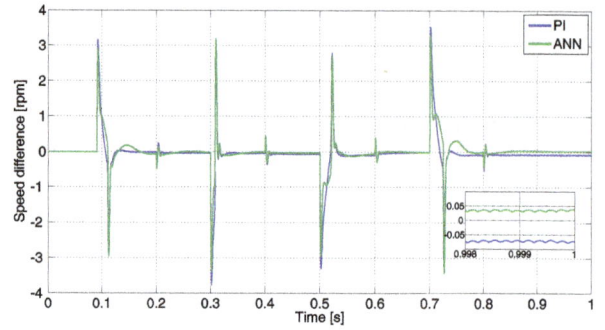

Fig. 5: RF-MRAS with PI (blue) and ANN (green) adaptation mechanisms, difference between real speed and estimated speed.

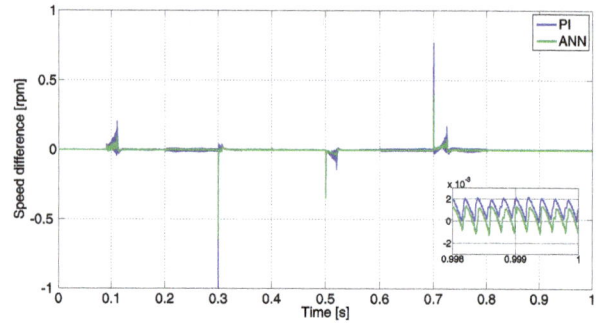

Fig. 6: CB-MRAS with PI (blue) and ANN (green) adaptation mechanisms, difference between real speed and estimated speed.

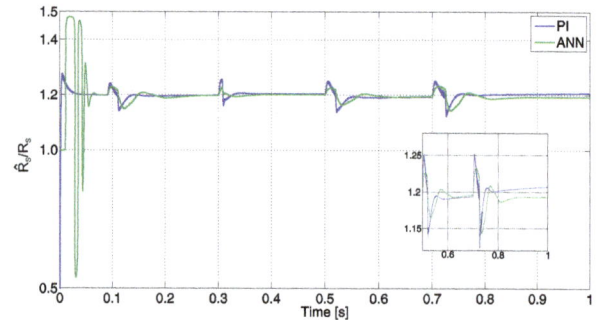

Fig. 7: RF-MRAS with PI (blue) and ANN (green) adaptation mechanisms, ratio between estimated value and known value of stator resistance.

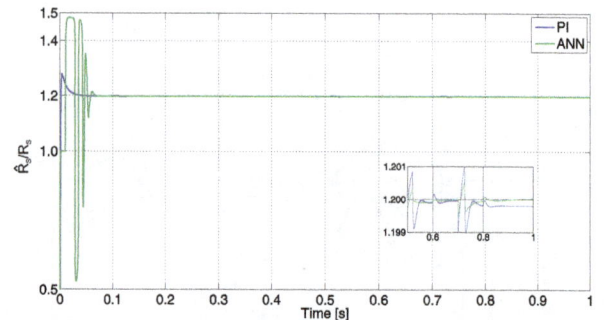

Fig. 8: CB-MRAS with PI (blue) and ANN (green) adaptation mechanisms, ratio between estimated value and known value of stator resistance speed.

The first rows of Tab. 1 and Tab. 2, Fig. 5 and Fig. 6 show that ANN mechanism gives smaller MSD than PI mechanism, MSD of CB-MRAS is smaller than that of RF-MRAS, and CB-MRAS with ANN mechanism (with only one-neuron in hidden layer) produces the smallest MSD. PI mechanism gives larger ESSR and smaller MESR (smaller overshoot) than ANN one does, and approximate-zero ESSR belongs to CB-MRAS with ANN mechanism as shown in Fig. 7 and Fig. 8 and the second and third rows of Tab. 1 and Tab. 2.

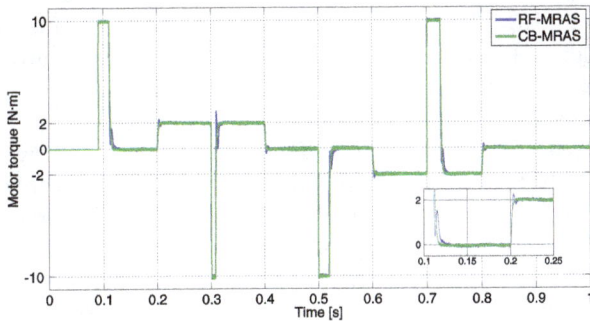

Fig. 9: RF-MRAS (blue) and CB-MRAS (green) with PI adaptation mechanism, motor torque.

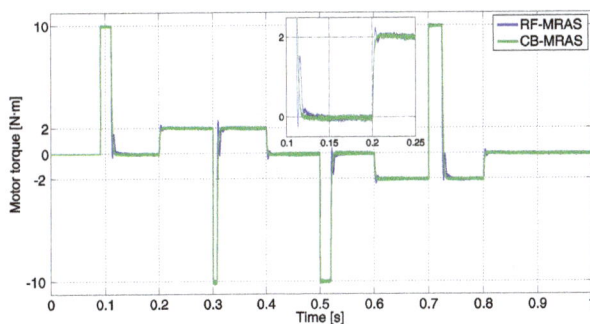

Fig. 10: RF-MRAS (blue) and CB-MRAS (green) with ANN adaptation mechanism, motor torque.

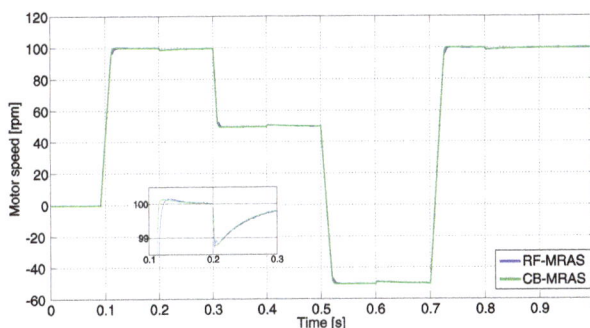

Fig. 11: RF-MRAS (blue) and CB-MRAS (green) with PI adaptation mechanism, motor speed.

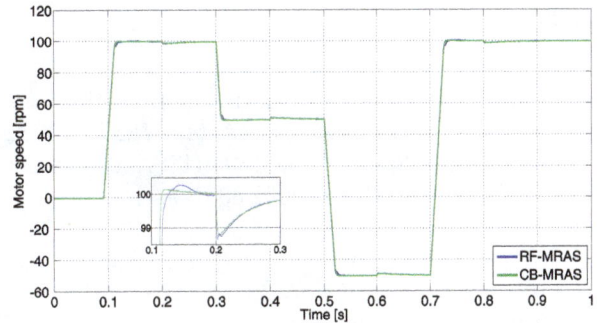

Fig. 12: RF-MRAS (blue) and CB-MRAS (green) with ANN adaptation mechanism, motor speed.

RMSET of both MRAS schemes with two SRAMs are almost equal (the fourth rows of Tab. 1 and Tab. 2), but motor torque and motor speed time courses of CB-MRAS are smoother than those of RF-MRAS (Fig. 9, Fig. 10, Fig. 11, Fig. 12).

4. Conclusion

The MRAS-based speed sensorless induction motor drive with the direct torque control was presented in the paper. The DTC drive with two MRAS estimators together with two MRAS-based SRAMs gave good dynamic responses and the estimation of the rotor speed was good in steady state and also in transient state. The CB-MRAS scheme gave higher accuracy of the rotor speed estimation than the RF-MRAS scheme, and the ANN mechanism gave higher accuracy of the stator resistance adaptation than the PI mechanism. However, response of the stator resistance estimation for ANN adaptation mechanism at beginning times is not good and should be improved. The CB-MRAS estimator with both MRAS-based SRAMs could be used for speed estimation with the uncertainty of stator resistance without speed encoder in the control system with digital signal processor. The simple MRAS-based ANN adaptation mechanism with only one-neuron hidden layer can be applied for estimating other parameters of induction motor.

Acknowledgment

The paper was supported by the projects: Center for Intelligent Drives and Advanced Machine Control (CIDAM) project, reg. no. TE02000103 funded by the Technology Agency of the Czech Republic, project reg. no. SP2016/83 funded by the Student Grant Competition of VSB–Technical University of Ostrava.

References

[1] HOLTZ, J. Sensorless Control of Induction Motor Drives. *Proceedings of the IEEE*. 2002, vol. 90, iss. 8, pp. 1359–1394. ISSN 0018-9219. DOI: 10.1109/JPROC.2002.800726.

[2] VAS, P. *Sensorless Vector and Direct Torque Control*. Oxford University Press, 1998. ISBN 978-0198564652.

[3] GADOUE, S. M., D. GIAOURIS and J. W. FINCH. Sensorless Control of Induction Motor Drives at Very Low and Zero Speeds Using Neural Network Flux Observers. *IEEE Transactions on Industrial Electronics*. 2009, vol. 56, iss. 8, pp. 3029–3039. ISSN 0278-0046. DOI: 10.1109/TIE.2009.2024665.

[4] SUTNAR, Z., Z. PEROUTKA and M. RODIC. Comparison of sliding mode observer and extended Kalman filter for sensorless DTC controlled induction motor drive. In: *14th International Power Electronics and Motion Control Conference (EPE/PEMC)*. Ohrid: IEEE, 2010, pp. 55–62. ISBN 978-1-4244-7856-9. DOI: 10.1109/EPE PEMC.2010.5606844.

[5] LASCU, C., I. BOLDEA and F. BLAABJERG. Comparative Study of Adaptive and Inherently Sensorless Observers for Variable-Speed Induction-Motor Drives. *IEEE Transactions on Industrial Electronics*. 2006, vol. 53, iss. 1, pp. 57–65. ISSN 0278-0046. DOI: 10.1109/TIE.2005.862314.

[6] GACHO, J. and M. ZALMAN. IM Based Speed Servodrive with Luenberger Observer. *Journal of Electrical Engineering*. 2010, vol. 61, iss. 3, pp. 149–156. ISSN 1335-3632. DOI: 10.2478/v10187-011-0021-8.

[7] ORLOWSKA-KOWALSKA, T. and M. DYBKOWSKI. Stator-Current-Based MRAS Estimator for a Wide Range Speed-Sensorless Induction-Motor Drive. *IEEE Transactions on Industrial Electronics*. 2010, vol. 57, iss. 4, pp. 1296–1308. ISSN 0278-0046. DOI: 10.1109/TIE.2009.2031134.

[8] BRANDSTETTER, P., P. CHLEBIS, P. PALACKY and O. SKUTA. Application of RBF Network in Rotor Time Constant Adaptation. *Electronics and Electrical Engineering*. 2011, vol. 113, iss. 7, pp. 21–26. ISSN 1392-1215. DOI: 10.5755/j01.eee.113.7.606.

[9] TOLIYAT, H. A., E. LEVI and M. RAINA. A Review of RFO Induction Motor Parameter Estimation Techniques. *IEEE Transactions on Energy Conversion*. 2003, vol. 18, iss. 2, pp. 271–283. ISSN 0885-8969. DOI: 10.1109/TEC.2003.811719.

About Authors

Hau Huu VO was born in Binh Thuan, Vietnam. He received his M.Sc. degree in Automation Engineering from Ho Chi Minh City of Technology, Vietnam in 2009. His research interests include robotics, control theory and modern control methods of electrical drives.

Pavel BRANDSTETTER was born in Ostrava, Czech Republic. He received his M.Sc. and Ph.D. degrees in Electrical Engineering from Brno University of Technology, Czech Republic, in 1979 and 1987, respectively. He is currently a full professor in electrical machines, apparatus and drives at Department of Electronics, VSB–Technical University of Ostrava. His current research interests are modern control methods of electrical drives, for example sensorless control of AC drives, applications of soft computing methods in control of AC drives.

Chau Si Thien DONG was born in Ho Chi Minh City, Vietnam. She received her M.Sc. degree in Automatic Control from Ho Chi Minh City of Technology, Vietnam in 2004. Her research interests include nonlinear control, adaptive control, robust control, neural network and modern control methods of electrical drives.

Thinh Cong TRAN was born in Da Nang City, Vietnam. He received his M.Sc. degree in Electrical and Electronic Engineering from Ho Chi Minh City of Technology, Vietnam in 1998. His research interests include microcontroller systems and modern control methods of electrical drives.

Appendix: Induction Motor Drive Quantities and Parameters

- $u_{s\alpha}, u_{R\beta}, i_{s\alpha}, i_{s\beta}, \hat{i}_{s\alpha}, \hat{i}_{s\beta}, \psi_{R\alpha}, \psi_{R\beta}, \hat{\psi}_{R\alpha}, \hat{\psi}_{R\beta}$: α, β components of stator voltage vector, stator current vector, estimated stator current vector, rotor flux vector, estimated rotor flux vector,

- $\omega_r, \hat{\omega}_r, T_e, T_L$: rotor speed, estimated rotor speed, motor torque, load torque,

- R_s: stator resistance value ($R_s = 1.115\ \Omega$),

- R_r: rotor resistance ($R_r = 1.083\ \Omega$),

- $\bar{R}_s, R_s$: real value and estimated value of stator resistance,

- L_m, L_s, L_r: magnetizing inductance ($L_m = 0.2037$ H), stator inductance ($L_s = 0.2097$ H), rotor inductance ($L_r = 0.2097$ H),

- T_r, J, σ, p: rotor time constant ($T_r = 0.1936$ s), moment of inertia ($J = 0.02$ kg·m^2), total leakage constant ($\sigma = 0.0562$), number of pole pairs ($p = 2$),

- ζ, K_P, K_I: adaptive signal, proportional coefficient ($K_P = 2000$), integral coefficient ($K_I = 1000000$) of MRAS speed estimators,

- ζ_{Rs}, K_{PRs}, K_{IRs}: adaptive signal, proportional coefficient ($K_{PRs} = 10$), integral coefficient ($K_{IRs} = 1000$) of stator resistance adaptation mechanisms.

7

Improvement in Performance of Cascaded Multilevel Inverter Using Triangular and Trapezoidal Triangular Multi Carrier SVPWM

Lokeshwar Reddy CHINTALA[1]*, Satish Kumar PEDDAPELLI*[2]*, Sushama MALAJI*[3]

[1]Department of Electrical and Electronics Engineering, CVR College of Engineering,
Ibrahimpatnam, Hyderabad 501 510, Telangana, India
[2]Department of Electrical Engineering, University College of Engineering, Osmania University,
Hyderabad 500 007, Telangana, India
[3]Department of Electrical and Electronics Engineering, College of Engineering, Jawaharlal Nehru
Technological University Hyderabad, Kukatpally, Hyderabad 500 085, Telangana, India

reddy.lokeshwar@gmail.com, satish_8020@yahoo.co.in, m73sushama@yahoo.com

Abstract. *In this paper a new Trapezoidal Triangular Multi Carrier SVPWM (TTMC-SVPWM) technique has been proposed for a cascaded H bridge multilevel inverter. This technique has been implemented for a 11-level cascaded H-bridge multilevel inverter to evaluate the performance parameters and also to compare it with other types of carrier based PWM techniques such as Phase Disposition (PD), Phase Opposition Disposition (POD), Alternative Phase Opposition Disposition (APOD) and Phase Shifted Carrier (PSC) PWM techniques. The simulation has been carried out for 11-level H-bridge multilevel inverter using MATLAB/Simulink. The detailed analysis of the results has been presented and studied in terms of fundamental component of output voltage and THD.*

Keywords

APODSVPWM, cascaded H-bridge, multilevel inverter, PODSVPWM, PSCSVPWM, TTMC-PDSVPWM.

1. Introduction

In recent years much work has been carried out on Multilevel Inverters (MLI) to reduce harmonics in the output current and voltage waveforms. These harmonics result in extra power loss. All undesired operating characteristics exhibited by converters can be overcome in multi-level converters using a proper PWM technique. Multilevel inverters are used more popularly for

generating higher voltage levels with reduced harmonics. This topic has been discussed in detail in [1], [2], [3], [4] and [5]. Each converter of MLI operates at a low switching frequency, thus reducing the stress on semiconductor devices, and the power loss due to switching is reduced [6] and [7]. Increase in number of levels in the inverter stage leads to the generation of staircase voltage waveform, which has a reduced harmonic distortion. On the other side higher number of levels increase the complexity in inverter control and present voltage inequality problems. Multilevel inverters are found to be suitable for static VAR compensators, active power filters and motor drive applications. Common mode voltages are the major issues in the MLI when they are interfaced with drives and they can be overcome with sophisticated modulation methods [8].

Multilevel inverters are classified into [9]:

- diode clamped or neutral point clamped,
- capacitor clamped or flying capacitor,
- cascaded multilevel inverters.

Due to modularity and simplicity in control Cascaded H-Bridge (CHB) MLI are used broadly in practical applications. Number of output phase voltage levels in cascaded inverter are $2n + 1$, where n is the number of H bridges used in one phase.

Several SPWM techniques like PDSPWM, PODSPWM and APODSPWM techniques have already been implemented on MLI. These techniques have high THD and low magnitude in output voltage. These can be improved by using the proposed techniques.

In this paper various carrier pulse width modulation methods are introduced and compared. In addition to this a new modulation method named Trapezoidal Triangular Multi Carrier (TTMC) SVPWM is introduced and implemented and compared with other methods. This modulation method gives higher output voltage and better harmonic distortion.

2. Multi Carrier SPWM Techniques

By implementing modulation technique, low frequency voltage harmonics are removed perfectly. This modulation technique produces nearly perfect sinusoidal waveforms, with lower THD. A very wide spread method in industrial applications is the classic carrier-based Sinusoidal PWM (SPWM) that uses the carrier shifting technique to condense the harmonics in voltages generated by inverters. The carriers used in multilevel inverter may be shifted vertically or horizontally.

The vertically shifted carrier scheme can be easily realizable on any digital controller. This scheme comes with three different techniques:

- All carrier signals are in Phase (Phase Disposition (PD)).

- Half of the carrier signals above are in same phase and half below carriers are in same phase but the phase difference between these two half's is 180° (Phase Opposition Disposition (POD)).

- All carriers are alternatvely in opposition (Alternate Phase Opposition Disposition (APOD)).

If the carriers are shifted horizontally corresponding PWM is Phase Shifted Carrier PWM (PSCPWM). The number of carriers required are $(m-1)$ for an m level inverter in all the variants.

In the APOD, the sideband harmonics corresponding to first set are centered around the carrier frequency. In the APOD and the POD, harmonics will not exist at pulse number m_f, due to odd symmetry of their PWM waveforms. The APOD and the POD strategies provide similar performance for three level converters [10]. In PD, the triplen harmonics of voltage will be removed because the waveforms are asymmetric and thus harmonics at m_f are removed if m_f is chosen as a multiple of three. So the PD is more expedient due to minute values of other harmonics. The PD strategy is now well recognised for attaining the lowest line-to-line harmonic voltage distortion.

The PSCSVPWM technique results in the termination of all carrier and connected sideband harmonics

up to $2N_c^{th}$ carrier group, where N_c is the number of H-bridges in each phase. Phase Shifted Carrier PWM (PSCPWM) is the common PWM for cascaded MLI. The switching transitions for PSCPWM are $2N$ times the number of switching transitions for APOD.

2.1. Modulation Index

The modulation index is the ratio of peak magnitudes of the modulating signal V_m and the carrier signal V_c.

$$m = \frac{V_m}{V_c} . \tag{1}$$

The modulation index in SPWM technique for cascaded multilevel inverter configuration is given by:

$$m = \frac{V_m}{(N-1)V_c}, \tag{2}$$

where N is number of levels. For undermodulation $0 < m < 1$. For overmodulation $m > 1$.

Generally, overmodulation is not desired because of the presence of the lower frequency harmonics in the output voltage and subsequent distortion in the load current.

2.2. Frequency Modulation Ratio or Pulse Number (m_f)

It is the ratio of frequency of the triangular carrier signal f_c to the frequency of sinusoidal reference signal f_s. It controls harmonics in the output voltage.

$$m_f = \frac{f_c}{f_s}. \tag{3}$$

1) For an Odd Integer Value of m_f:

To reduce the harmonics in output voltage, carrier signal would be synchronized with the reference signal. The carrier signal frequency f_c is an integer multiple of the reference sinusoidal signal frequency f_s, that is the pulse number m_f must be an exact integer. Sub harmonics will exist at the inverter output voltage, if m_f is not an integer [11]. The output voltage signal generated by implementing of SPWM technique has harmonics of several orders in the phase voltage waveform. The leading harmonics are of fundamental and other order of m_f and $m_f \pm 2$. Thus m_f would be an odd integer to diminish even harmonics. If m_f is not odd, DC component may exist and even harmonics will be present at output voltage [11].

2)   For m_f as a Multiple of 3:

The triplen harmonics of three-phase PWM inverter will be reduced by selecting m_f as multiple of 3. Selecting a multiple of 3 is also expedient to use the same triangular waveform as the carrier in all three phases, leading to some simplification in hardware control and implementation [12].

3)   For High m_f:

The PWM technique pushes the harmonics into the high frequency range nearby the carrier frequency and its multiples. Harmonic content at inverter output is reduced with larger number of pulses that is with high value of m_f or f_c. But, a high carrier frequency results in larger number of switchings per cycle and therefore the power loss increases.

In few circumstances the ratio of carrier signal frequencies and modulating signal frequencies cannot be very high but the pole voltage has a fundamental frequency component in-phase and proportionate to the modulating signal. The vital benefit of having very high carrier frequency, in contrast to the modulating wave frequency, is that the useful fundamental frequency component of pole voltage and unwanted harmonics are far apart on the frequency spectrum. We can virtually filter away the harmonic voltages without attenuating the magnitude of the fundamental frequency component by choosing a proper low pass filter. If the harmonics are of high frequencies, then the required filter size is to be small.

In AC motor drive application, the intrinsic low pass filtering characteristics of the motor load are enough to reasonably block the harmonic current flow to the load. In such cases requirement of filter may not arise. The switches used in high power applications can be switched on only at sub kilohertz frequency and hence the carrier frequency will not be high. The switching frequency losses should also be considered before deciding the carrier frequency of the sine-PWM inverter. The switching frequencies in this case are 1050 Hz ($m_f = 21$) and 1350 Hz ($m_f = 27$).

Quarter and half wave symmetry safeguards that even harmonics will not exist in the output voltage spectrum. This can be realised by choosing m_f odd. Significant even harmonic which is removed, is the DC component. The harmonic components below the fundamental frequency known as sub-harmonics will not exist. For different power system applications, the switching frequencies typically range from 2 kHz to 15 kHz [11].

3.   Generalized TTMC - Space Vector PWM for Cascaded Multilevel Inverter

In conventional SVPWM for multilevel inverters to find the switching time duration, for different inverter vectors, the mapping of the outer sectors to an inner sub hexagon sector is to be done. The switching inverter vectors corresponding to the concrete sectors are switched and the time periods premeditated from the mapped inner sectors. Implementing such a scheme in multilevel inverters will be very difficult, because higher number of sectors and inverter vectors are present. And in this method the computation time is increased for real time application.

In carrier based PWM scheme a proper offset voltage is added to sinusoidal references before comparing with carrier waves, to attain the performance of a SVPWM [13]. The offset voltage calculation is based on the modulus function depending on the DC link voltage, the number of levels and the phase voltage amplitudes.

One more modulation scheme is offered where sinusoidal reference phase voltages are added with common mode voltage of suitable magnitude all through the duration [14] and [15]. Addition of common mode voltage will not give SVPWM like performance, because middle inverter vectors will not be centered in a sampling interval [16]. Another modulation technique is offered in [17], where a fixed common mode voltage is added to the reference phase voltage all through the modulation range.

A simplified method is presented, where correct offset times are determined for centering the time durations of middle inverter vectors in a sampling interval. A procedure is given in [18] and [19] for finding the maximum probable peak amplitude of the fundamental phase voltage in linear modulation. The following equations are used to calculate offset time T_{offset}.

$$T_a = \frac{V_a \cdot T_s}{V_{dc}}, \tag{4}$$

$$T_b = \frac{V_b \cdot T_s}{V_{dc}}, \tag{5}$$

$$T_c = \frac{V_c \cdot T_s}{V_{dc}}, \tag{6}$$

where T_a, T_b and T_c are the time periods of imaginary switching, proportional to the instantaneous values of the reference phase voltages V_a, V_b and V_c. T_s is the sampling time period.

$$T_{offset} = \frac{T_0}{2} - T_{\min}, \tag{7}$$

(a) PDSVPWM.

(b) PODSVPWM.

(c) APODSVPWM.

(d) PSCSVPWM.

(e) Trapezoidal triangular PDSVPWM.

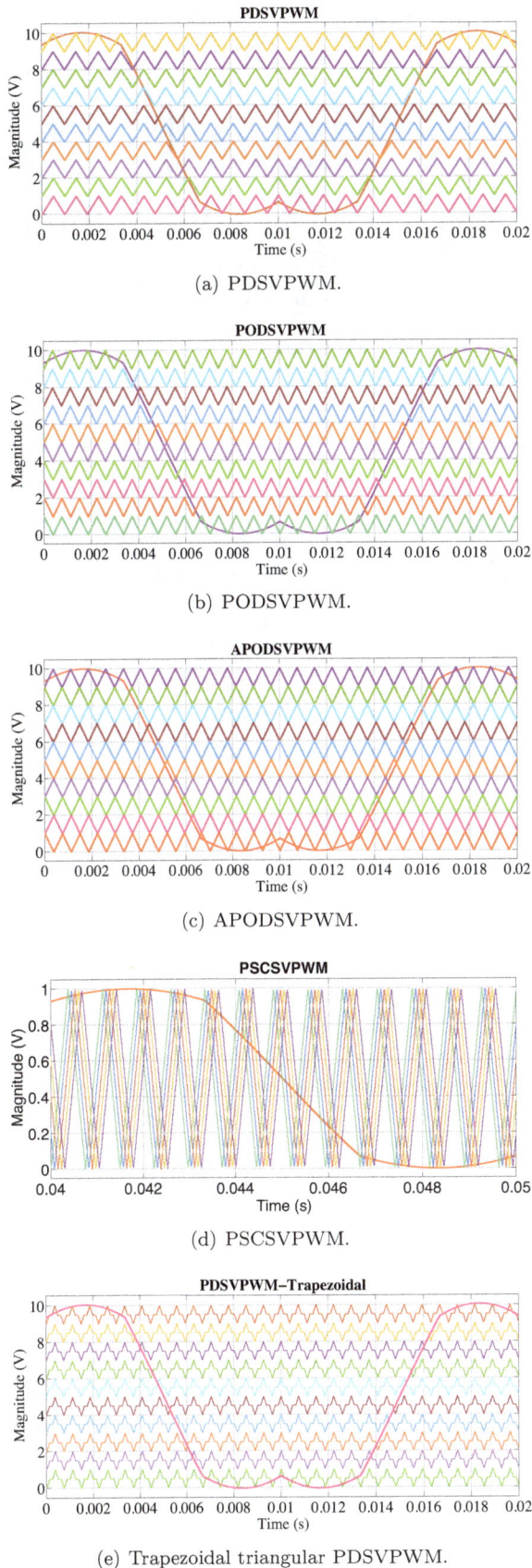

Fig. 1: Different modulation techniques.

$$T_0 = T_s - T_{effect}, \qquad (8)$$

$$T_{effect} = T_{\max} - T_{\min}, \qquad (9)$$

where $T_{\max}$ is the maximum magnitude of the three reference phase voltages, in a sampling interval and $T_{\min}$ is the minimum magnitude of the three reference phase voltages, in a sampling interval.

The inverter switching vectors are centered in a sampling interval by the addition of offset voltage to the reference phase voltages that equates the performance of SPWM technique with the SVPWM technique [20]. Figure 1 shows the different modulation techniques by the comparison of modulating wave generated in generalized SVPWM with triangular wave and trapezoidal triangular wave.

This proposed SVPWM signal generation does not involve look up table, sector identification, angle information and voltage space vector amplitude measurement for switching vector determination required in the conventional multilevel SVPWM technique. This scheme is more effective when compared with conventional multilevel SVPWM technique.

4. Simulation Results and Discussion

Simulations are carried out using Matlab/Simulink environment for 11-level cascaded H-bridge MLI by implementing proposed TTMC-SVPWM and TTMC-SPWM techniques. A 3 phase induction motor is con-

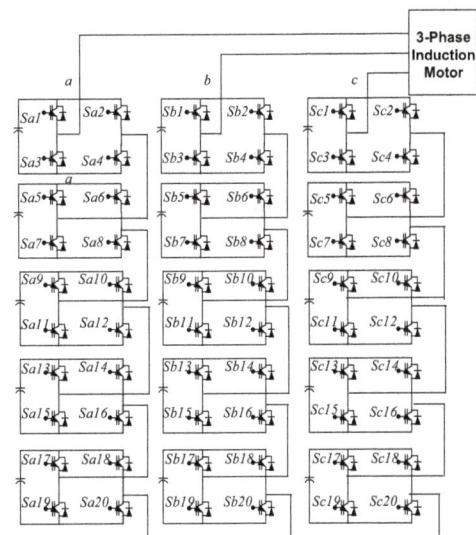

Fig. 2: 11-level cascaded H-bridge multilevel inverter.

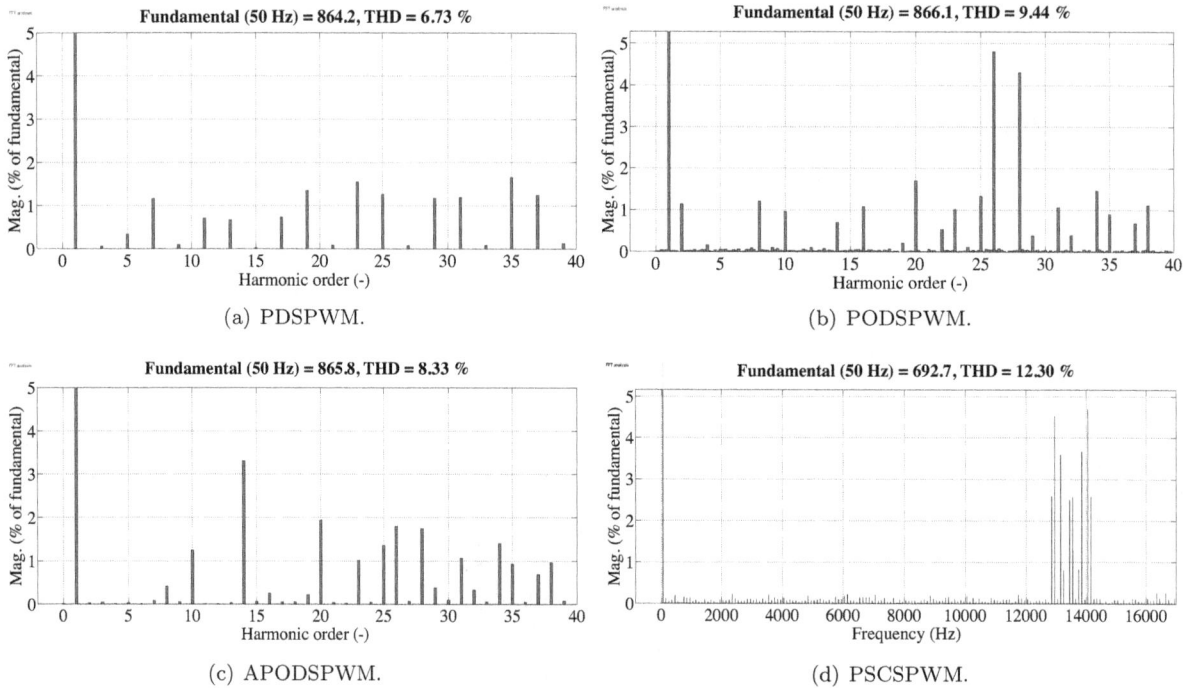

(a) PDSPWM.

(b) PODSPWM.

(c) APODSPWM.

(d) PSCSPWM.

Fig. 3: Harmonic analysis for different SPWM Techniques with triangular carrier and $m_f = 27$.

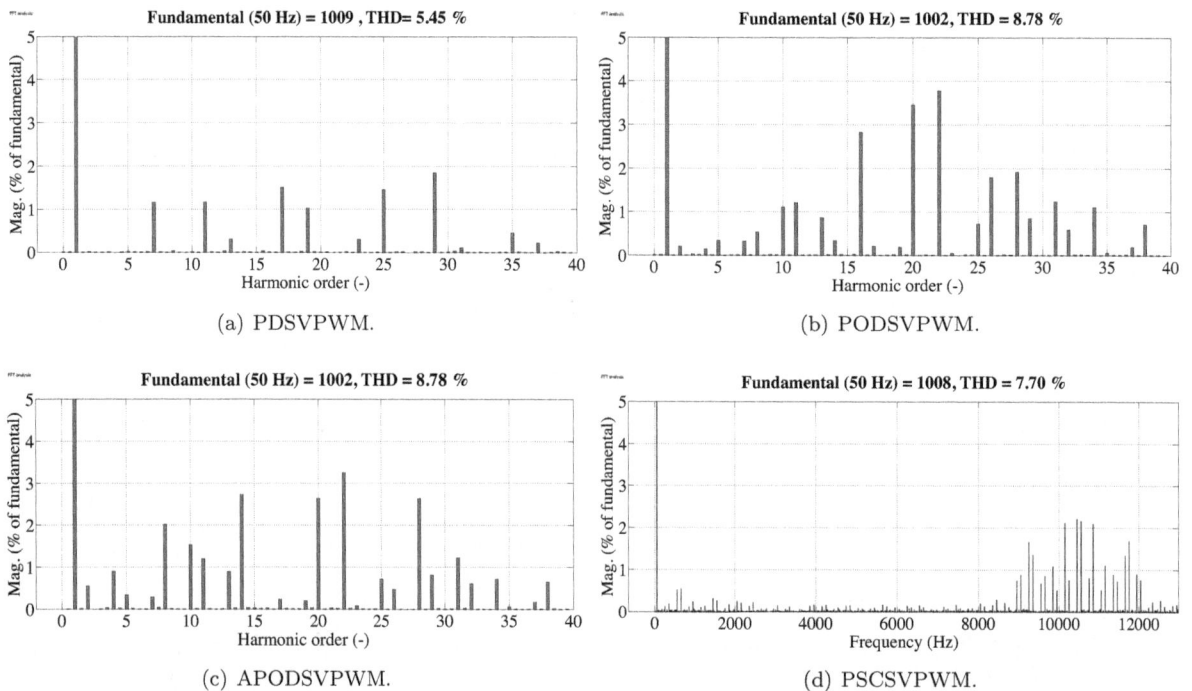

(a) PDSVPWM.

(b) PODSVPWM.

(c) APODSVPWM.

(d) PSCSVPWM.

Fig. 4: Harmonic analysis for different SVPWM techniques with triangular carrier and $m_f = 21$.

sidered as load for this scheme. The separate DC voltages sources are set to 100 V for each H-bridge and the switching frequencies are 1050 Hz and 1350 Hz considered for triangular and trapezoidal triangular carrier waves. The simulated 11-level cascaded H-bridge MLI connected to induction motor load is shown in Fig. 2.

Each phase consists of five H-bridges to generate 11-levels in phase voltage.

Figure 3 shows the harmonic analysis of inverter output line voltage for different SPWM techniques with triangular carrier having m_f of 27. The fundamental component magnitude in PDSVPWM, PODSVPWM

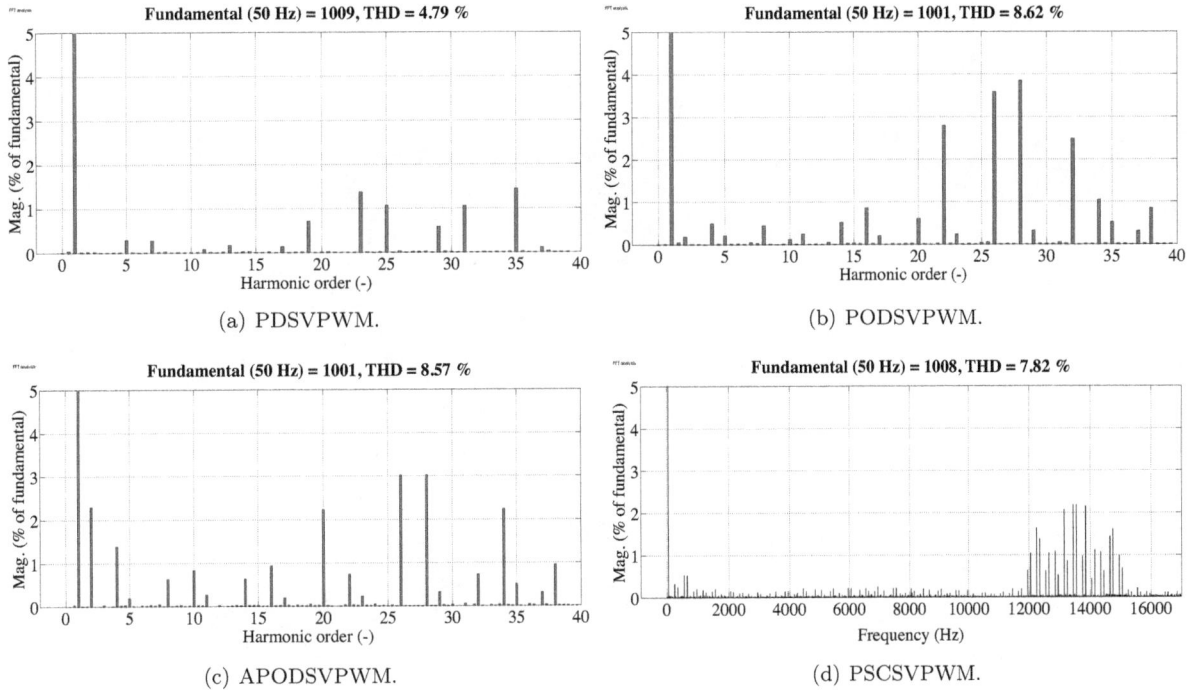

Fig. 5: Harmonic analysis for different SVPWM techniques with triangular carrier and $m_f = 27$.

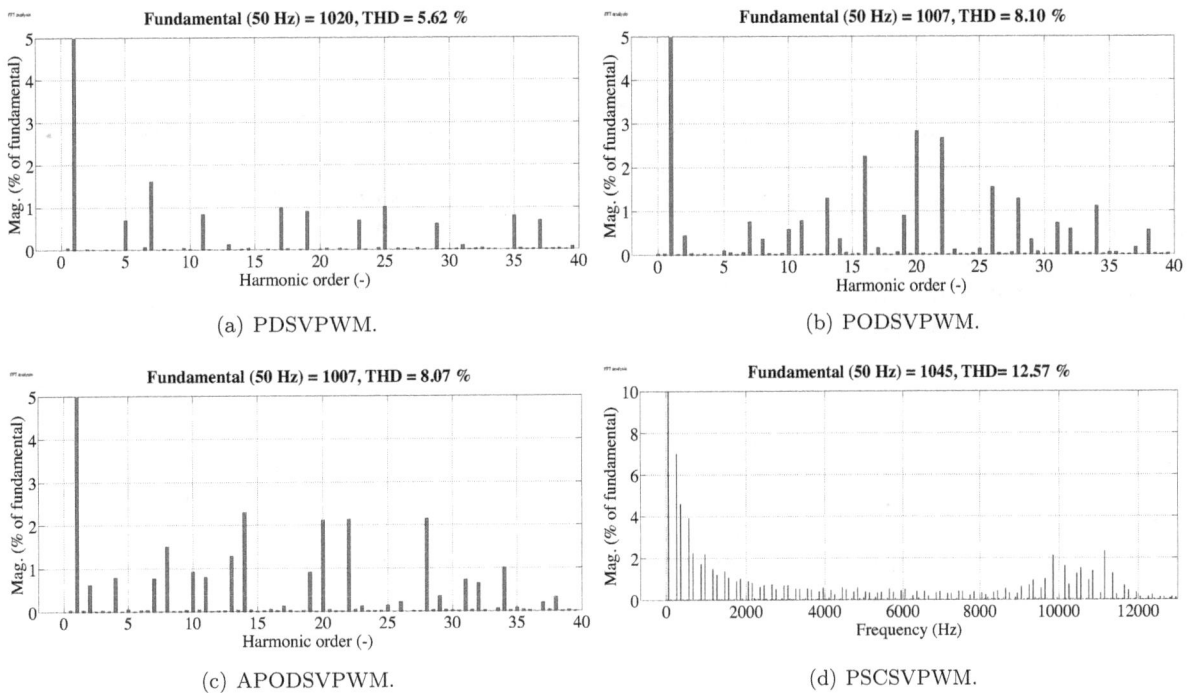

Fig. 6: Harmonic analysis for different SVPWM techniques with trapezoidal carrier and $m_f = 21$.

and APODSVPWM is the same and it is more than PSCPWM technique. In PSCSPWM technique no harmonics present up to $2N_c^{th}$ carrier group that is up to 13.5 kHz for carrier frequency of 1350 Hz. The PDSPWM technique gives better harmonic performance in comparison to the other SPWM techniques.

Figure 4 and Fig. 5 show the harmonic analysis of inverter output line voltage for different SVPWM techniques with the triangular carrier wave having m_f of 21 and 27. The fundamental component of output voltage in all the PWM techniques is almost the same but 15 % more than all the SPWM techniques except PSCPWM. The performance of PODSVPWM

(a) PDSVPWM.

(b) PODSVPWM.

(c) APODSVPWM.

(d) PSCSVPWM.

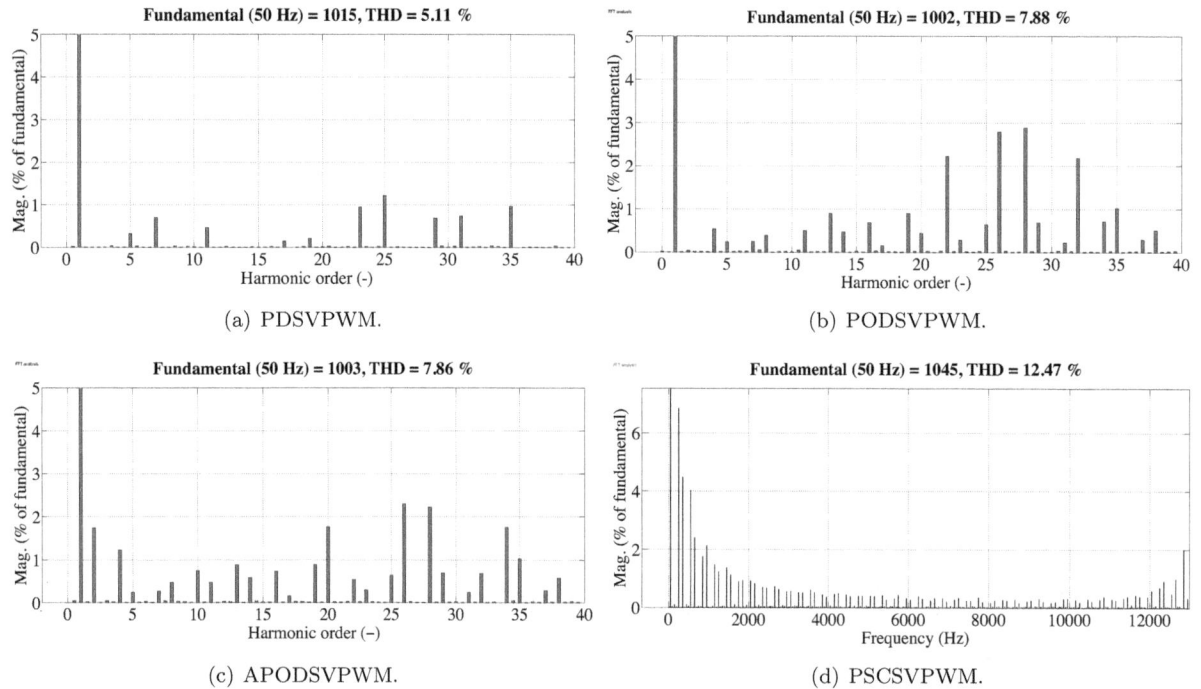

Fig. 7: Harmonic analysis for different SVPWM techniques with trapezoidal carrier and $m_f = 27$.

and APODSVPWM techniques is almost the same in terms of THD, but PSCSVPWM performance is better than the performance of these two methods. The PDSVPWM technique gives 5.45 % of THD, which gives better performance with respect to all other techniques. If the carrier frequency increases, the THD in PDSVPWM, PODSVPWM and APODSVPWM techniques reduces, whereas no change is observed in the magnitude of fundamental component. In PSCPWM technique the harmonics are shifted towards higher frequency.

Figure 6 and Fig. 7 show the harmonic analysis of output line voltages, when the inverter is controlled by different SVPWM techniques with trapezoidal triangular carrier waves having $m_f = 21$ and $m_f = 27$. In this SVPWM technique with $m_f = 21$, the magnitude of fundamental component is increased in all techniques but the THD is almost the same as triangular carrier wave except the PSCSVPWM. In PSCSVPWM the THD values are increased corresponding to the triangular carrier wave. If the carrier frequency is increased, there is a reduction in the THD in all the methods except PSCSVPWM. The performance of PSCSVPWM remains the same even though carrier frequency increased.

Table 1 shows the comparison of different SPWM techniques with trapezoidal triangular carrier and different SVPWM techniques with triangular carrier and trapezoidal triangular carrier waves. It is observed from tabulated simulation results that PDSVPWM technique with triangular carrier wave having m_f of

Tab. 1: Comparison of %THD and fundamental component of voltage for different PWM techniques.

S. no.	PWM technique	Fundamental component of line voltage	THD (%)
SPWM-triangular carrier wave with {$m_f = 27$}			
1	PDSPWM	864.2	6.73
2	PODSPWM	866.1	9.44
3	APODSPWM	865.8	8.33
4	PSCSPWM	692.7	12.30
SVPWM-triangular carrier wave with {$m_f = 21$}			
5	PDSVPWM	1009	5.45
6	PODSVPWM	1002	8.78
7	APODSVPWM	1002	8.78
8	PSCSVPWM	1008	7.70
SVPWM-triangular carrier wave with {$m_f = 27$}			
9	PDSVPWM	1009	4.79
10	PODSVPWM	1001	8.62
11	APODSVPWM	1001	8.57
12	PSCSVPWM	1008	7.82
SVPWM-trapezoidal triangular carrier wave with {$m_f = 21$}			
13	PDSVPWM	1020	5.62
14	PODSVPWM	1007	8.10
15	APODSVPWM	1007	8.07
16	PSCSVPWM	1045	12.57
SVPWM-trapezoidal triangular carrier wave with {$m_f = 27$}			
17	PDSVPWM	1015	5.11
18	PODSVPWM	1002	7.88
19	APODSVPWM	1003	7.86
20	PSCSVPWM	1045	12.47

27 gives better THD as 4.79 % and 1009 V fundamental voltage. Whereas the PDSVPWM technique with trapezoidal triangular carrier wave having m_f of 21 gives better THD of 5.11 % and 1020 V fundamental component of voltage.

5. Conclusion

In this paper new trapezoidal triangular multicarrier SVPWM techniques such as Phase Disposition (PD), Phase Opposition Disposition (POD), Alternate Phase Opposition Disposition (APOD), and Phase Shifted Carrier PWM (PSCPWM) have been implemented for 11-level cascaded MLI. The above results are compared among each other and with different trapezoidal triangular carrier SPWM techniques. It is observed that PDSVPWM technique with 1350 Hz triangular carrier wave gives 4.79 % of THD, which is the better performance for a 11 cascaded H-bridge MLI, whereas PDSVPWM with 1350 Hz trapezoidal triangular carrier gives 1015 voltage of high fundamental component and 5.11 % of THD value. It is concluded that triangular carrier PDSVPWM gives better THD, with trapezoidal triangular carrier wave, all the techniques give high fundamental component of line voltage, but PSCPWM with trapezoidal triangular carrier wave gives higher fundamental component voltage among all other techniques along with higher THD values. In all these methods if the carrier frequency is increased beyond 1350 Hz, the THD in the line voltage is increased.

References

[1] LAI, J.-S. and F. Z. PENG. Multilevel converters - a new breed of power converters. In: *IEEE Industry Applications Conference*. Orlando: IEEE, 1995, pp. 2348–2356. ISBN 0-7803-3008-0. DOI: 10.1109/IAS.1995.530601.

[2] RODRIGUES, J., J.-S. LAI and F. Z. PENG. Multilevel inverters: a survey of topologies, controls, and applications. *IEEE Transactions on Industrial Electronics*. 2002, vol. 49, no. 4, pp. 724–738. ISSN 1557-9948. DOI: 10.1109/TIE.2002.801052.

[3] MANJREKAR, M. and G. VENKATARA-MANAN. Advanced topologies and modulation strategies for multilevel inverters. In: *27th Annual IEEE Power Electronics Specialists Conference*. Baveno: IEEE, 1996, pp. 1013–1018. ISBN 0-7803-3500-7. DOI: 10.1109/PESC.1996.548706.

[4] CORZINE, K. and Y. FAMILIANT. A new cascaded multilevel H-bridge drive. *IEEE Transactions on Power Electronics*. 2002, vol. 17, iss. 1, pp. 125–131. ISSN 1941-0107. DOI: 10.1109/63.988678.

[5] RODRIGUEZ, J., L. MORAN, J. PONTT, P. CORREA and C. SILVA. A high-performance vector control of an 11-level inverter. *IEEE Transactions on Industrial Electronics*. 2003, vol. 50, iss. 1, pp. 80–85. ISSN 1557-9948. DOI: 10.1109/TIE.2002.804975.

[6] DIXON, J. and L. MORAN. Multilevel inverter, based on multi-stage connection of three-level converters scaled in power of three. In: *28th Annual Conference of the IEEE Industrial Electronics Society (IECON 02)*. Sevilla: IEEE, 2002, pp. 886–891. ISBN 0-7803-7474-6. DOI: 10.1109/IECON.2002.1185389.

[7] DIXON, J. W., M. ORTUZAR and F. RIOS. Traction Drive System for Electric Vehicles, Using Multilevel Converters. In: *19th Electric Vehicle Symposium (EVS-19)*. Busan: EDTA, 2002, pp. 19–23.

[8] CENGELCI, E., S. U. SULISTIJO, B. O. WOOM, P. ENJETI, R. TEODORESCU and F. BLAABJERG. A new medium-voltage PWM inverter topology for adjustable-speed drives. *IEEE Transactions on Industry Applications*. 1999, vol. 35, iss. 3, pp. 628–637. ISSN 1939-9367. DOI: 10.1109/28.767014.

[9] RODRIGUEZ, J., J.-S. LAI and F. Z. PENG. Multilevel inverters: a survey of topologies, controls, and applications. *IEEE Transactions on Industrial Electronics*. 2002, vol. 49, iss. 4, pp. 724–738. ISSN 1557-9948. DOI: 10.1109/TIE.2002.801052.

[10] MCGRATH, B. P. and D. G. HOLMES. Multicarrier PWM strategies for multilevel inverters. *IEEE Transactions on Industrial Electronics*. 2002, vol. 49, iss. 4, pp. 858–867. ISSN 1557-9948. DOI: 10.1109/TIE.2002.801073.

[11] ASPALLI, M. S. and A. WAMANRAO. Sinusoidal pulse width modulation (SPWM) with variable carrier synchronization for multilevel inverter controllers. In: *International Conference on Control, Automation, Communication and Energy Conservation (INCACEC 2009)*. Perundurai: IEEE, 2009, pp. 1–6. ISBN 978-1-4244-4789-3.

[12] AHIRRAO, D., B. GAWRE, P. KAKADE and S. CHAWDA. Analysis Of Single Phase Matrix Converter. *International Journal of Engineering Research and Applications*. 2014, vol. 4, iss. 3, pp. 856–861. ISSN 2248-9622.

[13] KUMAR, A. Direct torque control of induction motor using imaginary switching times with 0-1-2-7 & 0-1-2 switching sequences: a comparative study. In: *30th Annual Conference of IEEE Industrial Electronics Society (IECON 2004)*. Busan: IEEE, 2002, pp. 1492–1497. ISBN 0-7803-8730-9. DOI: 10.1109/IECON.2004.1431799.

[14] NADERI, R. and A. RAHMATI. Phase-Shifted Carrier PWM Technique for General Cascaded Inverters. *IEEE Transactions on Power Electronics*. 2008, vol. 23, iss. 3, pp. 1257–1269. ISSN 1941-0107. DOI: 10.1109/TPEL.2008.921186.

[15] CELANOVIC, N. and D. BOROYEVICH. A fast space-vector modulation algorithm for multilevel three-phase converters. *IEEE Transactions on Industry Applications*. 2001, vol. 37, iss. 2, pp. 637–641. ISSN 1939-9367. DOI: 10.1109/28.913731.

[16] LEE, D.-C. and G.-M. LEE. A novel overmodulation technique for space-vector PWM inverters. *IEEE Transactions on Power Electronics*. 1999, vol. 13, iss. 6, pp. 1144–1151. ISSN 1941-0107. DOI: 10.1109/63.728341.

[17] CHUNG, D.-W., J.-S. KIM and S.-K. SUL. Unified voltage modulation technique for real-time three-phase power conversion. *IEEE Transactions on Industry Applications*. 1998, vol. 34, iss. 2, pp. 374–380. ISSN 1939-9367. DOI: 10.1109/28.663482.

[18] REDDY, T. B., B. K. REDDY, J. AMARNATH and D. S. RAYUDU. New Hybrid SVPWM Techniques for Direct Torque Controlled Induction Motor Drive to Reduce Current and Torque Ripples without Angle and Sector Estimation. In: *IEEE International Conference on Industrial Technology (ICIT 2006)*. Mumbai: IEEE, 2006, pp. 1624–1629. ISBN 1-4244-0725-7. DOI: 10.1109/ICIT.2006.372482.

[19] REDDY, T. B., J. AMARNATH and D. SUBBARAYUDU. New Hybrid SVPWM Methods for Direct Torque Controlled Induction Motor Drive for Reduced Current Ripple. In: *International Conference on Power Electronics, Drives and Energy Systems (PEDES '06)*. New Delhi: IEEE, 2006, pp. 1–6. ISBN 0-7803-9771-1. DOI: 10.1109/PEDES.2006.344416.

[20] REDDY, C. L., P. S. KUMAR and M. SUSHAMA. Modified Modulation Techniques for Cascaded Multilevel Inverter Fed Induction Motor Drive. *Global Journal of Research in Engineering*. 2015, vol. 15, iss. 9, pp. 17–24. ISSN 0975-5861.

About Authors

Lokeshwar Reddy CHINTALA was born in Khammam, Andhra Pradesh, India in 1977. He obtained B.Tech. in Electrical and Electronics Engineering from Kakatiya University, Warangal in 1999 and M.Tech. in High Voltage Engineering in 2001 from Jawaharlal Nehru Technological University, Kakinada. He has 14 years teaching experience. He is working as Associate Professor in the Department of Electrical and Electronics Engineering, CVR College of Engineering, Hyderabad. He presented many research papers in various national and international conferences and journals. His research interests include Power Electronics Drives and Multilevel inverters.

Satish Kumar PEDDAPELLI was born in Karimnagar, Telangana, India in 1974. He obtained B.Tech. in Electrical and Electronics Engineering from JNTU College of Engineering, Kakinada, India in 1996. He obtained M.Tech. in Power Electronics in 2003 and Ph.D. in 2011 from Jawaharlal Nehru Technological University, Hyderabad. He has more than 19 years of teaching experience and at present he is an Assistant Professor in the Department of Electrical Engineering, University College of Engineering, Osmania University, Hyderabad, India. His research interests include Power Electronics, Special machines, Drives and Multilevel inverters and guiding eight research scholars.

Sushama MALAJI was born in Nalgonda district, Telangana state, India in 1973. Obtained her B.Tech. in 1993 and M.Tech. in 2003 with specialization Electrical Power Systems from JNTU, India. She obtained her Ph.D. from JNTUH College of Engineering, Hyderabad, in 2009 in the area of "Power Quality" using Wavelet Transforms. She has more than 20 years of teaching experience and at present she is working as Professor in the Department of Electrical and Electronics Engineering, JNTUH College of Engineering, Hyderabad, India. Her research interests includes green energy, power systems, power electronics, FACTs controllers and power quality and harmonics.

8

Ensuring the Visibility and Traceability of Items through Logistics Chain of Automotive Industry based on AutoEPCNet Usage

Pavel STASA[1], Filip BENES[1], Jiri SVUB[1], Yong-Shin KANG[2], Jakub UNUCKA[1], Lukas VOJTECH[3], Vladimir KEBO[1], Jong-Tae RHEE[4]

[1]Institute of Economics and Control Systems, Faculty of Mining and Geology,
VSB–Technical University of Ostrava, 17. listopadu 15, 708 33 Ostrava, Czech Republic
[2]Department of Systems Management Engineering, College of Engineering, Sungkyunkwan University,
300 Cheoncheon-dong, Jangan-gu, Suwon, Republic of Korea
[3]Department of Telecommunication Engineering, Faculty of Electrical Engineering,
Czech Technical University in Prague, Technicka 2, 166 27 Prague 6, Czech Republic
[4]Department of Industrial and Systems Engineering, College of Engineering, Dongguk University,
26 Pil-dong 3-ga, Jun-gu, Seoul, Republic of Korea

pavel.stasa@vsb.cz, filip.benes@vsb.cz, jiri.svub@vsb.cz, yskang7867@skku.edu, jakub.unucka@vsb.cz,
vojtecl@fel.cvut.cz, vladimir.kebo@vsb.cz, jtrhee@dgu.edu

Abstract. *Traceability in logistics is the capability of the participants to trace the products throughout the supply chain by means of either the product and/or container identifiers in a forward and/or backward direction. In today's competitive economic environment, traceability is a key concept related to all products and all types of supply chains. The goal of this paper is to describe development of application that enables to create and share information about the physical movement and status of products as they travel throughout the supply chain. The main purpose of this paper is to describe the development of RFID based track and trace system for ensuring the visibility and traceability of items in logistics chain especially in automotive industry. The proposed solution is based on EPCglobal Network Architecture.*

Keywords

AutoEPCNet, Auto-ID, logistics, RFID, SAP R/3, supply chain, visibility and traceability.

1. Introduction

The main subject of the logistics is the management of material flow and the associated flow of information and financial flows. The aim of this procedure is to achieve a competitive service level as well as minimizing logistics costs. Instrument for achieving that objective is a corporate logistics system which is one of the support functions of the company.

The Service Level (SL) is the company's ability to respond to demands and it can be generally expressed as a percentage, for example, between the actual fulfilment of customer requirements and their original content.

Logistics costs can be defined as all funds that we have to spend in order to achieve a given service level.

The quality suppliers are needed for effective supply chain management. These suppliers have to deliver goods to a Distribution Centres in time and in the required quantity. For the evaluation of suppliers, so called Service Level is used, which testifies to the quality of these suppliers' services. It can be expressed as follows [1]:

$$\text{SL}(\%) = \frac{\text{NCD}}{\text{NWD}} \cdot 100, \qquad (1)$$

where NCD is number of correct deliveries and NWD is number of wrong deliveries.

It is expressed in percentage and indicates how many deliveries were correct in defined period of time. The SL is expressed for the month, and if it stays bad for a long time, various sanctions are usually applied. SL of less than 80 % is considered as unsatisfactory. The SL value of 95 % represents an exclusive supplier that is very desirable to have deeper relations with [1].

The complex structure of the automotive industry puts high demands on logistics and associated service level. The automotive industry is characterized by a large number of different kinds of products and high volumes of fluctuating production. Each supply chain has to meet high standards of quality and flexibility. One of the most dynamic trends that fundamentally affects the operation of companies in the automotive industry is the gradual substitution of original forms of communication by the new ones. Globalization, variability of demand, production on order, sequential deliveries and many other processes reveal shortcomings of traditional methods of communication and they are being replaced by new technologies, including already established Electronic Data Interchange (EDI) standard the EPCglobal Network begins to assert.

The primary impetus for the use of the EPCglobal Network for a communication between partners in the automotive industry is to reduce costs and streamline processes in the delivery of goods. If the logistics chain should work as a complex system and bring benefits to all stakeholders, it is assumed that most vendors will not only be able to supply in the JIT mode, but also that they will share their data and reports electronically with their own suppliers as well as with their customers. The shared data can then be used to print logistic labels, to compare different generations of order references, to generate order references for own suppliers, etc. Until recently, the high investment requirements were an obstacle to extend this model to all levels of the supply chain for the acquisition of an integrated system which would provide the mentioned functionality.

2.   Related Work

Standards for Electronic Product Information Services (EPCIS) systems have been available for quite a long time. EPCIS could be considered as a key enabling technology to be primarily implemented in such areas where it is for any reason necessary to show the history of the production, transportation, ensure authenticity and higher level of safety of the products.

Typical areas corresponding to the above description, are pharmaceutical industry, transportation, processing and storage of drugs and preparation of consumer units of medicines in order to clearly demonstrate their pedigree and traceability of individual consumer packages.

Dozens of countries are trying to implement drug supply chain regulatory requirements nowadays. Most of the products are under a government reporting requirement. For example, USA order the compliance information exchange between trading partners in pharmaceutical industry. Rather than dictating the data formats and mechanism for the data interchange, the FDA has left it up to industry to decide.

One of the logistic data exchange options is the precise use of the systems based on the EPCIS and other EPCglobal standards and interconnection of the logistic data. Another option is to use Advanced Shipping Notice (ASN) that is a part of the EDI system.

The EPCIS is one of the most interesting possibilities for interchange of serialized data. EPCIS has not yet been much used for exchanging lot-level compliance data. Most of the wholesale distributors have accepted it in the past. Most of them prefer to receive ASNs that deliver logistical information in addition to compliance data instead. In 2014 pharmaceutical companies started the implementation process of the HDMA ASN 856 Guideline for lot based Drug Supply Chain Security Act (DSCSA) requirements and this soon after that became an industrial standard. As drug stores and hospitals are connecting to the system, the adoption of the ASN has continued [2].

EPCIS is not the official format for exchange of serialization data in the US, but supply chain partners may still choose to exchange data in this format. As we mentioned above, EPCIS was especially designed to exchange large volumes of serialized data. If anyone chooses EPCIS standard implementation, they have to decide if they would use EPCIS to exchange compliance data together with ASNs, or would the logistical data be included in the EPCIS communication avoiding the use of ASN. Many within the industry feel that the two should co-exist, with EPCIS serving purely a compliance purpose and ASNs managing logistical data, which was their first purpose.

Because there are already verification requirements for the entire supply chain in regards to returns, some manufacturers have begun to express that they would like to send consumer unit level serialized data prior to 2023, when they are required to do so. The serialization itself should start in 2017.

It is a beneficial to think of EPCIS adoption beyond simply meeting the compliance date. Since sooner or later they would like to offer the traceability service of each product, the investments in system covering both tasks would be smaller than investment in both systems one by one. According to a recent Acsis survey, researchers cautioned manufacturers taking a "compliance now" approach that they may get caught with long-term costs and rework of initiatives to build a truly traceable, serialized warehouse. In general, unit level serialization can protect brands from risk in a variety of ways, including improved recall readiness, and providing the ability to document chain of custody which helps eliminate opportunities for counterfeit drugs to circulate through the supply chain. By failing to recognize the long term pros, companies can risk lowering productivity through disintegrated systems and additional manual processes time and costs [3].

3. EPCglobal Network Architecture

In order to implement delivery of individual products and state of their lifecycle in the logistic chain, to ensure the information flow across information systems of the individual parties who are involved, it is necessary to use the Electronic Product Code Information Services (EPCIS) standard and other EPCglobal standards.

The EPCglobal Network is a set of technologies, enabling business partners to monitor product movement in the logistic chain. EPCglobal Network, supports improvement of organizations efficiency by enabling dynamic and accurate information distribution (i.e. information about the product movements) in real time.

EPCglobal Network comprises:

- Object Naming Service (ONS),

- EPC Discovery Services (DS),

- EPC Security Services (SS),

- EPC Information Services (EPCIS).

3.1. EPC Information Services

The goal of EPCIS (EPC Information Services) is to enable applications to create and share visibility event data, both within and across enterprises. Ultimately, this sharing is aimed at enabling users to gain a shared view of physical or digital objects within a relevant business context. EPCIS is a GS1 standard [5] that enables trading partners to share information about the physical movement and status of products as they travel throughout the supply chain of trading partners - from business to business and sometimes even to consumers.

The EPCIS Capture Interface acts as an interface between the "Capture" and "Share" data. The EPCIS Query Interface provides visibility event data both to internal applications and for sharing with trading partners.

The Back-end Application is software system consisting of databases and information systems, addressing the different business areas, from regular monitoring and reporting process of reading up to complex solutions of logistics chain traceability. These systems play a crucial role in implementation of traceability and visibility services because there is a necessity to develop interface for data exchanging between RFID system and Enterprise Resource Planning (ERP) systems. Section 5 deals with this issue.

Generally, EPCIS processes two kinds of data: event data and master data. Event data arises in the course of carrying out business processes, and are captured through the EPCIS Capture Interface and are available for query through the EPCIS Query Interfaces. Master data are additional data that provide the necessary context for interpretation of the event data.

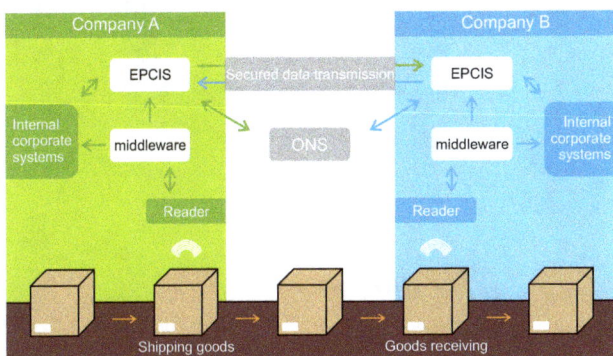

Fig. 1: Example of EPCglobal architecture.

The EPCglobal Architecture Framework includes software standards at various levels of abstraction, from low-level interfaces to RFID reader devices all the way up to the business application level [4].

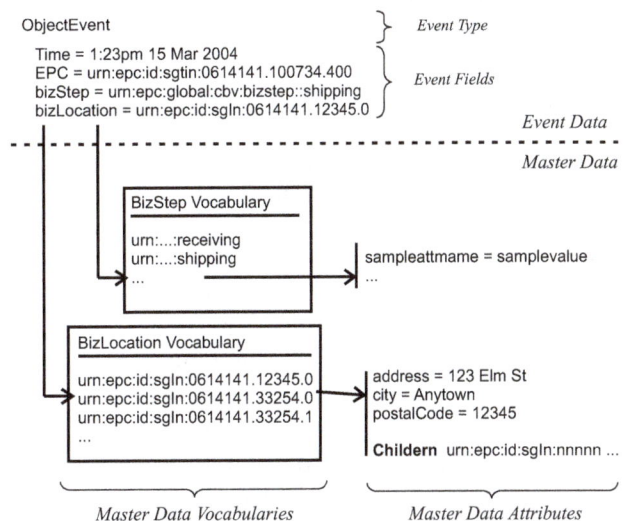

Fig. 2: Event and Master Data [5].

The basic element of the whole EPCIS system is event-driven data processing. As an event we can understand the real event usually associated with the process in the logistic chain accompanied by RFID identification. The event must be presented in a form that is understandable for software components, so it is realized through a standardized XML document. Events may relate to the identification of one or more objects which are identified by the EPC, the aggregation of objects into higher logistic units, business transaction or a simple inventory quantity of objects of that type.

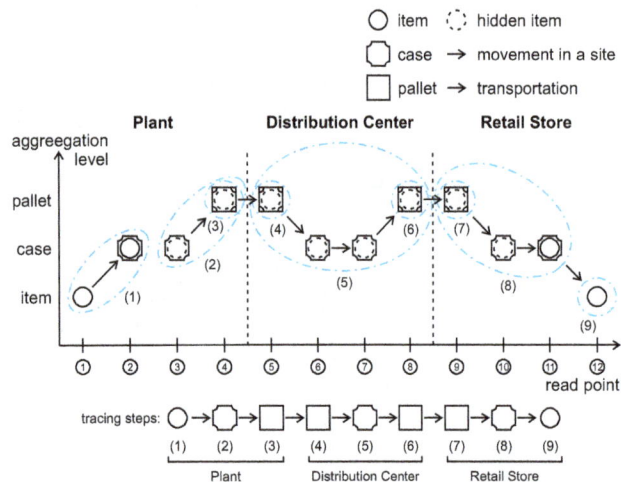

Fig. 3: Business process showing generation of EPCIS data.

The information in an EPCIS event records the essentials of what happened during a step of a business process in which physical or digital objects were handled, expressed via the four dimensions of **what**, **where**, **when**, and **why**.

4.   Developed Solution

The developed solution fully respects the rules concerning the standardized communication within EPCIS that was discussed in the previous sections. The vision of an integrated management system and visualization of lifecycle was primary overall concept of this system purpose. The essence of the project was to increase the efficiency of the process of tracking and tracing objects in order to maximize efficiencies in the supply chain. On the basis of observed information, it will be possible to create substantial added value in terms of providing information services to its business partners.

Based on the analysis of communication standards of the best-selling UHF RFID readers, it was decided to use the LLRP standard [7] defined by the GS1 organization in the EPC Architecture of the Framework System EPC Global.

There are various traceability needs in industry, which defines the different functional requirements of the traceability services, as was mentioned by Kang in his article [8]. Although traceability requirements are different for various industries, there is a common fundamental requirement - traceability for serial-level (or item-level) product [9], [10] and [11]. Based on this serial-level product traceability, companies in the manufacturing industry emphasize the traceability for manufacturing lots or batches, and the traceability for operation conditions, as they are to identify potential sources of quality problems of defective products.

Large manufacturing industries, such as automobile industry, require packaging information about an item such as bill-of-material information. Similarly, logistics companies have to obtain aggregation information between items and logistics units such as pallets and cases [12] and [13].

4.1.   AutoEPCNet Conceptual Scheme

AutoEPCNet is an information system to track and trace items marked with RFID tags in logistic chain. The main function can be characterized as follows:

- processing large-scale RFID event data, such as in-bound and out-bound logistics data,

- enabling history trace services tailored to the auto industry logistics system,

- building an information service hub by harnessing RFID technology.

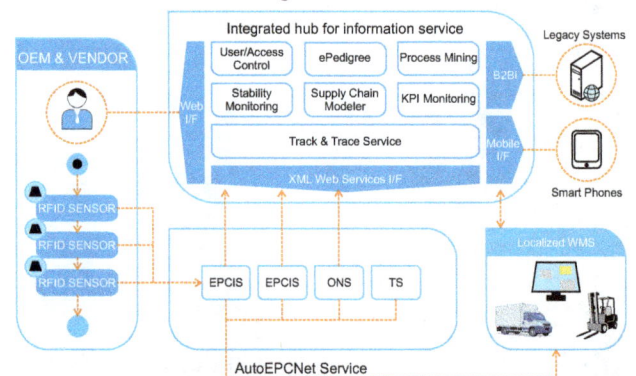

Fig. 4: Conceptual scheme.

As can be seen in Fig. 4, the system architecture is composed of the following crucial layers. The first important layer is called "interface" layer and it is responsible for receiving and interpreting users' requests. The second layer, called "business logic" layer, has the core functions for the queries that will be described in the

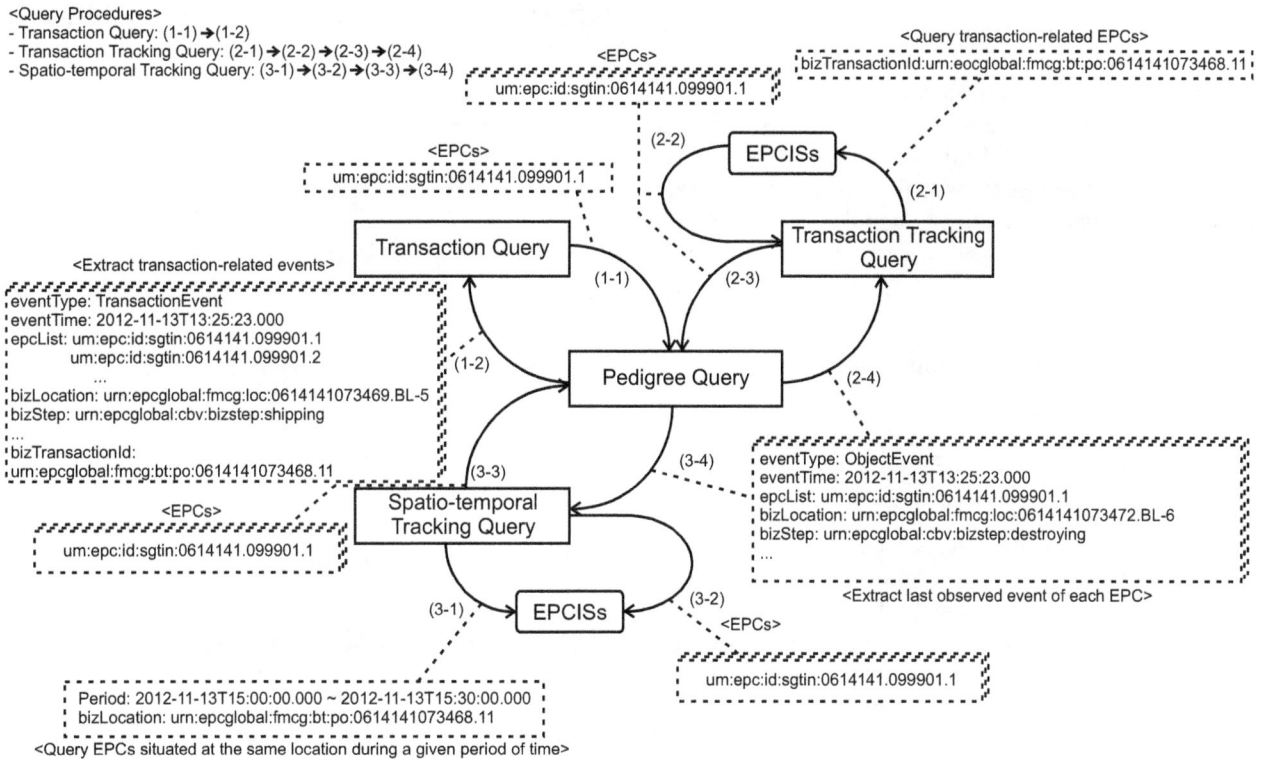

<Query Procedures>
- Transaction Query: (1-1)➜(1-2)
- Transaction Tracking Query: (2-1)➜(2-2)➜(2-3)➜(2-4)
- Spatio-temporal Tracking Query: (3-1)➜(3-2)➜(3-3)➜(3-4)

<EPCs>
um:epc:id:sgtin:0614141.099901.1

<Query transaction-related EPCs>
bizTransactionId:urn:eocglobal:fmcg:bt:po:0614141073468.11

<EPCs>
um:epc:id:sgtin:0614141.099901.1

EPCISs

Transaction Query

Transaction Tracking Query

<Extract transaction-related events>
eventType: TransactionEvent
eventTime: 2012-11-13T13:25:23.000
epcList: um:epc:id:sgtin:0614141.099901.1
         um:epc:id:sgtin:0614141.099901.2
         ...
bizLocation: urn:epcglobal:fmcg:loc:0614141073469.BL-5
bizStep: urn:epcglobal:cbv:bizstep:shipping

bizTransactionId:
urn:epcglobal:fmcg:bt:po:0614141073468.11

Pedigree Query

<Extract last observed event of each EPC>
eventType: ObjectEvent
eventTime: 2012-11-13T13:25:23.000
epcList: um:epc:id:sgtin:0614141.099901.1
bizLocation: urn:epcglobal:fmcg:loc:0614141073472.BL-6
bizStep: urn:epcglobal:cbv:bizstep:destroying
...

<EPCs>
um:epc:id:sgtin:0614141.099901.1

Spatio-temporal Tracking Query

EPCISs

<EPCs>
um:epc:id:sgtin:0614141.099901.1

Period: 2012-11-13T15:00:00.000 ~ 2012-11-13T15:30:00.000
bizLocation: urn:epcglobal:fmcg:bt:po:0614141073468.11
<Query EPCs situated at the same location during a given period of time>

Fig. 5: Information flows among the TS queries.

next subsection. And the last one, called "EPC client" layer allows the core functions of traceability service to invoke EPCglobal services such as EPCIS, DS and ONS. Thanks to use of multi-threading it was possible to implement the EPCIS invocation that enables that multiple EPCIS's can be called simultaneously. The separation of business logic and EPC client allows developers to accommodate any possible changes of the EPCglobal network services. The web-based Graphic User Interface (GUI) for supporting users' accessibility in the form of an RFID information portal system was implemented by the Korean colleagues.

Based on the traceability requirements analysis and based on earlier studies and industry survey, we have defined a set of essential RFID traceability requirements which are as follows.

There was developed a set of efficient algorithms for implementing the Traceability Service (TS) queries. The algorithms are based on the following two assumptions on the EPCglobal network operation, both are reasonable ones:

- Every EPC must be reported to DS at the first observation and the last observation within a site (e.g., a warehouse or a manufacturing plant).

- Once a given (child) EPC is aggregated to other (parent) EPC, the EPC can have neither a new child EPC nor a new parent EPC until disaggregated.

The event information (called EPCIS event) plays an important role in the query algorithms. In order to process the queries, TS needs to collect DS records, each of which contains the information on the EPCIS that a specific EPC has visited, and EPCIS events of different types (*ObjectEvent, AggregrationEvent* and *TransactionEvent*) [8].

1) Pedigree Query

The pedigree query is to reconstruct the complete history of individual products or other logistic units [13]. The typical usage of this query is to obtain an answer to the following questions: What is the full history of detections of item 123? Where and when was the item 123 lastly detected?

The procedure to retrieve the entire history for a given EPC is composed of two steps. The first step is to request all the event data of the EPC of interest (e.g., an item) to EPCIS's. A second step is to retrieve all the parents (e.g., pallets, containers) information if the EPC had parent(s).

2) Transaction Query

The transaction query is used to find the business transaction identifiers (e.g., purchase order, invoice and package) that were related to an item of interest during the lifecycle of an entire supply chain.

The algorithm of a transaction query is simple compared to a pedigree query algorithm. TS first executes a pedigree query with a target EPC, and then extracts business transaction-related events, which has a non-null bizTransactionList field. The query users can easily list up the business transactions in which the target EPC has involved through the supply chain by checking the bizTransaction field of each event.

3) Transaction Tracking Query

The transaction tracking query is to find out the current location of items which were related earlier to a specific business transaction.

In this query algorithm, TS first collects all the EPC's (including items and containers) that are associated with a specific business transaction using the input parameters (epcisAddress, bizTransactionType, and bizTransaction), and stores them on the EPCList. Then, TS executes a pedigree query for all the EPC's on the EPCList and it stores the last events (informing of the current locations) of all the resultant events.

4) Spatio-temporal Tracking Query

The spatio-temporal tracking query is to search for the last locations of products which were once situated at the same location during a given period of time.

Using the algorithm of the spatio-temporal query, TS first executes a simple event query to a given EPCIS in order to collect all the ObjectEvent and Aggragation-Event type events that occurred during a specific time period at a specific location defined by the readPoint. Then, TS executes a series of pedigree queries for all the EPC's extracted from the result of the simple event query, stores the last events of all the resultant events.

5) Aggregation Tracking Query

This query provides all the aggregation information of a product at a given period of time.

This query makes recursive function calls to search for the root EPC of a given (input) EPC and all the children of this root EPC at a given time (*dateTime*). First, TS determines the first EPCIS to be searched for the root EPC. This EPCIS is the last visited one before the given *dateTime*. Then the function *getRootEPC()* is used to search the root EPC of a given EPC. Finally, *getChildEPCs()* function is used to build up an , which defines the whole family pedigree of the input EPC in a tree data structure.

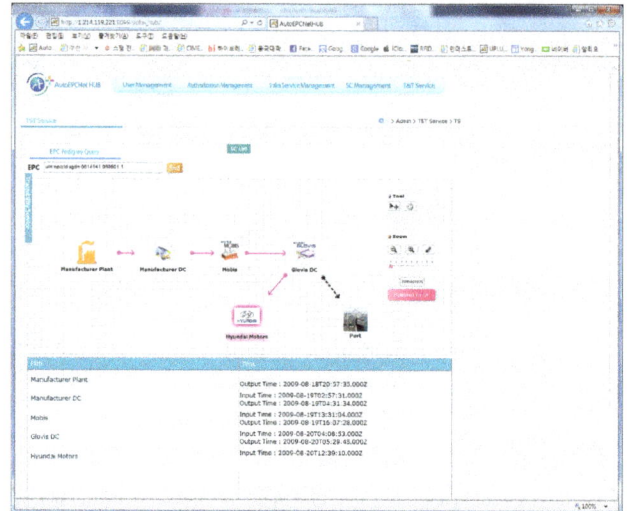

Fig. 6: Supply chain modeller.

4.2. Supply Chain Modeller

The first part of developed solution is called Supply Chain Modeller. This is a module that is used for modelling the logistic chain and any linkages that take place therein.

This tool facilitates handling and logistical follow-up activities that take place in the chain and gives these activities in a clear and mostly coherent unit. The Fig. 6 shows the main screen where the user defines all the elements involved in the supply chain and integrate these elements by oriented edges for marking a sequence of logistic data. Software has a predefined range of graphic elements but for the user's purposes it is possible to create and upload their own graphics which will be more appropriate to their needs.

4.3. Visibility and Analytics Service

This part of the system offers to user information about the status of all the participants and creates their clear graphical structure. The whole process of the supply chain can be monitored in a form of graphs and thanks to them it is possible to see and detect downtime and

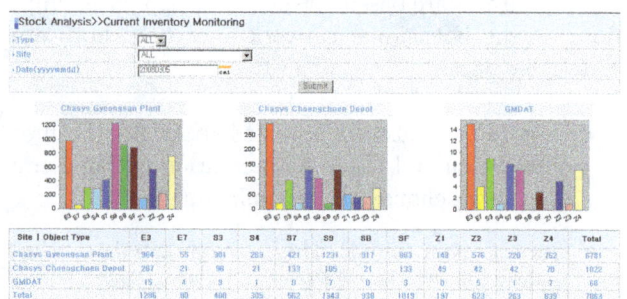

Fig. 7: Stock analysis - Current Inventory Monitoring.

loss of time in different parts of the chain, or you can view data on stocks or manufactured products.

The data can be interpreted in a synoptic charts, the program allows you to prepare any set of data needed for the application.

4.4.   e-Pedigree Service

One of the most important and most used parts of AutoEPCNet is an e-Pedigree tool which describes all necessary data about the product. The information about all phases of the product development are recorded into the pedigree, such as all identification data of the producer (name, address, state, country) and of course product information (name, description, product code) and further information about the location where the product is sent.

As was mentioned before, various types of industry need various types of requirements on the logistics chain. The great advantage of this system is that it can display data from external sensors besides EPC reading in an individual reading place. It is also possible to visualize other additional information that includes data from sensors, temperature, humidity, altitude, shock, sensors, opening mail, etc.

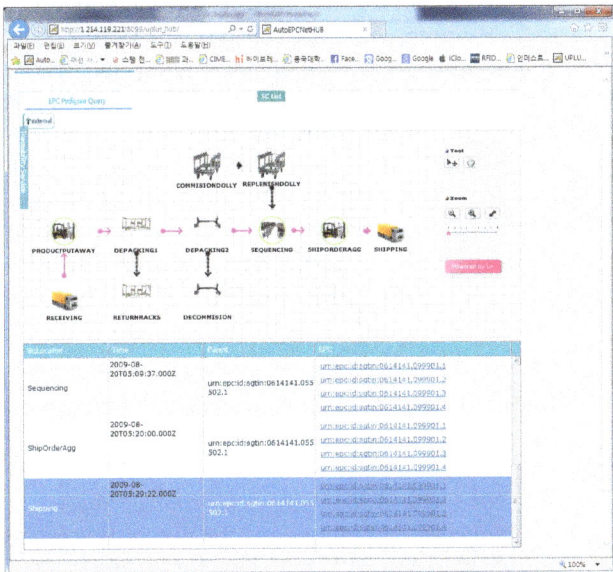

Fig. 8: e-Pedigree.

In every part of the supply chain there are the genealogy records information about the status and progress of delivery and thus the reliable and transparent functioning of the entire system is ensured. Through the genealogy, customer or one of the participating members can verify the authenticity of the product and make sure that the product has passed through all the stages that are necessary for its full value.

4.5.   Technical Parameters

The whole tested system is run on two hardware identical servers - Intel Xeon CPU E5 2665 @ 2.40 GHz with 8 GB of RAM. Both have the Oracle 10g database and Tomcat 6.0.45 with Java (JDK 7u55) installed.

These are the minimum hardware requirements. When system runs continuously, the memory usage is around 75 % of its capacity.

There are servlets implementing the services of AutoEPCNet Hub, DS, TS and ONS with necessary databases (AutoEPCNet Hub and DS database) on the first server (epcis1). On the second server (epcis2), there are servlets for 6 instances of EPCIS implementation including the Capture and Query Interface and Core Business Logic along with EPCIS databases that are essential for deployment.

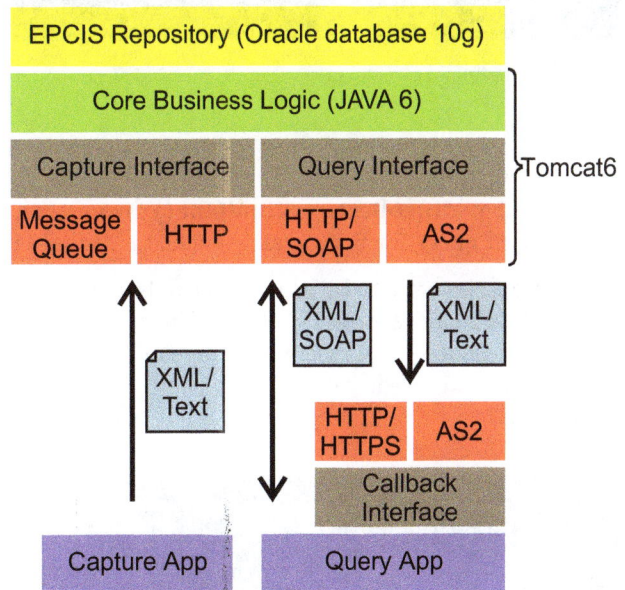

Fig. 9: AutoEPCNet architecture.

Functionalities of Capturing Application are:

- receives event data form Middleware or RFID reader,
- makes XML format of EPCIS event data based on pre-defined information,
- sends EPCIS event data to EPCIS capture interface (HTTP).

5.   Mass Data Throughput Experiment

The developed solution has to process and record big amount of data from whole logistic chain. The objective of this test is to confirm whether the database storage can handle 1 TB of data.

In laboratory environment, we inserted event data approximately 1 TB (1 086 402 562 560 Byte). Using command '**db.stats()**', we can easily check the volume as follows. According to Fig. 10, the number of event objects is 990 720 040.

```
"objects" : 990720040,
"avgObjsize" : 10960,
"dataSize" : 1086102562560,
"storageSize" : 1123344417120,
"numExtents" : 75,
"indexes" : 1,
"indexSize" : 32143535200,
"fileSize" : 1179878686720,
"extentFreeList" " {
"num" : 0,
"totalSize" : 0
},
"ok" : 1
}
```

Fig. 10: Checking the volume.

Using the '**db.collections.findOne()**' command, we repeated query for finding an event data 30 times.

All of queries returned within 30 seconds as follows. Each object includes eventType, eventTime, readPoint and so on. Using just single query, we can confirm all of event data we need.

```
mongos> db.Event.findOne();
{
    "_id" : ObjectId("566e2d21e4b03c99f5065a69"),
    "eventtype" : "Object",
    "event_id" : NumberLong(1),
    "eventTime" : ISODate("2012-12-10T09:00:00Z"),
    "recordTime" : ISODate("2012-12-10T09:00:00Z"),
    "eventTimeZoneOffset" : "+9:00",
    "action" : "Add",
    "bizStep" : "urn:epcglobal:cbv:bizstep:commissioning",
    "disposition" : "urn:epcglobal:cbv:disp:active",
    "readPoint" : "urn:epc:id:sgln:0614141150001.RP-1",
    "bizLocation" : "urn:epc:id:sgln:0614141150001.BL-1",
    "bizTransactionList" : {
        "type" : "urn:epcglobal:cbv:btt:po",
        "value" : "urn:epcglobal:cbv:btt:po:0614141150010.123"
    },
    "epcList" : [
        "urn:epc:id:sgtin:0614141.100000.1"
    ],
    "extension" : [
        {
            "fieldname" : "http://schema.hls.com/extension#temperature",
            "prefix" : "hls",
            "value" : [
                24.6295094935151
            ]
        }
    ]
}
```

Fig. 11: Confirmation of Event Data.

Using this test, we proved that our storage can handle 1 TB mass data and the Query Interface can provide visibility event data both to internal applications and for sharing with trading partners.

6. The Interface for Exchanging Data

The important aspect related to the implementation of RFID technology in the automotive industry was also to design and deploy a suitable additional simple system for the receiving gate in the car manufacturer warehouse - an interface for exchanging data between the AutoEPCNet system and corporate information system SAP R/3 for checking the received material. The system had to be connected to the SAP ERP interface and had to adopt GS1 standards with RFID reader communicating according to LLRP as well as EPCglobal RFID tags.

Applications for the reading gate include the following modules:

- **GB_GATE** - a module for reading data from the RFID reader. It provides continuous communication with RFID reader and sending and filtering data read from RFID tags and store them in a database.
- **GB_VIEW** - a module for data display and interaction with the operator. It provides information to the AT_INTF module when storing data in the database, sending information to module AT_INTF, saving the state change (completion of ASN filling, ASN manual termination). This module also provides data processing and status reporting of data from AT_INTF module (a list of items, the confirmation of processing ...), visualization of current data from the database. There is a submit button for sending reads into SAP and button for manual termination and confirmation of the storage command. Summary and statistics of activities can be also found here (in administrator mode).
- **AT_INTF** - a module for communication with SAP R/3. This module launches the RFC function for obtaining ASN items based on storing the first record with the new ASN into the database and store new ASN items into the database. It also launches RFC functions for accounting. The module launches RFC functions for termination of ASN processing or based on manual intervention. The module sends a confirmation about identified handling units into SAP and sends the request to accounting for incoming material.

7. Summary and Discussion

The system was tested at university environment [14] and in local automotive companies and brought

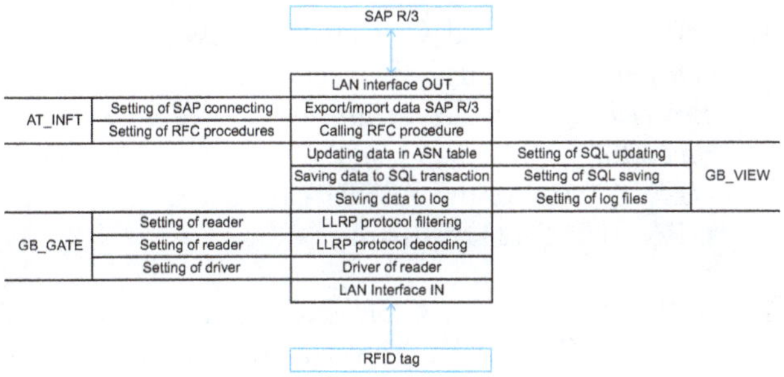

Fig. 12: Modules and theirs functions.

promising results. Together with the Koreans we are working on a gradual implementation to the automobile industry companies supply chain.

We deployed the simplified version of developed solution in logistic chain of car producer in Czech Republic. During the test period we can say that the main benefits of the solution are time savings and increase of Service Level (SL). The intended time savings relate to the time of delivery of material to the warehouse from unloading at the gate. Based on one-month testing process the time savings have been 29.71 minutes. This is due to fact that if the material is unloaded it can be directly moved to the warehouse. When the identifier of received material is in form of label, the delivered material has to be unloaded near the gate and when all incoming palettes are unloaded from truck, the worker in the warehouse is equipped with PDA containing application and he has to read all items one by one and send a command to SAP for receipt (GR) after reception end. After this, the material can be moved to warehouse.

Based on observation and evaluation of one-month system testing in real environment and according to Eq. (1), there was an increase of SL of 0.87 % compared to previous month. To respect the confidentiality, we cannot say the exact number.

8.   Conclusion

In comparison with the automotive the serialization requirements occurred a little bit sooner in the pharmaceutical industry. So we can presume the future development of traceability services in automotive industry will be analogous. Sooner or later, it will be necessary to make each important part of car traceable and with compliance ensured by pedigree standards.

The automotive industry is one of the most complicated, the fastest growing and most turbulent, which is associated with high costs and investments. The automotive industry is a very strong competitive environment and any hesitation entails high costs. The current state of the automotive industry requires faster

deliveries of increasingly complex products, throughout the supply chain from all manufacturers and subcontractors.

Due to cost reduction the automotive industry was the first, where the approach to organizing and managing processes (Lean Production) has begun to apply. The biggest losses of producers arise from overproduction, stock holding of materials, semi-finished products or finished products and last but not least also the loss due to suspended production because of delayed deliveries of goods.

The developed solution can help to meet strict requirements of automotive industry, ensures the traceability and visibility through logistics chain as well as helps to find bad habits of employees.

When the system was deployed at the producer of headlights we discovered very interesting phenomena during the first day of testing. When the pallets with headlights were despatched, it was found that the number of lights on a pallet disagrees with order. When all boxes were opened it was found that some headlights weren't marked by RFID tag. This situation was caused by employees who have hidden headlights and at a time when they are not able to meet the standards they take the headlights from their stocks. This is obviously undesirable condition since the date of production and batch does not correspond to reality.

Our system allows the user to model arbitrarily complex logistic chain geographically throughout the world. It is compatible with most of the UHF RFID readers, antennas and tags utilizing standards. The system allows you to visualize the flow of processes in real time and enables traceability of each EPC system. Provides clear graphical reports and outputs allows you to display data from sensors of physical quantities (temperature, humidity, shocks, etc).

Facts mentioned above show promising results for implementing of developed RFID-based track and trace solution that can be effectively deployed in various industrial and business areas such as asset management, supply chain optimization/visibility, order and delivery management, inventory management, and manufacturing optimization.

Acknowledgment

This contribution was supported by the Ministry of Education, Youth and Sport as a part of an Eureka international project titled "RFID Technologies in Logistics Networks of Automotive Industry" - identification number LF13005, international project code: E!7592 AutoEPCIS.

References

[1] BAZALA, J. Logistika v praxi: prakticka prirucka manazera logistiky. Prague: Dashoefer, 2003. ISBN 80-86229-71-8.

[2] Serialization: EPCIS for DSCSA: Why and When it Matters, and How to PrepareEP-CIS for DSCSA: Why and When it Matters, and How to Prepare. *Tracelink: Life Sciences Cloud* [online]. 2015. Available at: http://www.tracelink.com/resources.

[3] STURTEVANT, P. Standards Support: The Pharma Industry's Top Questions about EPCIS. *Pharmaceutical Comliance Monitor* [online]. 2016. Available at: http://www.pharmacompliancemonitor.com.

[4] TRAUB, K. The GS1 EPCglobal Architecture Framework: GS1 Standard Version 1.1. *GS1: The Global Language of Business* [online]. 2014. Available at: http://www.gs1.org.

[5] EPC Information Services (EPCIS) Version 1.1 Specification: GS1 Version 1.7. *GS1: The Global Language of Business* [online]. 2015. Available at: http://www.gs1.org.

[6] EPCIS and CBV Implementation Guideline: Using EPCIS and CBV standards to gain visibility of business processes. *GS1: The Global Language of Business* [online]. 2015. Available at: http://www.gs1.org.

[7] EPCglobal: Low Level Reader Protocol (LLRP). *GS1: The Global Language of Business* [online]. 2010. Available at: http://www.gs1.org.

[8] KANG, Y. S. and Y. H. LEE. Development of generic RFID traceability services. *Computers in Industry*. 2013, vol. 64, iss. 5, pp. 609–623. ISSN 0166-3615. DOI: 10.1016/j.compind.2013.03.004.

[9] MOE, T. Perspectives on traceability in food manufacture. *Trends in Food Science and Technology*. 1998, vol. 9, iss. 5, pp. 211–214. ISSN 0924-2244. DOI: 10.1016/S0924-2244(98)00037-5.

[10] VAN DORP, K. J. Tracking and tracing: a structure for development and contemporary practices. *Logistics Information Management*. 2002, vol. 15, iss. 1, pp. 24–33. ISSN 0957-6053. DOI: 10.1108/09576050210412648.

[11] CHENG, M. J. and J. SIMMONS. Traceability in manufacturing systems. *International Journal of Operations and Production Management*. 1994, vol. 14, iss. 10, pp. 4–16. ISSN 0144-3577. DOI: 10.1108/01443579410067199.

[12] MURTHY, K. and C. ROBSON. A model-based comparative study of traceability systems. In: *Proceedings of the International Conference on Information Systems, Logistics and Supply Chain (ILS 2008)*. Madison: EPFL, 2008, pp. 1–12. ISBN 978-1-4244-2708-6.

[13] AGRAWAL, R., A. CHEUNG, K. KAILING and S. SCHOENAUER. Towards traceability across sovereign, distributed RFID databases. In: *Proceedings of the 10th International Database Engineering and Applications Symposium (IDEAS 2006)*. Delhi: IEEE, 2006, pp. 174–184. ISBN 978-076952577-8.

[14] KODYM, O., L. KAVKA and M. SEDLACEK. Logistic chain data processing. In: *15th International Multidisciplinary Scientific Geoconference and EXPO*. Albena: SGEM, 2015, pp. 183–190. ISBN 978-619-7105-34-6. DOI: 10.5593/SGEM2015/B21/S7.024.

About Authors

Pavel STASA was born in Ostrava. He finished his doctoral studies at 2012 and he focuses his attention to modelling and simulation of flow using Fluent SW. He spent one semester at Dongguk University in Seoul, it was focused on RFID technology. His research area is modelling and simulation of flow using CFD program Fluent, RFID, NFC technology and IoT. Since 2015 he has been a head of RFID laboratory at VSB–Technical University of Ostrava.

Filip BENES was born in Bohumin. He received Ph.D. at 2012, his thesis was focused on RFID in retail. He has 7-year experience with research work on projects at the ILAB RFID. Experience in the fields of AutoID, Augmented Reality and IoT. His abilities include, above all, system analysis of a problem, testing, analysis and design of HW and SW equipment, research of RFID potential in perspective fields.

Jiri SVUB was born in Uherske Hradiste. He has 7-year experience with scientific work on projects at the ILAB RFID. His experience covers fields of automation, AutoID, technical equipment. His abilities include, above all, system analysis, analysis and design of HW and SW equipment, lab testing and experiments with RFID technology, statistic data processing, research of RFID potential in perspective fields.

Yong-Shin KANG was born in Suwon, South Korea. He worked as a researcher associate in u-SCM

Research Center at Dongguk University-Seoul in Republic of Korea. Since March 2016 he has been a research professor at Sungkyunkwan University. He received his B.Sc., M.Sc., and Ph.D. degree in Industrial and Systems Engineering from Dongguk University-Seoul, Korea. His current research interests include RFIDenabled supply chain traceability, sensor-integrated RFID application in manufacturing and logistics, modelling & simulation for u-supply chain and manufacturing processes, and prediction of product quality from real-time sensor data.

Jakub UNUCKA was born in Ostrava. He received his M.Sc. degree at Faculty of Mechanical Engineering at VSB–Technical University of Ostrava. He finished his MBA studies at 2012 at the Faculty of Economics, at the same university. Since 2014 he has studied doctoral degree program at Faculty of Mining and Geology at VSB–Technical University of Ostrava. He worked as head of the company AutoCont IPC. Since October 1998 he has been an executive director of the company Gaben. His research area is Auto-ID technologies.

Lukas VOJTECH was born in Nachod. He received the M.Sc. degree in electrical engineering from the Czech Technical University in Prague, Faculty of Electrical Engineering, in 2003. In 2005, he received the bachelor degree engineering pedagogy from the Masaryk Institute of Advanced Studies in Prague. In 2010, he received the Ph.D. degree from Czech Technical University in Prague. Currently, he works as an assistant professor at the Department of Telecommunication Engineering. His research interests include wireless technologies, technology RFID and mainly EMC in area of shielding materials.

Vladimir KEBO was born in Ostrava. He works at the University since 1984, since 2015 he is a researcher at CAIT, he used to work at the positions of vice-chancellor for research, development and international cooperation, vice-chancellor for cooperation with industry since 2003. Since 2009 he has been a presidium member of the Technology Agency of the Czech Republic (TA CR). His research area includes IoT, automatic control, Industry 4.0 and multi-agent and holonic systems.

Jongtae RHEE received the B.Sc. degree from Seoul National University, the M.Sc. degree from the Korea Advance Institute of Science and Technology, and the Ph.D. degree from the University of California, Berkeley. He is currently a Professor of Industrial and Systems Engineering and the Director of the Ubiquitous Supply Chain Management Research Centre with Dongguk University, Seoul, Korea. His current research interests are RFID, deep learning, image processing, artificial intelligence, fuzzy control and car sharing.

9

MODEL OF ELECTRIC AND HEAT BALANCE OF BIOGAS STATION

Jiri JANSA, Zdenek HRADILEK

Department of Electrical Power Engineering, Faculty of Electrical Engineering and Computer Science,
VSB–Technical University of Ostrava, 17. listopadu 15, 708 33, Ostrava, Czech Republic

jiri.jansa@vsb.cz, zdenek.hradilek@vsb.cz

Abstract. This paper deals with the model of a biogas station. The introduction provides a description of principle employed in this model. The model works with data reflecting the quantity and type of the input biomass to determine the potential energy and offer a selection from various co-generation units. Other important input parameters comprise fermentors and mainly the material composition of their walls to help with determination of heat losses. The input parameters are accompanied by the energetic balance model for a Biogas Station (BPS) concerned. The next section of this paper provides verification of the model with respect to an actual BPS subject to a series of measurements. These measurements focus on electric and heat parameters. The outcome from these measurements was then used to obtain the energetic balance figure. The comparison of measurement results against the model produces slight deviations of the model from the real biogas station only. This paper has been processed pursuant to a long-term research on biogas stations. The biogas station model will be developed further to obtain a more detailed energetic balance for the biogas station.

Keywords

Biogas station, energetic balance, heat losses, model, power losses.

1. Introduction

Each biogas station is a combined heat and power producing unit. Most of the current operators are focusing on the maximum utilisation of electric power. The heat producing remains mostly in the background with minor usage only. It would be mainly employed for the internal consumption within the BPS, that is warming up the fermentors and possibly even heating in the adjacent buildings. Biogas stations with contemporary design actually make use of some heat, since this parameter constituted one of the pre-conditions for award of subsidies. There are still options being explored to achieve a better efficiency. Using a greater amount of heat seems appropriate. To quantify the amount of heat available for further use, one needs to determine the internal consumption of heat within the BPS. That was the reason why we had opted to develop the model of BPS, which enables the operator to establish the overall energetic balance of the BPS.

2. Model Principle

The model processes the annual figures detailing individual input substrates to calculate the theoretical quantity of biogas produced. The total available power in the fuel is then calculated from this amount of biogas and biogas calorific value. The next step will be to select the type of one or more co-generation units by choosing from the drop-down menu with individual parameters. The model shows us whether the input power of Cogeneration Units (CGU) has exceeded the energy input from fuel. The next step requires entering of parameters for calculation of heat balance in the biogas station. That is the location of BPS – the options available show individual regions. Further details require to include parameters of fermentors, i.e. their height and radius. These values will be processed in the model to calculate the volume and individual areas for determination of heat losses. The list enables a selection of various materials that form the walls, floors and ceilings of fermentors respectively. The list is equipped with the database of most common materials with their heat transfer coefficients. The widths of individual layers are also required. The last important parameter is then represented by the length and char-

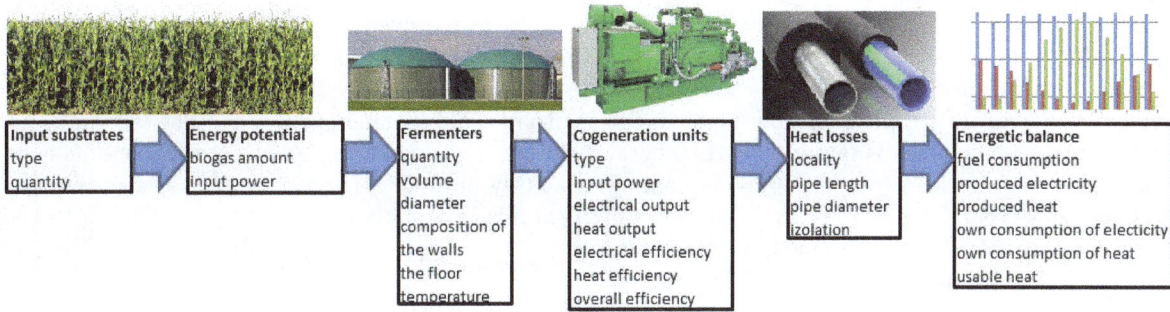

Fig. 1: Model block diagram.

acteristics of pipeline for transporting the heat from the CGU to fermentors. These parameters are used for calculation of heat losses and the internal consumption for warming the fermentors.

3. Operating the BPC Model

Using the values obtained from materials of biogas yield from various types of biomass, the amount of the biogas produced for the individual input substrates is calculated. It is merely a simple multiplication of the input amounts of biomass and the appropriate constant of biogas yield. Subsequently, the individual amounts of biogas produced from the individual ingredients are added up. The energy value of biogas from the biomass feedstocks is calculated from the amount of biogas produced per year as the product of this amount of biogas and the biogas calorific value. As the calorific value of biogas depends on the quantity of contained methane, it ranges from 18 to 25 MJ·m^3. This corresponds to the methane content of $48 - 68$ %. The biogas calorific value 18.61 MJ·m^{-3} was chosen, corresponding to the methane concentration of 50 %. This value was chosen deliberately for the lowest concentration which is normally present during the operation of biogas plants. The sum of all energy values of biogas produced from various types of biomass is the total energy value of bio-

gas obtained from the biomass feedstocks. This value is used to calculate fuel input as a share of the total energy of biomass of the input biogas and the number of hours per year.

An important part of BPS are bioreactors. The basic parameters of the fermenter are calculated according to the desired height and radius of the base. According to the chosen composition of the walls of the bioreactor, the heat transfer coefficient is calculated in (W·m^{-2}·K^{-1}).

$$U = \frac{1}{\dfrac{l_1}{\lambda_1} + \dfrac{l_2}{\lambda_2} + \dfrac{l_3}{\lambda_3} + \dfrac{l_4}{\lambda_4} + \dfrac{l_5}{\lambda_5}}, \qquad (1)$$

where l_i is the thickness of each layer in meters (m), λ_i is the thermal conductivity coefficient (W·m^{-1}·K^{-1}) of the respective layer. During the next steps, the set temperature values will be stored in the bioreactor for the subsequent calculation of heat loss. The calculation will be terminated upon the completion of all the required parameters in other questionnaires.

The form is used to select types of cogeneration units. To facilitate their selection, the calculated value of the value of the total fuel input is already loaded. It was already calculated after entering the first form, i.e. the amount of the biomass feedstocks. After selecting cogeneration units from the list, the user can check the

Fig. 2: Form cogeneration units.

input usage of the fuel. This button is called "Recalculation". After its activation, the value of the differential output will appear in the text field "Remaining output". This gives information on how much input power remains, or how much is missing for the operation of the selected cogeneration units. The remaining differential input is equal to the difference between the total input power of the biogas obtained from biomass, and the sum of the input power required for each of the selected cogeneration units:

$$P_{pz} = P_p - \sum_{i=1}^{5} P_p i, \qquad (2)$$

where P_{pz} is the differential input (kW), P_p is the total input power in the fuel (kW) and P_{pi} is the input power of the relevant cogeneration unit.

The subsequent step after the optimization of the selected cogeneration units is the choice of the service interval of a regular service outage associated with the maintenance of cogeneration units. The length of the service interval is fixed for 3 hours. This period was determined based on the real biogas experience. This period gives the operator enough time for maintenance, which includes checking operational fillings, cleaning or the replacement of filters. In the case of the cogeneration unit, using a spark to ignite the mixture in the cylinder, also for cleaning or replacing the spark plugs. This value is used to calculate the number of monthly hours of the cogeneration unit operation.

The next step after selecting the cogeneration unit is the transition to calculate the heat loss. First, we calculate the heat loss from the pipe leading from the cogeneration unit to the bioreactors. After the selection of standard pipe diameters from the database, appropriate parameters are selected – outside diameter, inside diameter and wall thickness. Another important parameter is the thickness of the insulation, which is

again selected from the used standard sizes. To calculate the heat loss in the pipe, it is necessary to enter the pipe length and the temperature of the medium in the output of the cogeneration unit. The last entry is the site selection, which assigns the corresponding average monthly outdoor temperature after the corresponding region is selected [1].

Consequently, the heat loss of the pipe Q_{ptr} (W) can be calculated:

$$Q_{ptr} = U_0 \cdot l \cdot (t_{in} - t_{out}), \qquad (3)$$

where l is the length of the pipe (m), tin the medium temperature on the output of the cogeneration units (°C), t_{out} outdoor temperature (°C) – it corresponds to the selected location and the appropriate month.

After selecting the location, the heat loss of the bioreactors Q_f (W) can be calculated:

$$Q_f = Q_p + Q_s + Q_{st}, \qquad (4)$$

where Q_p is the heat loss through the floor (W), Q_s the wall heat loss (W), Q_{st} the heat loss through the ceiling (W).

This is the total loss of the bioreactor, but the decomposition of biomass leads to the production of heat, this heat Q_b (W) is determined by the following formula:

$$Q_b = \beta \cdot V_f, \qquad (5)$$

where β is the coefficient of heat production by the decomposition of biomass (W· m^{-3}), V_f is volume of fermenter (m^3). This coefficient was determined from practical measurement when the heat flowing for heating the bioreactor at known outdoor temperature was measured. The heat loss was calculated using the known structure composed of the bioreactor. The difference between the heat flowing to the bioreactor and

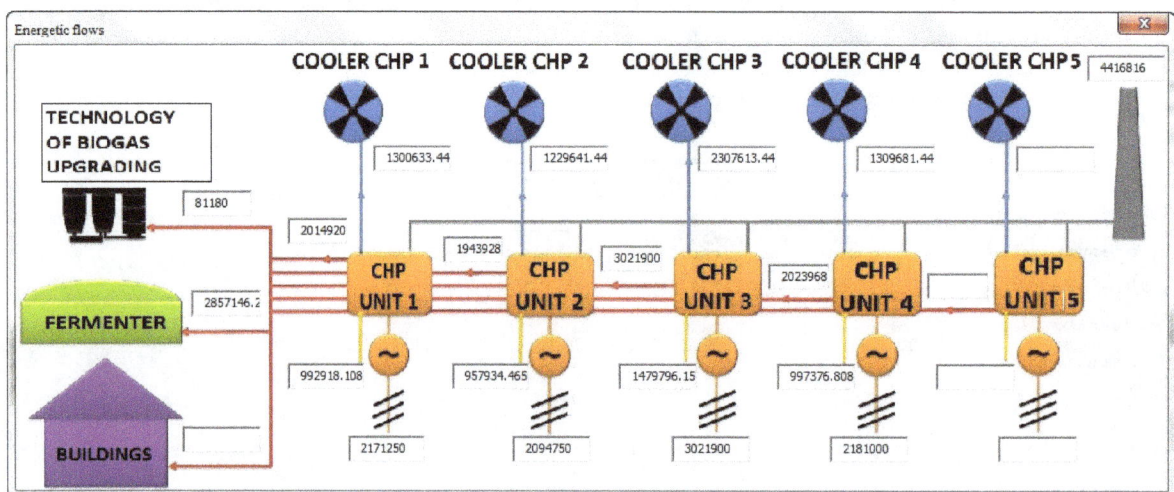

Fig. 3: Energetic flow scheme.

the heat loss of the fermenter, the value of the missing heat was obtained. The heat inside the bioreactor must be formed by decomposition of biomass, by dividing this value by the bioreactor volume, we obtain the value of the coefficient $\beta = 85$ W·m^{-3}.

The heat required for heating the bioreactor Q_v (W) is then given as the difference of the heat loss Q_f (W) and the heat resulting from biomass degradation Q_b (W):

$$Q_v = Q_f - Q_b. \tag{6}$$

For the total heat loss of the biogas plant Q_{bps} (W) the following formula then applies:

$$Q_{bps} = Q_{v1} + Q_{v2} + Q_{v3} + Q_{bp}, \tag{7}$$

where Q_{v1} (W) is the heat required for heating the first bioreactor, Q_{v2} is the heat required for heating the second bioreactor, Q_{v3} is the heat required for heating the after-bioreactor, Q_{bp} (W) the heat needed for the biogas treatment – to remove moisture (dehumidification):

$$Q_{bps} = V_{bp} \cdot \mu_{bp}, \tag{8}$$

where V_{bp} is the biogas consumption (m^3), μ_{bp} the coefficient of thermal energy consumption for dehumidification (W·m^{-3}), 1 (m^3) of raw biogas $\mu_{bp} = 18$ W·m^{-3}.

This value of the coefficient μ_{bp} was determined based on measurements carried out on real BPS, wherein the amount of heat flowing into the device for biogas treatment was measured while the value of biogas flowing into all the cogeneration units was read.

The result of all the specified parameters is the summarizing diagram of energy flows of the whole BPS per year.

- Energy required for biogas treatment – it is the energy that is needed to modify the biogas for the entire BPS according to the consumption of all cogeneration units.

- Energy for bioreactors – it is the energy that must be supplied for the bioreactors to cover the heat loss.

- Input energy to the cogeneration unit – it is the energy contained in the consumed biogas by the corresponding cogeneration unit according to the consumption H and operating hours per year.

- Electricity – it is the amount of electricity produced by the cogeneration unit according to the electrical output P_{el}, electrical efficiency η_{el} and operating hours.

- Thermal energy for own consumption – it is the energy supplied by the cogeneration unit for its own consumption BPS, it is calculated from its own heat consumption BPS divided by the proportional output of the relevant cogeneration unit.

- Cooling energy – it is the energy which must be diverted into the environment due to the cooling of the cogeneration unit – it is determined as the thermal energy produced by the cogeneration unit according to the heat output P_t, thermal efficiency η_t and operating hours, from which the heat, which the cogeneration unit supplies for its own consumption, is subtracted.

- Lost energy – it is the energy that is not utilized and its use is very problematic – it is the energy contained in flue gases – losses by incomplete combustion, heat losses of flue gas, because it cannot be cooled down to ambient temperature due to condensation in the exhaust system and insufficient thrust from the combustion chamber, the energy radiated by the surface of the cogeneration unit into the environment – every solid hotter than the environment radiates some heat. It is calculated as the difference of the input energy into the cogeneration unit and the output energy of the cogeneration unit, basically it is the remainder after deducting the electrical and thermal efficiency from 100 percent.

4. Model Verification

The model has been compared to the actual situation measured on the real biogas station. The Biogas Station (BGS) subject to measurement is located within the territory of Moravian-Silesian Region. This BPS is situated right in the premises of a pork farming enterprise processing mainly maize silage and pig's slurry. The main reason for this location is the source of pig's slurry for supply of fluid processed by the wet fermentation technology. Other reasons include the opportunity to use the waste heat in the heating system of the adjacent pigsty, the office building and the newly built harvest processing line in summer months. The biogas station comprises two fermentors, each offering the available volume of 1630 m^3 and the secondary fermentor with the capacity of 2090 m^3. The installed power capacity is equal to 1090 kW and the heat output is 1080 kW. Transformation of biogas into electric power is handled by four co-generation units. There are three identical compression ignition units delivering the output of 250 kWe and one spark ignition unit with the output of 340 kWe [2].

4.1. Measurement of Electric Parameters

The measurement was performed using an automatic digital measuring device Grid Analyser ENA 500 made by ELCOM, working in one-minute increments to measure and save effective values of phase voltages, currents per individual phase and power factors. The remaining values, i.e. effective, reactive, and apparent outputs were completed by calculation run in the device automatically.

Measurements of voltage were taken right at bus bars inside the distribution board RH1 and currents were measured with jaw currents converters MT-UNI using transformers 1500/5A already installed [2].

Fig. 4: Plant room and biogas processing unit.

4.2.   Measurement of Heat and Internal Consumption of Heat Energy

The measurement taken by the contact flow meter produced mass flow rates of heat transfer fluids. Thermal element was used for measurement of temperature at the outlet and the return pipe, the specific heat capacities were determined and established per type of the flowing fluid and the temperature values obtained by measurement were processed using the Engineering Equation Solver software. This data served for calculation of heat output carried by individual pipelines [3].

Since the flow of biogas into CGUs was measured under conditions different to the particular standards, these values had to be converted with respect to the temperature and pressure of gas to establish the so called 'normal conditions, i.e. the temperature of 0 °C and the pressure of 101,325 Pa. Energy inputs from biogas to co-generation units were calculated using the converted biogas flow with respect to normal conditions and the calorific capacity of biogas determined by analysis of chemical composition of biogas. The chemical constitution of biogas was previously conducted per order from the biogas station operator [4].

The energy input from biogas entering the CGU and production of electric power on the generator in CGU

were used to determine the electrical efficiency of CGU [4].

The BPS was run at almost the full nominal electric power, i.e. 1068 kW. The Zspark ignition unit was operated at the level of 340 kWe, which is the normal operating power. However, this is a unit with the nominal electric power equal to 350 kW. Yet this unit is operated at the output reduced by 10 kW here. That was because the contractor strove towards meeting the requirement for permitted installed capacity specified by the distribution grid operator. Two of the identical compression ignition units rated for 250 kW were operated at the output level of 249 kW, while the last CGU was operated at the level of 230 kW. The deviation of output from the nominal values ranges between 0 %, through 0.4 % up to the highest deviation of 4 %. The overall deviation from the nominal output from the entire BPS is then equal to 2 % [6].

Fig. 5: Schnell ZV250-V5 co-generation unit.

Fig. 6: Agrogen BGA222 co-generation unit.

The energy input from fuel has been obtained by conversion of the data reflecting the biogas flow into individual CGUs, its temperature and pressure. The electric input for the spark ignition unit was 864 kW, which is 22 kW below the level stated by its manufacturer. However, the value presented by the manufacture applies to CGUs operated at the full nominal output level. The energy input from fuel on two compression ignition units ranged around 565 kW, whereas the energy input for the last unit working at the lowest output level then approximated 545 kW. The deviation established in energy input from fuel then oscillated around 2.5 % in both the positive and negative terms. The manufacturer states the consumption may vary within 10 %. The heat output was measured using a flow meter and a thermal element at the outlets from individual CGUs; the values obtained were then converted. The output from compression ignition units ranged from 203 kW, through 219 kW, up to 220 kW, where the latter corresponds with the nominal output value. The compression ignition unit worked with the heat output of 359 kW that exceeds the value stated by the manufacturer. That was caused by operation of the CGU at a lower output level, since the machine is not working within the top efficiency region. The total heat output delivered by the BPS during the measurement period was equal to 1005 kW. These and further parameters can be found in the Tab. 1.

5. Energy Balance of BGS

The BPS model works with the values obtained from the manufacturer of individual co-generation units. As already mentioned in the introduction, the BPS model contains a database of biogas-fuelled CGUs from leading suppliers currently available on the market. To enable comparison of the values measured, the model

had to be extended with older versions of co-generation units used within the specific BPS subject to measurement. These units have been undergoing continuous improvements untill now and the parameters of existing units would not match the actual units employed. This extension included specifically the parameters of units SCHNELL type ZV250-V5 [5] and AGROGEN type BGA 222 [6]. These parameters are also presented in the synoptic Tab. 2.

It shall be mentioned once again that the AGRO-GEN BGA 222 CGU is defined by these parameters for nominal electric power of 350 kW. The manufacturer states this unit may be operated at a reduced output rate, yet there are not detailed parameters supplied for it.

6. Comparison of Model with Measurement Results

The values produced by identical SCHNELL ZV250-V5 co-generation units were averaged for better comparison. The results of comparison can be viewed in the following synoptic tables for particular types of CGUs.

These tables show that the difference between this model and the actual measurement within the scope of several per cent only. The lowest deviation is 0.4 % and the highest one equals 5.2 %. This large deviation is caused by operation of the CGU at a level different from its nominal parameters. The average deviation is 2.4 % [7].

Tab. 1: CGU parameters measured.

	Schnell ZV250-V5	Schnell ZV250-V5	Schnell ZV250-V5	Agrogen BGA222
Energy input from fuel	566 kW	564 kW	545 kW	864 kW
Electric power	246 kW	249 kW	230 kW	340 kW
Electric efficiency	44.1 %	44.3 %	42.2 %	39.4 %
Heat output	220 kW	219 kW	203 kW	359 kW
Heat efficiency	38.9 %	38.9 %	37.3 %	41.5 %

Tab. 2: Theoretical parameters of CGUs.

	Schnell ZV250-V5	Agrogen BGA222
Energy input from fuel	549 kW	886 kW
Electric power	250 kW	350 kW
Electric efficiency	45.5 %	39.5 %
Heat output	220 kW	350 kW
Heat efficiency	40 %	39.5 %

Tab. 3: Comparison of parameters of Schnell ZV250-V5 CGU.

Schnell ZV250-V5	Measurement	Model	Difference in values	Difference in percentage
Energy input from fuel	560 kW	549 kW	11 kW	2.003643
Electric power	249 kW	250 kW	-1 kW	-0.4
Electric efficiency	44.5 %	45.5 %	-1 %	-2.35643
Heat output	217 kW	220 kW	-3 kW	-1.36364
Heat efficiency	38.75 %	40 %	-1.25 %	-3.30114

Tab. 4: Comparison of parameters of agrogen BGA222 CGU

Agrogen BGA222	Measurement	Model	Difference in values	Difference in percentage
Energy input from fuel	864 kW	886 kW	-22 kW	-2.48307
Electric power	340 kW	350 kW	-10 kW	-2.85714
Electric efficiency	39.4 %	39.5 %	-0.1 %	-0.3836
Heat output	359 kW	350 kW	-9 kW	2.571429
Heat efficiency	41.5 %	39.5 %	-2 %	5.183201

7. Conclusion

This paper deals with the BPS model and its verification against the actual parameters measured. The model has been operated with the CGUs selected to match the ones in the real BPS. Those are co-generation units Schnell ZV250-V5 and Agrogen BGA222. To enable comparison of the parameters measured with respect to the electric power, the heat output and the energy input from fuel against the data in the model, we have also compared the calculated parameters based on such values obtained by measurement; those are details of the electric power, the heat output and the overall efficiency. The values measured show a slight deviation from values of the model based on the nameplate parameters advised by the equipment manufacturer. This deviation progresses in both directions, i.e. towards the positive (the value measured gains higher levels exceeding the model parameters) as well as the negative values respectively. However, the deviation experienced in the best scenario is 0.4 %, while the worst one is then equal to 5.2 %. The average deviation is then approximately 2.4 %. We were dealing mainly with the verification of parameters inherent to the model and used as background data to establish the energy balance of biogas station. This model still is and will remain under development towards a more detailed energy balance of BPS, i.e. hours of operation, the amount of electric power produced, the amount of heat produced, the internal consumption of electric power, the internal heat consumption. The model should further assess the option to use any unused heat produced by co-generation units emitted to the surrounding environment, which is to improve the overall efficiency of the biogas station.

Acknowledgment

This research was partially supported by the SGS grant from VSB–Technical University of Ostrava (No. SP2016/95).

References

[1] Map browser of CHMI outputs. *Czech Hydro Meteorogical Institute* [online]. Available at: http://portal.chmi.cz/files/portal/docs/poboc/OS/OMK/mapy/prohlizec.html?map=T_M.

[2] JANSA, J., Z. HRADILEK and J. JANSA. Energy Balance of Biogas Station. In: *8th International Scientific Symposium on Electrical Power Engineering.* Stara Lesna: Technical University of Kosice, 2015, pp. 168–171. ISBN 978-80-553-2187-5.

[3] JANSA, J. Analysis of Measuring the Thermal Part of the Biogas Plant. In: *Ph.D. Workshop of Faculty of Electrical Engineering and Computer Science.* Ostrava: VSB–Technical University of Ostrava, 2015, pp. 30–35. ISBN 978-80-248-3787-1.

[4] JANSA, J. *Heat Flow Measurement Report from BGS Lodznice.* Ostrava, 2015.

[5] Technical data Gensets BGA. *AGROGEN gasmotoren.* 2010.

[6] Produktubersicht 2012/2013. *SCHNELL motor.* 2012.

[7] JANSA, J. and Z. HRADILEK. Energy Balance Model of Biogas Station. In: *12th Electrical Networks Workshop.* Ostrava: VSB–Technical University of Ostrava, 2015, pp. 7–14. ISBN 978-80-248-3858-8.

About Authors

Jiri JANSA was born in Ostrava, Czech Republic. He is a Ph.D. student at VSB–Technical University of Ostrava, Faculty of Electrical Engineering and Computer Science, Department of Electrical Power

Engineering. The topic of his thesis focuses on Energetic problems of biogas power plants.

Zdenek HRADILEK was born in Brno. After graduation of college education at Faculty of Electrical Engineering and Computer Science Brno University of Technology in 1962 he worked as a technician in company Southern Moravian power plants in Brno, than he worked as a major power-supply director in Heat-supply Ostrava and from 1966 until now he is at the VSB–Technical University Ostrava. His scientific preparation graduated by his candidate dissertation defending at the Brno University of Technology in 1972. He defended his doctoral thesis at the Czech Technical University in Prague in 1988 and was appointed as professor.

10

THE DETECTION PROPERTIES COMPARISON FOR THE SNO₂ GAS SENSOR IN DIFFERENT HEATING REGIMES AND THE STEP CHANGE IN CONCENTRATION

Libor GAJDOSIK

Department of Telecommunications, Faculty of Electrotechnical Engineering and Computer Science,
VSB–Technical University of Ostrava, 17. listopadu 15, 708 33 Ostrava-Poruba, Czech Republic

libor.gajdosik@vsb.cz

Abstract. *Non-periodical heating regimes and also periodical heating regimes were applied for the sensors TGS 813 and SP 11. Ethanol vapour in air was used as a detected substance. It was found that the sensitivity is better in the periodical square-wave heating regime among the tested regimes. The stabilisation of the sensor conductivity, when the concentration abruptly varies its value, takes longer compared to the constant heating regime at maximum heating voltage.*

Keywords

Dynamic response, semiconductor gas sensor, temperature modulation, thermal cycling, tin dioxide.

1. Introduction

Tin dioxide gas sensors have been used as detectors of the occurrence of gaseous substances in the air in numerous applications for years. In each of these applications, the occurrence means the change of the concentration of a detected substance in the sensor surrounding. Sensors are often operated in the heating regime dependent on time, because of better detection properties compared the constant voltage heating regime [1], [2], [3], [4], [5], [6], [7], [8], [9], [10], [11] and [12]. The overview of the often used heating time-dependent regimes is presented in [17]. It follows, that detection properties of the sensor are affected by its thermal inertia. The comparison of the detection properties of the sensor in different heating regimes when concentration is changed has not been carried out yet and this is performed here.

2. Experiments

The sensor was tested in the glass chamber of the volume 2700 ml. In the tested chamber, the atmosphere of clean air of 31 % relative humidity was used either with or without the concentration of 100 ppm of ethanol vapour. The concentration of 100 ppm was formed by the injection of the certain volume of the saturated ethanol vapour into the tested chamber filled with clean air of 31 % relative humidity. The saturated ethanol vapour was taken above the liquid in the closed bottle at the ambient temperature. The Antoine equation was used for the calculation of the saturated vapour concentration. The sensors were inside the chamber, heated by the heating voltage and the responses of the sensors were sensed. The response of the sensor is a term used here to refer to the electrical conductivity. The heating voltage was realized by the following manners: the time independent constant value, the step change from one value to the next one, the non-periodical consecutively increasing linear function, denoted here the ramp function, and periodical functions. The periodical functions were formed by the following formulas and used in the programmable voltage source, which heated the sensors: the sine-wave function was formed according to Eq. (1) and Eq. (2):

$$U(t) = U_S + U_A \sin(\omega t), \quad (1)$$

$$U_S = \frac{U_M + U_m}{2}, \qquad U_A = \frac{U_M - U_m}{2}, \quad (2)$$

where t is the time, ω is the angular frequency, U_M is the maximum value of the heating voltage a U_m is the minimum value of the heating voltage, the periodical symmetrical square-wave function was formed according to Eq. (3) and Eq. (4),

$$U(t) = U_m \quad \text{for} \quad 0 \le t < \frac{T}{2}, \quad (3)$$

$$U_t = U_M \quad \text{for} \quad \frac{T}{2} \leq t \leq T, \qquad (4)$$

and the symmetrical triangle-wave function according to Eq. (5), Eq. (6) and Eq. (7),

$$U(t) = U_S + \frac{4U_A}{T}t \quad \text{for} \quad 0 \leq t < \frac{T}{4}, \qquad (5)$$

$$U(t) = U_S + 2U_A - \frac{4U_A}{T}t \quad \text{for} \quad \frac{T}{4} \leq t < \frac{3}{4}T, \qquad (6)$$

$$U(t) = U_S - 4U_A + \frac{4U_A}{T}t \quad \text{for} \quad \frac{3T}{4} \leq t < T. \qquad (7)$$

The values $U_M = 5$ V, $U_m = 2$ V, 2.5 V and 3 V were chosen respectively. The time for setting next value of the heating voltage was chosen considering practical experiences $t_s = 1$ s. The response was sensed just before setting the next value of the heating voltage. The sensor response was sensed in 32 equidistant points during one period T of the heating voltage. It leads to the frequency of the principal harmonic component of the heating voltage $f = 1/(32 \cdot t_s) = 0.0312$ Hz. The ramp function was formed according to Eq. (5). Its value was changed in time sufficiently slowly, the value of the response at the given value of the heating voltage can be considered independent of time. The concentration of ethanol vapour in the air was kept either constant equal 100 ppm or its step change in time either from 0 to 100 ppm or from 100 ppm to 0 was used. The step change of the concentration was carried out by the following manner. In the case of the step from 0 to 100 ppm, the certain volume of the ethanol vapour was injected into the gas chamber to form 100 ppm concentration while the sensors were inside. In the case of the step from 100 ppm to 0, the sensors were taken out of the chamber and immediately put into the second chamber filled with clean air of 31 % relative humidity.

3. Theory

The sensor SP 11 is formed on the alumina substrate on which the measuring electrodes are printed and covered by a metal oxide detection layer. A heater is printed on the reverse side of the substrate [13].

The sensor TGS 813 consists of the heating system realized by a resistive wire coil, which is located inside an alumina ceramic tube. The surface of the ceramic tube is covered with a pair of electrodes positioned to measure the electrical conductivity of the detection metal oxide layer deposited on the electrodes [14].

A detected substance in gaseous surrounding of the sensor must first diffuse into the detection layer to be detected. The detection layer in both cases is a polycrystalline porous structure of small pellets which touch each other and the space between them is filled with the air. A chemical substance in the gas phase can be detected, if the substance diffuses into the space between the pellets.

The electrical voltage heats the resistive conductor and heat penetrates through the ceramic layer into the detection layer. The detection of the gas became after the specific temperature of the detection layer was reached, which was manifested by the change of the electrical conductivity of the detection layer. It is possible to simplify the solution for transient heat conduction in the sensitive element according the literature [15]. The alumina substrate in the case of SP 11 is a plate of 2×2 mm dimensions and of 0.3 mm thickness. The alumina substrate in the case of TGS 813 is a cylinder of the similar size - the wall of the thickness 0.3 mm and of the length 2 mm. In both cases heat penetrates through 0.3 mm thick layer. The thickness is considerably smaller than the length and we consider transient heat conduction for relative short time after an abrupt temperature change at the beginning. According the literature[15] it does not matter on the shape of the object and we can consider it as an infinite plate of the finite thickness. It follows, that the Fourier Kirchhoff equation can be solved in one dimension. The sensing element of the sensor is surrounded by quiet air. The heat transfer coefficient of the air can be estimated $h = 8$ W $\cdot$ m$^{-2} \cdot$ K^{-1} [15] and alumina thermal conductivity $k = 20$ W $\cdot$ m$^{-1} \cdot$ K^{-1} [16]. It follows that air acts as a good thermal insulation of alumina substrate and during the transient state practically no heat penetrates from the plate into the air. Thus, it is possible to do further simplification. The alumina substrate can be replaced by the infinite half-space $x \geq 0$. The plane $x = 0$ represents the site of the heating system. At first the system is at uniform temperature ϑ_0 for all x. In the time $t = 0$ the step change of the heating is carried out to the temperature ϑ_1.

Transient heat conduction in the sensor can be solved by the Fourier Kirchhoff equation. We consider heat conduction in the half-plane $x \leq 0$:

$$a\frac{\partial^2 \vartheta}{\partial x^2} = \frac{\partial \vartheta}{\partial t}, \qquad (8)$$

where a is the thermal diffusivity of the material (m$^2 \cdot$s^{-1}), ϑ is the temperature and t is the time. The heating system is supposed to be in $x = 0$. Let us assume that the condition is given by the following equation:

$$\vartheta(x, 0) = \vartheta_0, \qquad (9)$$

where ϑ_0 is the initial temperature at the time $t = 0$ equal in every point of x. In the time $t = 0$ the step change of the temperature is carried out in the line $x = 0$ to the temperature ϑ_1. Then it is true:

$$\vartheta(0, t) = \vartheta_1. \qquad (10)$$

We calculate the distribution of the temperature in arbitrary point $x \neq 0$ at the time t. The Laplace transform can be applied on both sides of Eq. (8). The real variable of time t is transformed into the complex variable p. Let us designate the Laplace transform of the temperature $\Theta(x, p)$. We obtain the Laplace transform of Eq. (8) in the following form:

$$p\,\Theta(x, p) - \vartheta_0 = a\frac{\mathrm{d}^2\Theta(x, p)}{\mathrm{d}x^2}. \qquad (11)$$

Equation (11) is an ordinary differential equation, because the derivative of $\Theta(x, p)$ with respect to x does not depend on p. The associated homogeneous equation is:

$$\frac{\mathrm{d}^2\Theta(x, p)}{\mathrm{d}x^2} - \frac{p}{a}\Theta(x, p) = 0. \qquad (12)$$

The characteristic equation of Eq. (12) is:

$$\lambda^2 - \frac{p}{a} = 0. \qquad (13)$$

The roots of Eq. (13) are:

$$\lambda_{1/2} = \pm\sqrt{\frac{p}{a}}. \qquad (14)$$

The general solution of Eq. (12) is in the form:

$$\Theta(x, p) = K_1(p)\,\mathrm{e}^{\lambda_1 x} + K_2(p)\,\mathrm{e}^{\lambda_2 x}, \qquad (15)$$

where $K_1(p)$ and $K_2(p)$ are proper coefficients, which are independent of x but they are dependent on p. The temperature must be finite $\Theta(x, p) < \infty$ at $x \to \infty$, because in the half-plane $x \leq 0$ there are no additional sources of heat. This is why the complementary solution of the homogeneous equation is considered only in the following form containing the negative root λ_2:

$$\Theta(x, p) = K_2(p)\,\mathrm{e}^{-\sqrt{\frac{p}{a}}x}. \qquad (16)$$

The particular solution of non-homogeneous equation Eq. (11) can be found by the undetermined coefficient method. The particular solution describes the steady state which is independent on x. It can be supposed a constant designated $\Theta(x, p) = B$. After substitution into Eq. (11) we obtain:

$$pB - \vartheta_0 = a\frac{\mathrm{d}^2B}{\mathrm{d}x^2} = 0. \qquad (17)$$

As B is independent on x, its derivative with respect to x must be zero. Then it is true, according to Eq. (17), that the constant B equals:

$$B = \frac{\vartheta_0}{p}. \qquad (18)$$

The summation of the Eq. (16) and Eq. (18) makes the general solution of Eq. (11):

$$\Theta(x, p) = K_2(p)\,\mathrm{e}^{-\sqrt{\frac{p}{a}}x} + \frac{\vartheta_0}{p}. \qquad (19)$$

We determine the coefficient $K_2(p)$ in Eq. (19) using the Laplace transform of condition Eq. (10):

$$L\{\vartheta_1\} = \frac{\vartheta_1}{p}. \qquad (20)$$

We substitute Eq. (20) into Eq. (19). It must be true for $x = 0$, that:

$$\frac{\vartheta_1}{p} = K_2(p)\cdot 1 + \frac{\vartheta_0}{p}. \qquad (21)$$

It follows from Eq. (21), that the coefficient $K_2(p)$ equals:

$$K_2(p) = \frac{\vartheta_1}{p} - \frac{\vartheta_0}{p}. \qquad (22)$$

Then the solution of Eq. (19) has the following form:

$$\Theta(x, p) = \left(\frac{\vartheta_1}{p} - \frac{\vartheta_0}{p}\right)\mathrm{e}^{-\sqrt{\frac{p}{a}}x} + \frac{\vartheta_0}{p}. \qquad (23)$$

The solution must be re-transformed into the time domain by using the inverse Laplace transform. According to [18], if:

$$L^{-1}\left\{\frac{\mathrm{e}^{-\alpha\sqrt{p}}}{p}\right\} = 1 - \frac{2}{\sqrt{\pi}}\int_0^{-\frac{\alpha}{2\sqrt{t}}}\mathrm{e}^u\cdot\mathrm{d}u =$$

$$= \mathrm{erfc}\left(\frac{\alpha}{2\sqrt{t}}\right), \qquad (24)$$

where α represents:

$$\alpha = \frac{x}{\sqrt{a}}, \qquad (25)$$

and if:

$$\mathrm{erfc}\left(\frac{\alpha}{2\sqrt{t}}\right) = 1 - \mathrm{erf}\left(\frac{\alpha}{2\sqrt{t}}\right), \qquad (26)$$

then it is possible to express Eq. (23) by using Eq. (24) and Eq. (26) in the following form:

$$\vartheta(x, t) = \vartheta_1\,\mathrm{erfc}\,(z) + \vartheta_0\,\mathrm{erf}\,(z),$$

$$\text{where } z = \frac{x}{2\sqrt{at}}. \qquad (27)$$

If $\vartheta_0 < \vartheta_1$, then Eq. (27) describes the increase of the temperature. For example $\vartheta_0 = 150$ and $\vartheta_1 = 350$. If $\vartheta_0 > \vartheta_1$, then Eq. (27) describes the decrease of the temperature. For example $\vartheta_0 = 350$ and $\vartheta_1 = 150$. If we consider the temperature range between $(150 - 350\,°\mathrm{C})$, the thickness $x = 0.3$ mm, the thermal diffusivity $a = 1.2\cdot 10^{-5}$ mm$^2\cdot$s^{-1} (alumina [16]), the temperature reaches 90% of the target temperature approximately in 0.08 seconds.

4. Results

4.1. Test 1.

The sensor heating was maintained at the constant heating voltage U. The sensor response was sensed after the step change of the concentration from zero to 100 ppm of ethanol vapour in the air was carried out. The obtained results were similar for both used sensors. The results for the sensor TGS813 are presented in Fig. 1.

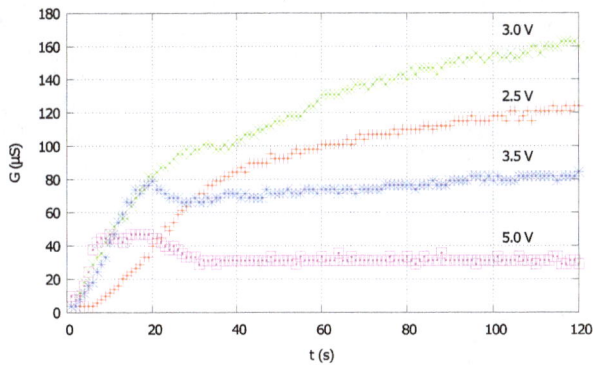

Fig. 1: The electrical conductance G of the sensor TGS813 when the step change of the concentration from 0 to 100 ppm of ethanol vapour was carried out. The constant heating voltage U is a parameter of the curves.

The experiment was repeated for different values of the heating voltage U. It is evident, that the response is dependent on the value of the heating voltage. The response reaches the greatest value for $U = 3$ V, but reaching the steady state value takes longer compared to the case for the voltage $U = 5$ V. The peak (with the duration approximately 25 s) appears in the responses for the voltages $U > 3.5$ V. The system of consecutive reactions between the detected compound and oxygen adsorbed in various forms [20] and [21] occurs to form reaction products during the detection. These phenomenons are dependent on temperature. The peak can be caused by a type of chemical reactions which occur at higher temperatures. It emerges from the experimental results, that it is proper to use higher heating voltage $U = 5$ V to obtain faster response despite the response reaches lower values here compared to the response for the heating voltage $U = 3$ V.

4.2. Test 2.

The sensor was tested in the atmosphere of 100 ppm of ethanol vapour in the air. The heating voltage was kept at $U = 2$ V. The step change of the heating voltage was carried out from $U = 2$ V to $U = 5$ V and the sensor response started to be sensed. The value of 2 V was chosen experimentally, the sensor temperature is

too low to detect the substance, but high enough not to adsorb substances like water which can affect the detection process [21]. The value of 5 V was chosen considering the previous results, to reach quickly the steady state. Measured values of the response for the sensors TGS813 and SP11 are in Fig. 2.

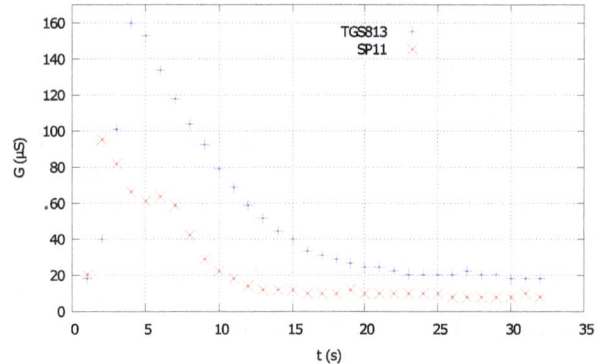

Fig. 2: The electrical conductance G of the sensors when the constant concentration of ethanol vapour in air and the step change of the heating voltage U from 2 V to 5 V were used.

4.3. Test 3.

The sensor was kept in the atmosphere of 100 ppm of ethanol vapour in the air. The heating voltage was increased from $U = 2$ V to $U = 5$ V slowly by the step 20 mV per 10 seconds. The response was sensed 10 seconds after the change of the heating voltage and just before setting the next value of the heating voltage. The step 20 mV per 10 seconds was chosen experimentally, because the responses at given heating voltage are practically independent of time. Measured values of the response for the sensors TGS813 and SP11 are shown in Fig. 3.

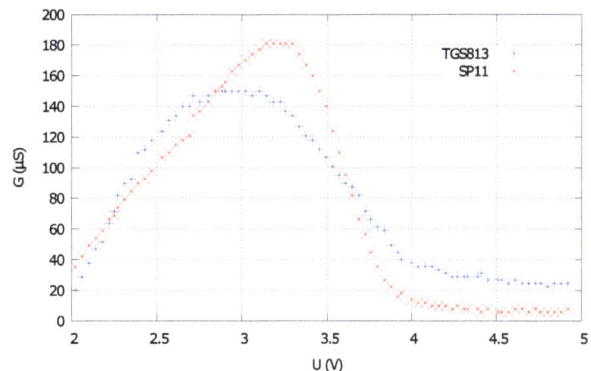

Fig. 3: The electrical conductance G of the sensors when ramp function for the heating voltage U and the constant concentration of the ethanol in the air were used.

It can be seen that the maximum lies in the range of heating voltage $2.5 < U < 3.5$ V of the electrical conductance. This phenomenon can be used for explanation of the peak occurrence in the response in Fig. 2. When the step change of the heating voltage from 2 V to 5 V is considered, see Fig. 2, the temperature of the sensor quickly moves across the value associated with the maximum of the electrical conductivity in Fig. 3.

4.4. Test 4.

The sensors were tested in the periodical steady state heating regime. This means, that the sensors are heated by given periodical voltage in the tested chamber filled with the clean air until the response is periodically stable. It took approximately two minutes. An example of the response sensed in periodical steady state in the clean air is demonstrated in Fig. 4.

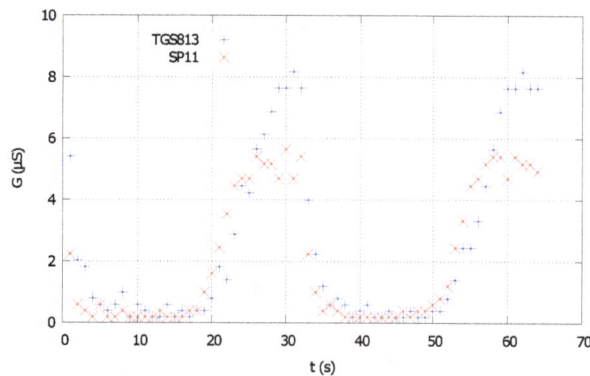

Fig. 4: The electrical conductivity G of the sensors when the periodical square-wave heating voltage of $U_m = 2$ V and $U_M = 5$ V and clean air were used. The record corresponds to two periods of the heating voltage.

At the beginning of the experiment, the concentration was abruptly changed from 0 – 100 ppm and then the response started to be sensed. An example of such a response is in Fig. 5. The change of the concentration caused a transient state and the peaks of the response exceeded 120 μS. It is 2.4 times more then the peak of 50 μS in Fig. 1. This follows better sensitivity of the sensor in the periodical heating regime compared to the constant voltage heating regime.

When the periodical steady state of the response had been restored, see Fig. 6, the peak values of the response decreased to 55 μS. The peak values in Fig. 6 correspond to the peak values in Fig. 1, about 50 μS, where the constant heating voltage of 5 V was used. Then the experiment continued. The step change of the concentration was carried out from 100 ppm to 0 and the response was sensed. The time dependency of the response is exhibited in Fig. 7.

It follows, that the peak values of the response remain almost equal and decrease slowly in time.

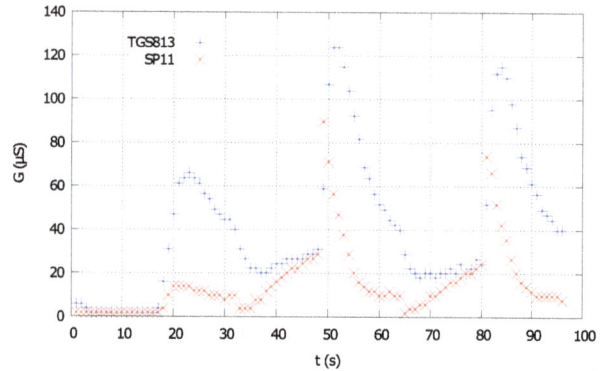

Fig. 5: The electrical conductivity G of the sensors when the periodical square-wave heating voltage of $U_m = 2$ V and $U_M = 5$ V was used. The step change of the concentration from $0 - 100$ ppm was carried out at the beginning of the record. The transient state caused by the step change occurs here. The record corresponds to three periods of the heating voltage.

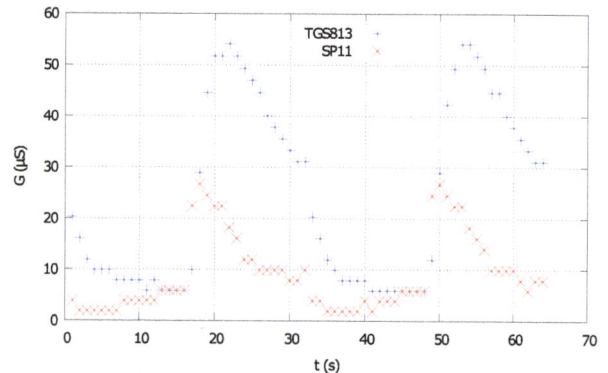

Fig. 6: The electrical conductivity G of the sensors when the square-wave heating voltage of $U_m = 2$ V and $U_M = 5$ V and the constant concentration 100 ppm were used. The record of the periodical steady state corresponds to two periods of the heating voltage.

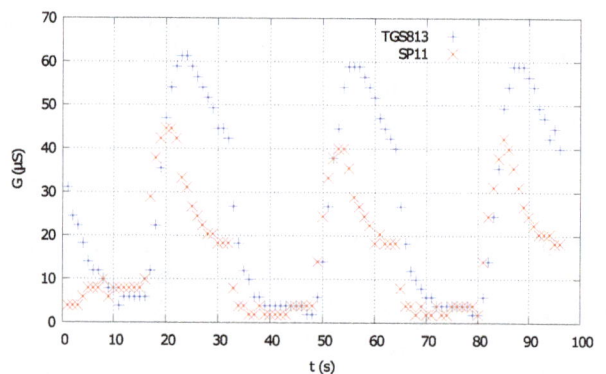

Fig. 7: The electrical conductivity G of the sensors when the periodical square-wave heating voltage of $U_m = 2$ V and $U_M = 5$ V and the step change of the concentration from 100 ppm – 0 were used. The transient state caused by the step change occurs here. The record corresponds to three periods of the heating voltage.

4.5.   Test 5.

The sensors were heated by the constant heating voltage $U = 5$ V in the atmosphere of 100 ppm ethanol vapour in the air until the time independent response was reached. Then the abrupt change of the concentration was carried out to zero and the responses of the sensors started being measured. An example of the responses is presented in Fig. 8.

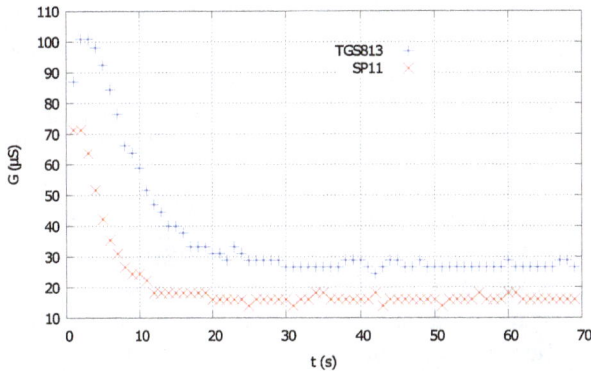

Fig. 8: The electrical conductivity G of the sensors when the constant heating voltage of $U = 5$ V and the step change of the concentration from 100 ppm – 0 were used.

It is demonstrated here, that the response reached its stability after 20 seconds or 10 seconds for TGS 813 or SP 11 respectively, which is considerably shorter time compared to the case in Fig. 7, where the response did not reach the value corresponding to zero concentration of the substance even after 100 seconds. It is possible to explain this phenomenon in the following way: the process of the consecutive chemical reactions, i.e. the rate constant and types of reaction products with different adsorption properties to the surface of SnO_2 [20] [22], [23] and [24], are affected by the variation of the temperature during a periodical heating regime.

4.6.   Test 6.

The sensors were tested in the steady state periodical heating regime. The voltage U_M was maintained at 5 V and the voltage U_m was maintained at values 2 V, 2.5 V and 3 V respectively. U_m was kept constant during given test. The concentration of ethanol vapour was used as the step change from 0 – 100 ppm, constant of 100 ppm and the step change from 100 ppm – 0 respectively for the measurement at given value of U_m.

Different range of heating voltages leads to different range of the temperature of the detection layer. It was found, that when the values of $U_m > 2$ V were used, the peak values of the responses associated to the step change of the concentration from 0 – 100 ppm are lower

compared to the responses at $U_m = 2$ V. The overview of the peak values is presented in Tab. 1.

Tab. 1: The peak values of the electrical conductivity G of the sensors. The square-wave heating voltage of $U_M = 5$ V and different values of U_m were used. The responses were sensed in the steady state periodical heating regime. G_r is the conductivity at the step change of the concentration from 0 – 100 ppm, G_u is the conductivity at the constant concentration of 100 ppm, G_f is the conductivity at the step change of the concentration from 100 ppm – 0.

	G_r (μS)	G_u (μS)	G_f (μS)
	$U_m = 2$ V		
TGS813	120	55	60
SP11	90	28	40
	$U_m = 2.5$ V		
TGS813	65	70	50
SP11	45	60	35
	$U_m = 3$ V		
TGS813	70	50	50
SP11	50	35	40

Identical experiments were carried out when the sine-wave and the triangle-wave heating voltage were applied to the heating system of the sensor. The same frequency ($f = 0.0321$ Hz) as in the square-wave case was used. The results are summarized in Tab. 2.

Tab. 2: The peak values of the electrical conductivity G of the sensors when different types of the heating voltage of the parameters $U_M = 5$ V and $U_m = 2$ V were used. The responses were sensed in the steady state periodical heating regime. G_r is the conductivity at the step change of the concentration from 0 – 100 ppm, G_u is the conductivity at the constant concentration of 100 ppm, G_f is the conductivity at the step change of the concentration from 100 ppm – 0.

	G_r (μS)	G_u (μS)	G_f (μS)
	triangle		
TGS813	90	50	160
SP11	45	30	100
	sine		
TGS813	80	55	140
SP11	35	35	80

It follows from Tab. 2 and Tab. 1, that the peak values of the response are lower in the sine-wave case or in the triangle-wave case compared to the square-wave case when the range of the heating voltage of $U_m = 2$ V , $U_M = 5$ V is used. It can be concluded, that the square-wave heating regime is more advantageous, due to the higher sensitivity. The longer stabilisation time is the disadvantage of the tested heating regimes. The stabilisation time is a term used here to refer to the time, when the sensor response decreases to 110 % or increases to 90 % of the value of the response related to the steady state. But it is possible to modify the periodical heating regime by setting the constant voltage regime of $U = 5$ V after the detection of the occurrence of the substance until the response corresponds to zero concentration. This could shorten the stabilisation time.

5. Conclusion

The experiments were carried out with the commercial sensors TGS813 and SP11. It was found, that the periodical square-wave heating regime is the most advantageous among the tested heating regimes. The stabilisation time of the sensor (time, when the concentration abruptly changes its value) is longer at the periodical heating regime compared to the constant heating regime at maximum heating voltage. It was suggested that a combination of both heating regimes, i.e. periodical and constant voltage, could shorten the stabilisation time and keep higher sensitivity of the sensor. Further research would be target to optimisation of the heating regime of the sensor.

Acknowledgment

This study was enabled due to Project TA03020439 entitled The Security of Optical Transferral Networks and the Development of Optoelectronics Devices for Power Energetics and Energy Piping Systems provided by TACR (The Technology Czech Science Foundation) and also due to Project CZ.1.07/2.3.00/20.0217 entitled The Developement of Research Team of Telecommunication Technics with Relationship to Foreign Cooperation provided by MSMT (The Ministry of Education).

References

[1] XINGJIU, H., J. LINHUAI, S. DONGLIANG, P. ZONXIN and Y. ZENGLIANG. Rectangular mode of operation for detecting pesticide residue by using a single SnO_2 -based gas sensor. *Sensors and Actuators B: Chemical*. 2003, vol. 96, iss. 3, pp. 630–635. ISSN 0925-4005. DOI: 10.1016/j.snb.2003.07.006.

[2] XINGJIU, H., M. FANLI, P. ZONGXIN, X. WEIHONG and L. JINHUAI. Gas sensing behavior of a single tin dioxide sensor under dynamic temperature modulation. *Sensors and Actuators B: Chemical*. 2004, vol. 99, iss. 2–3, pp. 444–450. ISSN 0925-4005. DOI: 10.1016/j.snb.2003.12.013.

[3] HERRERO-CARON, F., D. J. YANEZ and F. De BORJA RODRIGUEZ. An active, inverse temperature modulation strategy for single sensor odorant classification. *Sensors and Actuators B:*

Chemical. 2015, vol. 206, iss. 1, pp. 555–563. ISSN 0925-4005. DOI: 10.1016/j.snb.2014.09.085.

[4] NAKATA, S., H. NAKAMURA and K. YOSHIKAWA. New strategy for the development of a gas sensor based on the dynamic characteristics: principle and preliminary experiment. *Sensors and Actuators B: Chemical*. 1992, vol. 8, iss. 2, pp. 187–189. ISSN 0925-4005. DOI: 10.1016/0925-4005(92)80179-2.

[5] NAKATA, S., K. TAKEMURA and K. NEYA. Non-linear dynamic responses of a semiconductor gas sensor: Evaluation of kinetic parameters and competition effect on the sensor response. *Sensors and Actuators B: Chemical*. 2001, vol. 76, iss. 1–3, pp. 436–441. ISSN 0925-4005. DOI: 10.1016/S0925-4005(01)00652-9.

[6] NAKATA, S., H. OKUNISHI and Y. NAKASHIMA. Distinction of gases with a semiconductor sensor under a cyclic temperature modulation with second-harmonic heating. *Sensors and Actuators B: Chemical*. 2006, vol. 119, iss. 2–7, pp. 556–561. ISSN 0925-4005. DOI: 10.1016/j.snb.2006.01.009.

[7] JAEGLE, M., J. WOELLENSTEIN, T. MEISINGER, H. BOETTNER, G. MUELLER, T. BECKER and C. BOSCH-V.BRAUNMUEHL. Micromachined thin film SnO_2 gas sensors in temperature-pulsed operation mode. *Sensors and Actuators B: Chemical*. 1999, vol. 57, iss. 1–3, pp. 130–134. ISSN 0925-4005. DOI: 10.1016/S0925-4005(99)00074-X.

[8] HEILIG, A., N. BARSAN, U. WEIMAR, M. SCHWEIZER-BERBERICH, J. W. GARDNER and W. GOEPEL. Gas identification by modulating temperatures of SnO_2-based thick film sensors. *Sensors and Actuators B: Chemical*. 1997, vol. 43, iss. 1-3, pp. 45–51. ISSN 0925-4005. DOI: 10.1016/S0925-4005(97)00096-8.

[9] VERGARA, A., E. LLOBET, J. BREZMES, P. IVANOV, C. CANE, I. GRACIA, X. VILANOVA and X. CORREIG. Quantitative gas mixture analysis using temperature-modulated micro-hotplate gas sensors: Selection and validation of the optimal modulating frequencies. *Sensors and Actuators B: Chemical*. 2007, vol. 123, iss. 2, pp. 1002–1016. ISSN 0925-4005. DOI: 10.1016/j.snb.2006.11.010.

[10] ILLYASKUTTY, N., J. KNOBLAUCH, M. SCHWOTZER and H. KOHLER. Thermally modulated multi sensor arrays of SnO_2/additive/electrode combinations for enhanced gas identification. *Sensors and Actuators B: Chemical*. 2015, vol. 217, iss. 1, pp. 2–12. ISSN 0925-4005. DOI: 10.1016/j.snb.2015.03.018.

[11] HOSSEIN-BABAEI, F. and A. AMINI. A breakthrough in gas diagnosis with a temperature-modulated generic metal oxide gas sensor. *Sensors and Actuators B: Chemical*. 2012, vol. 166–167, iss. 1, pp. 289–425. ISSN 0925-4005. DOI: 10.1016/j.snb.2012.02.082.

[12] IONESCU, R. and E. LLOBET. Wavelet transform-based fast feature extraction from temperature modulated semiconductor gas sensors. *Sensors and Actuators B: Chemical*. 2002, vol. 81, iss. 2–3, pp. 289–295. ISSN 0925-4005. DOI: 10.1016/S0925-4005(01)00968-6.

[13] FIS GAS SENSOR SP-11: for hydrocarbons gas detection. In: *FIS Inc.* [online]. 2006. Available at: http://www.fisinc.co.jp/en/common/pdf/ESP1100.pdf.

[14] Technical information for TGS813. In: *Figaro Engineering Inc.* [online]. 2006. Available at: http://www.figarosensor.com/products/813Dtl.pdf.

[15] CENGEL, Y. *Heat and Mass Transfer: A Practical Approach*. New York: McGraw-Hill, 2007. ISBN 978-0073250359.

[16] BROWN, A. I. and S. M. MARCO. *Introduction to Heat Transfer*. New York: McGraw-Hill Book Company, 1958.

[17] LEE, A. P. and B. J. REEDY. Temperature modulation in semiconductor gas sensing. *Sensors and Actuators B: Chemical*. 1999, vol. 60, iss. 1–2, pp. 35–42. ISSN 0925-4005. DOI: 10.1016/S0925-4005(99)00241-5.

[18] REKTORYS, K. *Prehled uzite matematiky II*. Praha: Prometheus, 1995. ISBN 80-85849-62-3.

[19] BRODKEY, R. S. and H. C. HERSHEY. *Transport Phenomena: A Unified Approach*. Columbus: Brodkey Publishing, 2001. ISBN 0-9726635-9-2.

[20] KOHL, D. Oxidic semiconductor gas sensors. *Gas sensors: principles, operation, and developments*. Dordrecht: Kluwer Academic Publishers, 1992, pp. 43–88. ISBN 0-7923-2004-2.

[21] WATSON, J., K. IHOKURA and G. S. V. COLES. The tin dioxide gas sensor. *Measurement Science and Technology*. 1993, vol. 4, iss. 7, pp. 711–719. ISSN 0957-0233. DOI: 10.1088/0957-0233/4/7/001.

[22] SINGH, R. C., N. KOHLI, N. P. SINGH and O. SINGH. Ethanol and LPG sensing characteristics of SnO_2 activated Cr_2O_3 thick film sensor. *Bulletin of Materials Science*. 2010, vol. 33, iss. 5, pp. 575–579. ISSN 0250-4707. DOI: 033/05/0575-0579.

[23] SCHMID, W., N. BARSAN and U. WEIMAR. Sensing of hydrocarbons with tin oxide sensors: possible reaction path as revealed by consumption measurements. *Sensors and Actuators B: Chemical*. 2003, vol. 89, iss. 3, pp. 232–236. ISSN 0925-4005.

[24] HAE-WON, C. and L. MAN-JONG. Sensing characteristics and surface reaction mechanism of alcohol sensors based on doped SnO_2. *Journal of Ceramic Processing Research*. 2006, vol. 7, iss. 3, pp. 183–191. ISSN 1229-9162.

About Authors

Libor GAJDOSIK graduated in telecommunication engineering from the Czech Technical University (Prague, Czech Republic) in 1983. He received his Ph.D. in 1998 from the Technical University of Ostrava (Ostrava, Czech Republic). His thesis was focused on the detection properties of tin dioxide gas sensors at the dynamic operating modes. He has been an assistant professor at the Technical University of Ostrava since 1989. His main interests are electronics, circuit theory and chemical sensors.

11

Assessment of Long Thermal Ageing on the Oil-Paper Insulation

Iraida KOLCUNOVA, Marek PAVLIK, Lukas LISON

Department of Electric Power Engineering, Faculty of Electrical Engineering and Informatics,
Technical University of Kosice, Letna 9, 040 01 Kosice, Slovak republic

iraida.kolcunova@tuke.sk, marek.pavlik@tuke.sk, lukas.lison@tuke.sk

Abstract. *Electric power equipment has complex construction. Therefore, it is very important to have enough information about the state of equipment. High voltage transformers play a very important role in the electric power system. One of the most important parts of electric power equipment is the insulation system. Insulation system must be in a good condition for reliable and safe operation of electrical devices. Insulation system of electrical equipment is exposed to various factors which could have negative influence on its condition. Oil impregnated insulation paper is one of the oldest insulation systems used in electrical power equipment. Mineral oils have been used for decades as transformer fluids because of their excellent dielectric properties and availability. However, performance of mineral oil starts to be limited due to environmental consideration. The aim of this paper is to simulate a real insulation system of transformer and to show the influence of accelerated thermal ageing on the insulation system. Properties such as relative permittivity, dissipation factor and the breakdown voltage will be described and analysed.*

Keywords

Breakdown voltage, dissipation factor, oil-paper insulating system, permittivity, thermal ageing.

1. Introduction

Proper operation of electrical equipment requires the use of high quality insulating materials. Power equipment as transformers are one of the vital and expensive elements in the industry of electrical energy [1]. The insulating system is composed of the combination of oil and Kraft paper. The oil is providing both electrical insulation and a means for transferring thermal losses to a cooling system, so as coolant. For more than one hundred years, the majority of liquid-immersed transformers have been filled with mineral oil. The significant use of this petroleum-based product has been justified until now by its wide availability, its good properties, its good combination with cellulose and its low cost [2]. The main oil properties can be divided to three categories: physical, chemical and electrical. Viscosity, flash point, pour point and interfacial tension are the main physical properties of oil. Water content, oxidation stability, total acid and sulphur corrosion are the most important chemical properties. Breakdown voltage, dissipation factor and permittivity belong to the group of important electrical properties. Mineral oil is mainly used in power transformers for its good oxidation stability and low water content. Disadvantages of this naphthenic product are poor biodegradable and low flash point. In recent years, the research thinks ahead for the esters as insulating fluids, which could replace the conventional used insulation liquids based on naphthenic oils [3] and [4]. These liquids are more environmentally friendly because they are almost fully biodegradable (95 %) in relation to the conventional mineral oils, and also they have higher fire point, about 300 °C [5]. This work investigates the influence of temperature on the high voltage transformer system composed of the insulating liquid and Kraft paper.

2. Measurement Methods

There were used some measurements for investigating the ageing mechanism. Dielectric properties as relative permittivity, dissipation factor and breakdown voltage were analysed.

2.1. Breakdown Voltage and Electrical Breakdown Strength

Electrical breakdown voltage U_b is one of the most important properties of the dielectric material. It presents a degree of the insulation ability to resist the electric stress [6]. It is the minimal voltage value, which causes growth of the electric conductivity to the level that initiates an electric breakdown. High value of the current flowing through the breakdown area causes mechanical, thermal and chemical processes that change dielectric characteristics [7]. Electric breakdown strength E_b is one of the basic qualitative characteristics of the dielectrics in addition to the breakdown voltage and the dielectric loss. In the electric field, the dielectric keeps its insulating characteristic only up to the specific value of the electric field intensity. After reaching the boundary of the critical field intensity, the dielectric resistance decreases rapidly to the resistance level of the conductive material. In the case of a homogeneous electric field, the field intensity is the same in the whole space between electrodes, therefore the electric breakdown strength can be calculated using the formula Eq. (1), where d is the thickness of the insulation material [8].

$$E_b = \frac{U_b}{d} . \qquad (1)$$

Figure 1 shows general dependence of electrical breakdown strength on thickness of the insulation material.

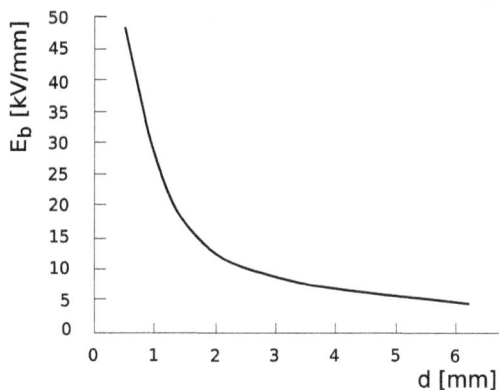

Fig. 1: Dependence of the electrical breakdown strength on the insulation thickness.

2.2. Dissipation Factor

Dielectric dissipation factor $(\tan \delta)$ is another of important diagnostic tools to monitor the condition of the solid/liquid insulation. Periodical measurement of $\tan \delta$ gives the rate of the deterioration of the insulation state [9]. The dissipation factor is a parameter which is a dimension for the quality of the dielectric losses in the insulation system and it gives the relation between the real and reactive components. The total dissipation factor of the whole insulating system depends on the dissipation factor and quality of each component. Higher values of the dielectric dissipation factor indicate presence of moisture and contaminating agents [10]. Tan delta of the insulation is dependent on the water content, impurities and the presence of partial discharges. Besides temperature and frequency, applied voltage has one of the strongest influence on the values of the dissipation factor. General characteristic of the dissipation factor versus applied voltage is shown on Fig. 2, where curve 1 responds to good insulation, curve 2 responds to insulation with partial discharges. The critical value of the applied voltage is the initial voltage of partial discharges.

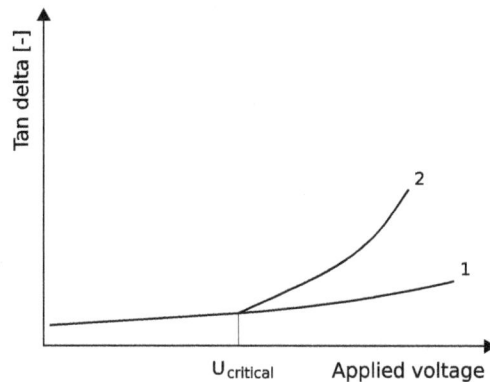

Fig. 2: Dissipation factor versus applied voltage.

2.3. Relative Permittivity

Relative permittivity ε_r is a characteristic material constant typical for every material, and describes the relationship of how much the capacitance of a capacitor changes when filled with a certain material C_x in relationship to a capacitor filled with air C_0. The relative permittivity expresses material skills to accumulation the electric charge. The following formula applies:

$$\varepsilon_r = \frac{C_x}{C_0} . \qquad (2)$$

Relative permittivity is a macroscopic characteristic of the substances that depends on the microscopic properties of structure units, their polarizability α and dipole moment p of the material molecules and their speed in the electrical field [2] and [10]. It depends on temperature, frequency and applied voltage. Relative permittivity of liquids and solid materials is always more than 1.

3. Experimental Setup and Samples

This experiment is focused on measurement of properties of the oil paper insulation system after thermal ageing. The temperature of accelerated ageing test was 90 °C and there were chosen two time intervals of ageing. The first measurement was realized with new samples, the second one was realized after 750 hours of ageing, and the third one after 1500 hours of ageing. A transformer paper with 0.06 mm thickness and two types of impregnating oils were used for these purposes. The first type was natural sunflower oil and the second one was inhibited mineral oil ITO 100. The insulating oil, transformer paper and 10 grams of the copper wire were placed in the glass vessel during the thermal aging process.

3.1. Breakdown Voltage Measurement

The AC breakdown voltage was measured using two electrodes which were 25 mm diameter brass hemispherical types. AC breakdown voltage of mineral oil paper insulation and sunflower oil paper insulation were measured by the instrument High-Voltage DTS-60D. For each type of oil, five oil impregnated paper samples were tested, and the average value was calculated. The voltage was applied at a rise rate of $2 \text{ kV} \cdot \text{s}^{-1}$ until breakdown. The paper insulation was changed after each breakdown. There was a one-minute break between two measurements and the oil sample was mixed. Values of the electric breakdown strength were calculated according to Eq. (1).

3.2. Dissipation Factor and Relative Permittivity Measurement

The relative permittivity in dependence on voltage were measured with automatic Schering Bridge TET-TEX AG. The measurement was realized at the room temperature of 20 °C and normal pressure. The applied voltage was increased from 0.2 kV to 2 kV with the step 0.2 kV at the frequency of 50 Hz.

4. Experimental Results and Discussion

The results of the measurement of AC breakdown voltage and the electrical breakdown strength are shown in Tab. 1 and Tab. 2. Then the average value was

calculated according to the Eq. (3).

$$U_b = \frac{\sum_{i=1}^{n} U_{bi}}{n} \ . \tag{3}$$

Tab. 1: The values of the breakdown voltage U_b and the electric breakdown strength E_b for sunflower oil paper.

Number of measurement	Before ageing	750 hours ageing at 90 °C	1500 hours ageing at 90 °C
	U_b [kV]	U_b [kV]	U_b [kV]
1	22	24.5	23.1
2	21.8	22.5	23.2
3	21.4	23.3	20.9
4	19.6	22.6	21.4
5	20.7	24	24.5
Average	**21.1**	**23.38**	**22.62**
Number of measurement	Before ageing	750 hours ageing at 90 °C	1500 hours ageing at 90 °C
	E_b [kV/mm]	E_b [kV/mm]	E_b [kV/mm]
1	61.11	68.06	64.17
2	60.56	62.50	64.44
3	59.44	64.72	58.06
4	54.44	62.78	59.44
5	57.50	66.67	68.06
Average	**58.61**	**64.94**	**62.83**

Tab. 2: The values of the breakdown voltage U_b and the electric breakdown strength E_b for mineral oil paper.

Number of measurement	Before ageing	750 hours ageing at 90 °C	1500 hours ageing at 90 °C
	U_b [kV]	U_b [kV]	U_b [kV]
1	18	19.4	21.3
2	18	20.1	22
3	18.9	20.4	22
4	17.9	20.6	20.4
5	18.2	21.1	21.7
Average	**18.2**	**20.32**	**21.48**
Number of measurement	Before ageing	750 hours ageing at 90 °C	1500 hours ageing at 90 °C
	E_b [kV/mm]	E_b [kV/mm]	E_b [kV/mm]
1	50.00	53.89	59.17
2	50.00	55.83	61.11
3	52.50	56.67	61.11
4	49.72	57.22	56.67
5	50.56	58.61	60.28
Average	**50.56**	**56.44**	**59.67**

The AC breakdown voltage of natural esters is generally higher than the breakdown voltage of mineral oils. This is because sunflower oil has relative permittivity greater than transformer mineral oil. The initial average values of the breakdown voltage were 21.1 kV for the sunflower oil paper and 18.2 kV for the mineral oil paper. 750 hours of the thermal stress caused the growth of the average values of breakdown voltage from 21.1 to 23.38 kV for the sunflower oil paper, and from 18.2 to 20.32 kV for the mineral oil paper. After 1500-hour thermal aging the breakdown voltage for the sunflower oil paper declined slightly to the value 22.62 kV. This phenomenon is probably caused by hydrolysis. The breakdown voltage for the mineral oil paper increased to the value 21.48 kV. The electrical breakdown strength for these specimens corresponds with the situation like the breakdown voltage.

The voltage dependence on relative permittivity and dissipation factor for the sunflower oil paper are shown in Fig. 3 and Fig. 5, and for the mineral oil paper are shown in Fig. 4 and Fig. 6. The initial values of ε_r for the sunflower oil paper, due to its polar nature, are higher than those for the mineral oil paper. The values of relative permittivity are decreasing for mineral oil paper insulation after each degradation cycle. At the beginning of the thermal test the relative permittivity for sunflower oil paper in comparison with mineral oil has higher value due to the ability of natural ester to absorb moisture. The final value of ε_r after 1500-hour ageing is 2.2 for sunflower oil paper and 2.5 for mineral oil paper. This may be due to the drying of the oil.

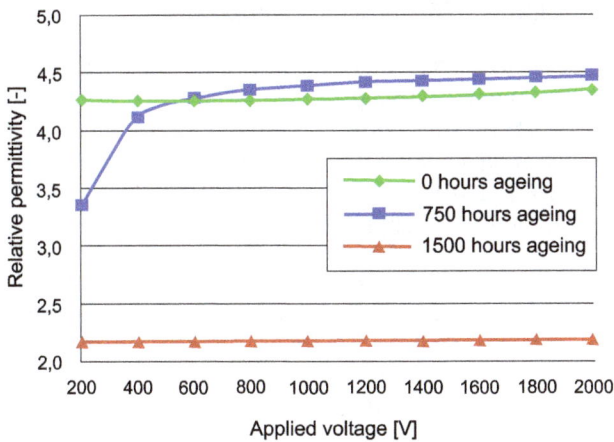

Fig. 3: Relative permittivity of the Kraft paper impregnated with a sunflower oil.

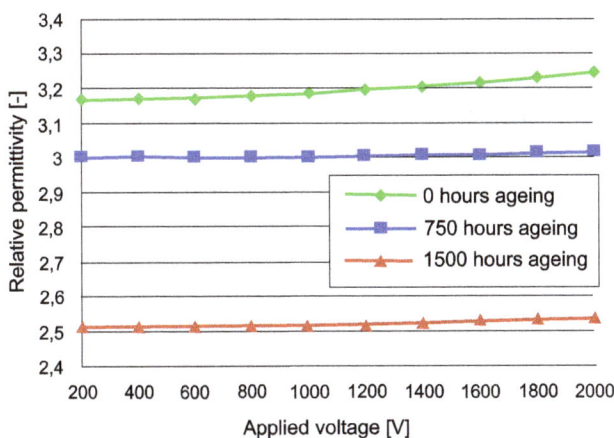

Fig. 4: Relative permittivity of the Kraft paper impregnated with a mineral oil.

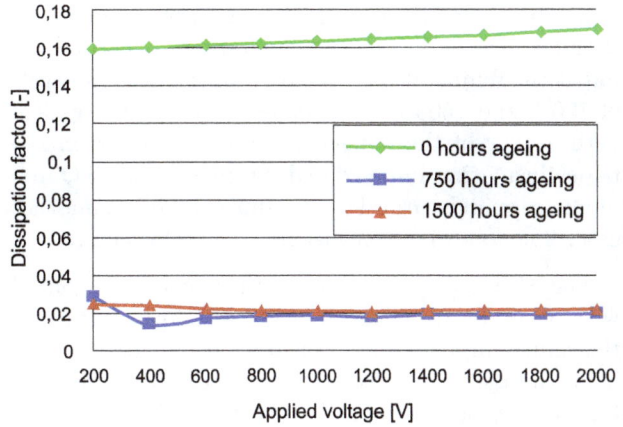

Fig. 5: Dissipation factor of Kraft paper impregnated with a sunflower oil.

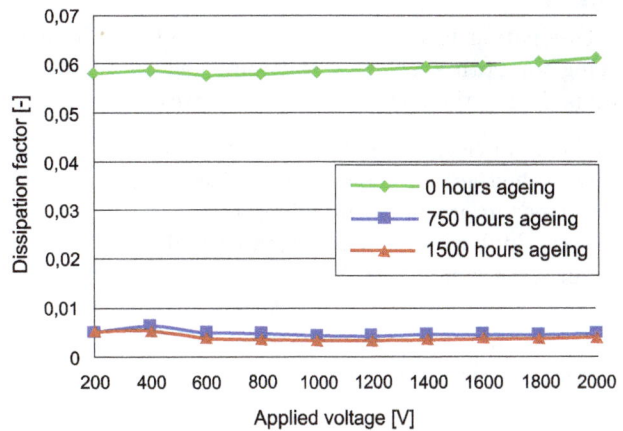

Fig. 6: Dissipation factor of Kraft paper impregnated with a mineral oil.

The dissipation factor has a very strong connection with the relative permittivity of the insulating material. The values of the dissipation factor significantly decreased for each sample. It can be noticed that the behaviour of samples at accelerated aging is very similar. This fact has a connection with moisture which is evaporated after thermal stress. The thermal ageing process caused that the dissipation factor of the samples decreased for each one. Higher dielectric losses were determined at the sunflower oil paper insulation. This status is caused by polar molecules which natural esters, like sunflower oil, contain. Initial values of tan delta were 0.16 for sunflower oil paper and 0.059 for mineral oil paper at 1 kV applied voltage. Temperature 90 °C and period of time 1500 hours caused that tan delta decreased to 0.02 and 0.004 respectively.

5.   Conclusion

The aim of this paper was to compare the dielectric properties of commonly used transformer oil with natural ester in combination with transformer paper. All samples were subjected to accelerating of thermal aging at the temperature T = 90 °C in the time intervals of 750 h and 1500 h. This paper provides the readers with a view of the influence of thermal stress on the paper insulation impregnated by mineral oil ITO 100 in one case and natural ester - sunflower oil in another case. The following conclusions may be formulated.

The AC breakdown voltage of the paper impregnated with natural ester is higher than breakdown voltage of the paper impregnated with mineral oil. After 1500-hour thermal aging electrical breakdown strength is higher than before thermal stress for both specimens.

The values of relative permittivity are decreasing gradually for both mineral and natural ester oil paper insulation after each degradation cycle.

Dissipation factor fell down after 750-hour thermal aging for both specimens and remained on the same value during the rest of the thermal stress.

It can be visible from this experiment that the influence of the thermal stress on the dielectric properties of the paper impregnated by mineral oil or natural ester is very similar. Therefore, research and development in this area is much needed.

Acknowledgment

The authors also wish to acknowledge Scientific Grant Agency of The Ministry of Slovak Republic and Slovak Academy of Science for funding experimental works in the frame of VEGA No. 1/0132/15, VEGA No. 1/0311/15 and APVV-15-0438.

References

[1] HARAMIJA, V., D. VRSALJKO and V. DURINA. Thermal properties of synthetic ester-based transformer oil during ageing in laboratory conditions. In: *International Conference on Liquid Dielectrics*. Bled: IEEE, 2014, pp. 1–4. ISBN 978-1-4799-2063-1. DOI: 10.1109/ICDL.2014.6893111.

[2] CIMALA, R., M. GERMAN-SOBEK and S. BUCKO. The Assessment of Influence of Thermal Aging to Dielectric Properties of XLPE Insulation Using Dielectric Relaxation Spectroscopy. *Acta Electrotechnica et Informatica*. 2015, vol. 15, iss. 3, pp. 14–17. ISSN 1335-8243. DOI: 10.15546/aeei-2015-0023.

[3] LU, G., Q. HUANG, H. SONG and P. WU. Impregnation Model and Experimental Investi-gation of Vegetable Insulating Oil-paper Insulation. In: *International Conference on High Voltage Engineering and Application*. Poznan: IEEE, 2014, pp. 1–4. ISBN 978-1-4799-6613-4. DOI: 10.1109/ICHVE.2014.7035471.

[4] DARWIN, A., C. PERRIER and P. FOLLOIOT. The use of natural ester fluids in transformers. In: *European Conference on HV & MV Substation Equipment*. Lyon: SEE, 2008, pp. 1–6. Available at: http://2011.matpost.org.

[5] MARTINS, M. A. G. and A. R. GOMES. Comparative study of the thermal degradation of synthetic and natural esters and mineral oil: effect of oil type in the thermal degradation of insulating kraft paper. *IEEE Electrical Insulation Magazine*. 2012, vol. 28, iss. 2, pp. 22–28. ISSN 0883-7554. DOI: 10.1109/MEI.2012.6159178.

[6] LISON, L., I. KOLCUNOVA and M. KMEC. Effect of Thermal Aging on the Oil-Paper Insulation. *Acta Electrotechnica et Informatica*. 2014, vol. 14, iss. 4, pp. 23–26. ISSN 1335-8243. DOI: 10.15546/aeei-2014-0036.

[7] MARCI, M. and I. KOLCUNOVA. Electric breakdown strength measurement in liquid dielectrics. In: *International Conference on Environment and Electrical Engineering*. Prague: IEEE, 2010, pp. 427–430. ISBN 978-1-4244-5370-2. DOI: 10.1109/EEEIC.2010.5489916.

[8] LISON, L., I. KOLCUNOVA and S. BUCKO. Vplyv sposobu impregnacie na prieraznu pevnost kombinovanej izolacie olej-papier. *Starnutie elektroizolacnych systemov*. 2014, vol. 9, iss. 1, pp. 9–11. ISSN 1337-0103.

[9] MALPURE, B. D. and K. BABURAO. Failure Analysis & Diagnostics of PowerTransformer Using Dielectric dissipation factor. In: *International Conference on Condition Monitoring and Diagnosis*. Beijing: IEEE, 2008, pp. 497–501. ISBN 978-1-4244-1622-6. DOI: 10.1109/CMD.2008.4580334.

[10] DEDINSKA, L. *Vplyv elektrotepelneho namahania na elektroizolacne vlastnosti rastlinnych olejov*. Kosice, 2012. Dissertation thesis, Technical University of Kosice.

About Authors

Iraida KOLCUNOVA was born in Kotlas, Russia, in 1955. She graduated from the Department of High Voltage Engineering, the Faculty of Power Engineering at the Moscow Power Engineering Institute in 1979. She received the Ph.D. degree from Slovak Technical

University in Bratislava in 1993. She became a Professor of Electric Power Engineering at the Faculty of Electrical Engineering and Informatics at the Technical University of Kosice in 2009. Since 1979 she has worked at the Technical University, first as a research worker in High Voltage Laboratory, then as a teacher at the Department of Electrical Power Engineering. She deals with degradation of insulating materials and measuring partial discharges in high voltage equipment. She is lecturing on Diagnostics of High Voltage Equipment and High Voltage Engineering.

Marek PAVLIK was born in 1985 and he is an assistant professor at the Department of Electric Power Engineering on the Faculty of Electrical Engineering and Informatics at Technical University in Kosice. He received a master degree in electric power engineering. His scientific research is mainly focused on a research of reflection, penetration and absorption of electromagnetic waves in layered materials.

Lukas LISON is a Ph.D. student in the Department of Electric Power Engineering on the Faculty of Electrical Engineering and Informatics at Technical University in Košice. He received a master degree in electric power engineering on the subject of thermal degradation of insulation oils. His scientific research is mainly focused on research of electrophysical properties of paper insulation impregnated by natural esters.

12

The Determination of the Asynchronous Traction Motor Characteristics of Locomotive

Pavel Grigorievich KOLPAKHCHYAN[1], Alexey Rifkatovich SHAIKHIEV[1],
Alexander Evgenievich KOCHIN[2], Konstantin Stepanovich PERFILIEV[3],
Jan OTYPKA[4], Andrey Valerievich SUKHANOV[2,4]

[1]Science and Production Association "Don Technology",
Mikhaylovskaya st. 164 a, 346414 Novocherkassk, Russian Federation
[2]Electrical Machines and Apparatus Department, Faculty of Power Engineering,
Rostov State Transport University, Rostovskogo Strelkovogo Polka Narodnogo
Opolchenya sq. 2, 344038 Rostov on Don, Russian Federation
[3]OJSC "VNIKTI", Oktyabrskoy Revolutcii st. 410, 140400 Kolomna, Russian Federation
[4]Department of Electrical Engineering, Faculty of Electrical Engineering and Computer Science,
VSB–Technical University of Ostrava, 17. listopadu 15/2172, 708 33 Ostrava, Czech Republic

kolpahchyan@mail.ru, alex1807@mail.ru, kochinalex@list.ru, otvsp-ksp@yandex.ru, jan.otypka@vsb.cz,
a.suhanov@rfniias.ru

Abstract. *The article deals with the problem of the locomotive asynchronous traction motor control with the AC diesel-electric transmission. The limitations of the torque of the traction motor when powered by the inverter are determined. The recommendations to improve the use of asynchronous traction motor of locomotives with the AC diesel-electric transmission are given.*

Keywords

AC traction motor, diesel-electrical locomotive, diesel-electrical transmission.

1. Introduction

Currently, the update of the locomotive fleet, including the fleet of freight locomotives is carried on the railways of the Russian Federation. The locomotives with asynchronous traction motors are considered to be promising in accordance with the technical policy of the industry [1]. The batch of the freight locomotives 2TE25A "Vityaz" is released. The first new experience of the locomotives operating has shown that some of the technical solutions require further, more in-depth elaboration.

The development of algorithms and principles of the control system, the improving of the pull and energy characteristics of locomotives to increase the effectiveness of their operation are the main directions of research in the field of improvement of the locomotives power transmission.

The optimization of the locomotives traction characteristics to better harness the advantages of the used traction motors is one of the most urgent issues.

The work in the field-weakening mode in a significant speed range is one of the work features of the locomotive asynchronous traction motor. The article examines the formation of characteristics of the asynchronous traction motor working in such conditions.

2. The Problem Statement

The use of the asynchronous traction motor on the traction rolling stock is related to the number of limitations. It is particularly true for diesel locomotives, where the diesel electrical transmission with AC machines is used.

One of the most promising locomotives is a new freight locomotive with asynchronous traction motor 2TE25A [1], [2] and [3]. It has the diesel generator set of 2500 kW capacity in each six-axle section. Each of the two three-phase stator windings of the synchronous traction generator feeds the appropriate traction converter. Each converter includes uncontrolled bridge rectifier and three autonomous voltage inverters. Each asynchronous traction motor receives the power from its self-commutated voltage inverter. Therefore, the individual (by each axis) regulation of traction is realized. It improves the locomotive haulage capacity.

The output voltage of autonomous voltage inverter is limited and depends on the voltage in the DC link. Permissible heating windings and inverter of the IGBT-modules limit the phase current of asynchronous traction motor. The coupling conditions in contact "wheel - rail" limit the torque of the traction motor. In the formation of the traction motor characteristics it is necessary to take into account the saturation of its magnetic system, limiting rotor flux linkage. The voltage on the DC link (at the inverter input) is not constant and depends on the speed of the diesel generator, which varies over a wide range.

The rotation of the rotor of the asynchronous traction motor varies widely when the locomotive works. It is necessary to take it into account when creating the System of the Automatic Control (SAC). Two modes of the traction motors can be defined: the torque limit mode and the power limit mode.

The torque limit mode is used at low speed. The rotor flux linkage is maintained at a constant level, it is independent of the rotor speed of the asynchronous traction motor. The torque value is limited to the conditions of the clutch and the amount of current consumed by the converter of the asynchronous traction motor.

The back-EMF of the motor is increasing with speed increase and there is an increase of the output voltage of autonomous voltage inverter. The torque limit mode is possible up to a speed at which the output voltage of the inverter becomes maximum possible with the used modulation method. The limitation of inverter voltage does not allow regulation at constant flux linkage and requires the use of field weakening.

With further speed increase, the stator voltage remains constant and equal to its maximum value, which the inverter limits. The stator current may not exceed the nominal one, which the current load on the power semiconductors of the inverter determines.

Thus, in these modes the regulation of asynchronous traction motor should be carried out at a constant voltage and a constant current of the stator, which corresponds to the implementation of the law of the constancy of the full power supplied to the stator. This law allows the best use of electrical equipment of the locomotive and it is optimal in terms of traction properties.

Therefore, according to the theory of electrical machines, the traction motor operation takes place in two zones: in a constant field zone and in a weak field zone.

3. The Mathematical Description of Processes in an Asynchronous Traction Motor

As known [4], [5] and [6], the predetermined electromagnetic torque on the asynchronous traction motor shaft can be obtained by various values of the rotor flux linkage. The flux linkage increase reduces the stator current; it reduces the current load on IGBT-modules of the inverter. However, the flux linkage growth above the nominal value leads to the saturation of the magnetic system of the asynchronous traction motor and to the significant increase of the magnetic losses. In addition, stator current component significantly increases on the d-axis, which may lead to the stator current increase, instead of its reduction. Therefore, in the control zone without field weakening, it is advisable to implement an asynchronous traction motor control with a nominal flux linkage.

The equations in the d - q coordinate system describe the processes in the stator winding of the asynchronous traction motor. The engine is operated with the use of vector control principles and the rotor flux linkage vector is directed along the d-axis: $\Psi_r = \Psi_{rd} = \text{const}$, $\Psi_{rq} = 0$. In this case, the equations system for the projection of the stator current vector on the axis of the rotating d - q coordinate system is following:

$$\begin{cases} L_s' \frac{dI_{sd}}{dt} = U_{sd} - R_s I_{sd} + L_s' p\omega\, i_{sq}, \\ L_s' \frac{dI_{sq}}{dt} = U_{sq} - R_s I_{sq} - L_s' p\omega\, i_{sd} - p\omega\, \frac{L_m}{L_r'}\, \Psi_{rd}, \end{cases} \quad (1)$$

where L_m – magnetizing inductance of the asynchronous traction motor; $L_s' = L_s \cdot \left(1 - \frac{L_m^2}{L_s \cdot L_r'}\right)$ - total leakage inductance from the stator; $L_{\sigma s}$, $L_s = L_m + L_{\sigma s}$ - leakage inductance and the total inductance of the stator; $L_{\sigma r}'$, $L_r' = L_m + L_{\sigma r}'$ - leakage inductance and the total inductance of the rotor, referred to the stator; R_s - stator resistance.

Given that $\Psi_{rd} = L_m \cdot I_{sd}$, the electromagnetic torque is determined using the expression:

$$M_{em} = \frac{3}{2} \cdot p \cdot \frac{L_m}{L_r'} \cdot \Psi_{rd} \cdot I_{sq} = \frac{3}{2} \cdot p \cdot \frac{L_m^2}{L_r'} \cdot I_{sd} \cdot I_{sq}. \quad (2)$$

Absolute slip (rotor current frequency) is determined by the formula:

$$\omega_r = I_{sq} \cdot \frac{L_m R_r'}{L_r' \cdot \Psi_{rd}}, \quad (3)$$

where $R_r^{'}$ - rotor resistance referred to the stator.

Figure 1 shows a vector diagram of asynchronous traction motor in the traction mode, where the following notations are given: θ - the angle of the rotor rotation in electric space; δ - the angle between the rotor flux linkage and the stator current (load angle); φ - the angle between the vectors of the stator voltage and current; $\vec{E} = p \cdot \omega \cdot \vec{\Psi}_r$ - EMF vector of the stator induced by the rotor flux.

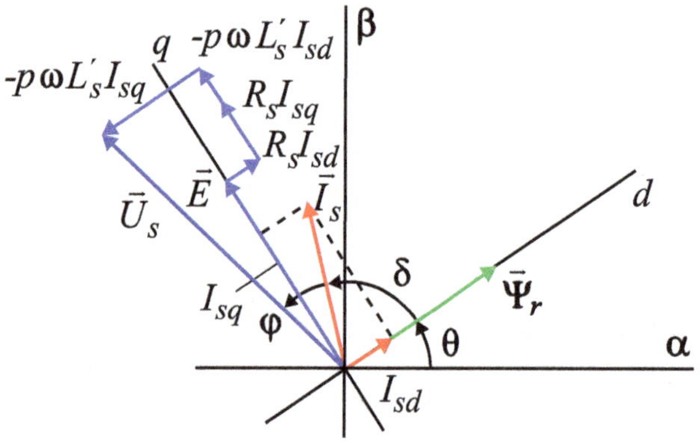

Fig. 1: The vector diagram of asynchronous traction motor in the traction mode.

In the locomotive under consideration, the asynchronous traction motor of the AD-917 type with the following parameters is installed:

- $p = 3$,
- $L_m = 0.01238$ H,
- $L_{\sigma s} = 0.001405$ H, $L_{\sigma r}^{'} = 0.000913$ H,
- $R_s = 0.03$ Ω, $R_r^{'} = 0.0274$ Ω.

The basic data of the traction motor AD-917 are following:

- $U_{ln} = 605$ V (1150 V) – the linear voltage for a long period (maximum),
- $I_{sn} = 440$ A (480 A) – the phase current of a long operating mode (starting),
- $f_{sn} = 18.6$ Hz (125 Hz) – stator current frequency for a long operating mode (maximum),
- $M_n = 10\,200$ Nm – shaft torque in a long operating mode,
- $\Psi_{rd} = 4.18$ Wb – the rotor flux linkage in a long operating mode.

The values of voltage and current of the stator, the rotor flux linkage are given as actual values.

4. Characteristics of the AC Traction Motors

The traction characteristic of the locomotive has two zones. In the first one the torque of asynchronous traction motor is limited by clutch conditions. In the second zone the torque is determined by the constancy of power. At the transition from the first to the second control zone, the speed value is determined by the torque value in the starting mode and the power of the diesel generator set.

In some cases, at this speed the output voltage of a diesel generator set reaches its maximum value, but this condition is not mandatory. On the locomotive under consideration, the diesel generator set voltage reaches the maximum value at the speed more than the speed of the transition to the power limit. Thus, three zones can be distinguished in the regulation characteristic of the asynchronous traction motor on the locomotive.

In the speed range from zero to the diesel generator set full power speed, the traction motor implements torque that is equal to starting one (the first control zone). The voltage applied to the stator winding rises to a value smaller then maximum. With further increase of the speed, there is the reduction of torque to maintain the power constancy law (the second control zone). The voltage applied to the stator winding continues to increase. When it reaches the maximum value that is limited by inverter, the motor goes into the field-weakening operation mode (the third control zone).

As it is known, the electromagnetic torque of the asynchronous traction motor can be obtained by various combinations of the stator current components along the d and q axes [3], [4], [5], [6], [7] and [8]. The specified torque is realized with the least current of the stator when these components of the current are equal [3], [4], [5] and [6]. The operation with a minimum of stator current leads to a considerable increase of the rotor flux linkage and causes strong saturation of the magnetic system at the starting mode of the asynchronous traction motor AD-917. The flux linkage is equal to the nominal value during operation with the long operating mode torque when calculating the characteristics of asynchronous traction motor.

At steady mode, taking into account Eq. (1) the vector projection, the stator voltage on the d and q axes are equal:

$$U_{sd} = R_s \cdot I_{sd} - L_s^{'} \cdot p \cdot \omega \cdot I_{sq},$$
$$U_{sq} = R_s \cdot I_{sq} + L_s^{'} \cdot p \cdot \omega \cdot I_{sd} + p \cdot \omega \cdot \tfrac{L_m}{L_r^{'}} \cdot \Psi_{rd}. \quad (4)$$

In the speed range from zero to the speed of transition to the power limit mode (the first control zone), the given torque and value of the rotor flux linkage determines the stator current, the formula determines the required voltage stator:

$$U_s = \sqrt{U_{sd}^2 + U_{sq}^2}. \qquad (5)$$

The torque of the asynchronous traction motor decreases when the speed exceeds the speed of transition in the power limitation mode, in the second control zone. The flux linkage value of the rotor is also reduced when a predetermined torque is realized with a minimum of stator current. The magnetic field of the motor slightly reduces.

The traction characteristic of the locomotive has two zones. In the first one the torque of asynchronous traction motor is limited by clutch conditions. In the second zone the torque is determined by the constancy of power. At the transition from the first to the second control zone, the speed value is determined by the torque value in the starting mode and the power of the diesel generator set.

In some cases, at this speed the output voltage of a diesel generator set reaches its maximum value, but this condition is not mandatory. On the locomotive under consideration, the diesel generator set voltage reaches the maximum value at the speed more than the speed of the transition to the power limit. Thus, three zones can be distinguished in the regulation characteristic of the asynchronous traction motor on the locomotive.

In the speed range from zero to the diesel generator set full power speed, the traction motor implements torque that is equal to starting one (the first control zone). The voltage applied to the stator winding rises to a value smaller then maximum. With further increase of the speed, there is the reduction of torque to maintain the power constancy law (the second control zone). The voltage applied to the stator winding continues to increase. When it reaches the maximum value that is limited by inverter, the motor goes into the field-weakening operation mode (the third control zone).

In the third control zone, the control of the asynchronous traction motor is carried out when the torque is maximum and when the current and the stator voltage are restricted. The load angle value is determined in accordance with this condition.

As it has been noted, the work with a deep field weakening is a feature of use of the asynchronous traction motor in the locomotive. In this mode it is necessary to control the position of the operating point on the mechanical characteristics of the asynchronous traction motor; it must be within the ascending branch (motor slip should be less than critical). It provides the stable motor work. The ratio of the critical points to

the current torque must not be less than a predetermined value. The ensuring of the sustainability of the asynchronous traction motor is relevant at high speeds in the third control zone. The Kloss specified formula can determine the rotor current frequency at which the required ratio of the critical moment to the current value is implemented [9]:

$$\frac{M_{emM}}{M_{em\,max}} = \frac{2 + 2 \cdot q}{\dfrac{\Omega_r}{\Omega_{r\,max}} + \dfrac{\Omega_{r\,max}}{\Omega_r} + q}, \qquad (6)$$

where $\Omega_{r\,max}$, $M_{em\,max}$ – critical absolute slip and torque; $q = 2 \cdot \frac{R_s}{R_r'} \cdot \frac{\Omega_{r\,max}}{\Omega+\Omega_r}$ - the coefficient of the stator active resistance.

After the transformation (Eq. (3)) the form is the following:

$$\Omega_r^2 - (2 \cdot k_m + k_m \cdot q - q) \cdot \Omega_r \cdot \Omega_{r\,max} + \Omega_r^2 = 0, \quad (7)$$

where $\frac{M_{em\,max}}{M_{em}} = k_m$ is stability margin of the asynchronous traction motor.

The absolute slip value is obtained by solving the equation Eq. (7) for the predetermined value k_m. Providing that $k_m > 1$ equation Eq. (6) has two real roots. The lowest of them corresponds to the work of the asynchronous traction motor within the ascending branch of the mechanical characteristics. This absolute slip value is maximum for a given stability margin value. The current value of the absolute slip that is determined by the formula Eq. (3), must not exceed this value. It is a condition for the steady work of the asynchronous traction motor.

The increasing voltage drop in the reactance of the stator and rotor is the reason for decrease of the stability margin with the speed increase of the asynchronous traction motor. Therefore, the performance of the asynchronous traction motor stability requires the reducing of the stator current value.

The asynchronous traction motor characteristics obtained with accepted principles of control were calculated taking into account the described approach. Figure 2 shows the dependence of the current value of the stator line voltage, the stator current and the electromagnetic torque on the rotor speed of the asynchronous traction motor that are obtained taking into account the above-mentioned restrictions and the importance of stability margin 1.1.

Figure 3 shows that the point A corresponds to the transition from the regulatory mode with the constant torque to the control mode with constant power (the boundary between the 1st and the 2nd zones of control). The point B shows the transition to voltage constraints of the stator and the engine limit under the continuous operation current and maximum current of the stator respectively (the transition between the 2nd and the 3rd control zone).

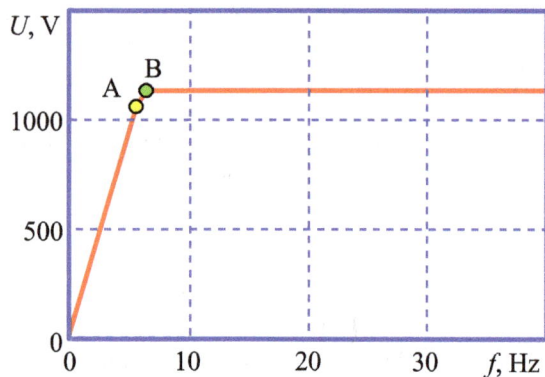

(a)

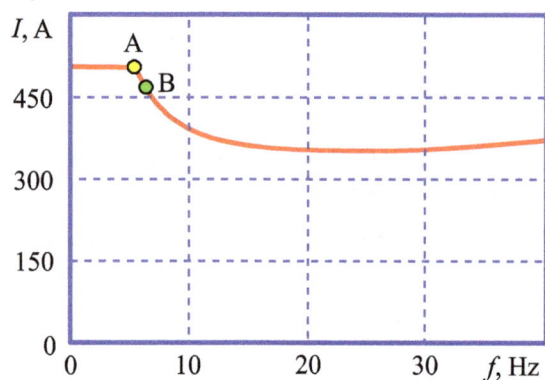

(b)

Fig. 2: The dependence of the stator line-to-line voltage (a) and stator phase current (b).

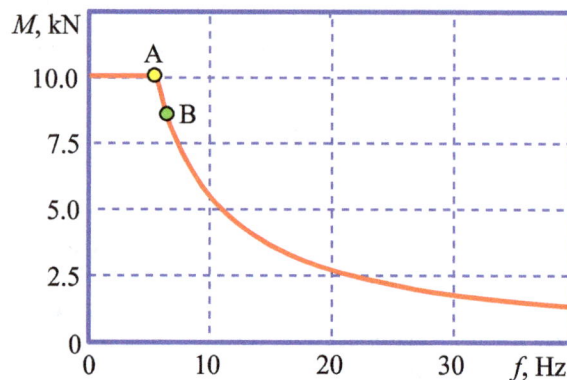

(a)

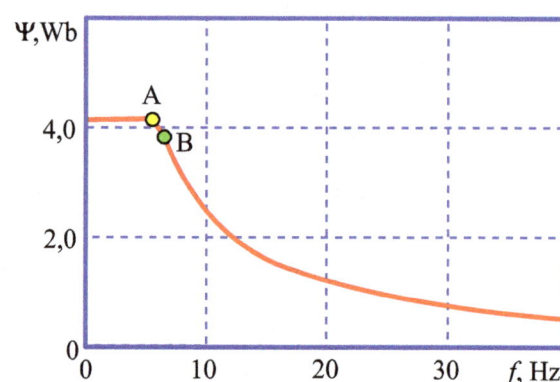

(b)

Fig. 3: The dependence of the electromagnetic torque (a) and rotor flux linkage (b).

The analysis of the obtained dependencies shows the achievement of the minimum stator current is possible with a slight weakening of the motor field in the constant power mode and with the voltage reserve. At the same time, the value of the stator current reduces in comparison with the 1st zone of regulation.

At first the electromagnetic shaft power increases due to the reduction of the reactive component of the stator current and then it reduces with the increasing speed and the decreasing rotor flux linkage. It is due to the voltage drop on the stator and rotor reactance increasing proportional to the rotor speed. As the rotor speed grows the influence of this effect increases. Also, stability margin reduces. When it reaches minimum allowed value, there is a need to reduce the stator current.

The described effects are particularly strong in the initial control positions because in this case the limit value of the stator voltage is much lower than the nominal value; and the voltage drop across the active and reactive resistances, depending on the current and speed of the rotation, are comparable to the motor EMF.

The increase of the stator current allows rising of the power of the asynchronous traction motor in the case of a full field. The stator current increase gives effect only until the current stability limit when the motor operates in the field weakening mode. It is a consequence of the rapid increase in the voltage drop across the reactance of the stator and the rotor, which are defined by the increase of these resistances value and by high current.

5. Conclusion

Analysis of the results leads to the following conclusions. The asynchronous traction motor as a part of the traction electric locomotive works mainly in the field weakening mode, due to the need to obtain a fixed power from a diesel generator set. In the field-weakening area, it is necessary to the stability margin, as the maximum engine rotation speed is much higher than the rotation speed of the continuous mode.

The control of the asynchronous traction motor on the locomotive is carried out with three control zones. It is possible when the speed of the transition from the torque restriction to the power restriction is lower than the speed when the voltage limit of the diesel-generator set is reached. In the 1st zone, the engine torque and rotor flux linkage are constant. In the 2nd zone, the locomotive power is constant. In the 3rd zone, the regulation is carried out with field weakening after reaching a voltage limit on the inverter output.

Stable operation of the asynchronous traction motor is possible, if the absolute slip does not exceed the value determined by the required stability margin. Its maintenance requires the reduction of the stator current. For the considered traction motor, that moment occurs when the rate of rotation is half or twice the nominal rate, depending on the stator current. Thus, the power of the asynchronous traction motor is determined by the steady work condition within the range of medium and high speeds.

Acknowledgment

The work was partly supported by Project SP2017/153 Research antenna systems, diagnostics and reliability of electric drives of the Student Grant System, VSB–Technical University of Ostrava.

References

[1] BABKOV, J. V., J. I. KLIMENKO, K. S. PERFILEV, P. L. CHUDAKOV and V. A. LINKOV. Main-line diesel-electric locomotive 2TE25A. *Locomotive*. 2012, vol. 7, iss. 1, pp. 28–32. ISSN 0869-8147.

[2] BABKOV, J. V., V. B. ZORIN, J. I. KLIMENKO, A. A. KURAEV and K. S. PERFILEV. Traction and auxiliary converters for locomotive. *Heavy Engineering*. 2006, vol. 8, iss. 1, pp. 31–33. ISSN 1924-7106.

[3] ANDRJUSHHENKO, A. A., J. V. BABKOV, A. A. ZARIFJAN, G. F. KASHNIKOV, P. G. KOLPAKHCHJAN, K. S. PERFILEV and V. P. JANOV. *Locomotive's asynchronous traction drive*. Moscow: UMC ZhDT, 2013. ISBN 978-5-89035-631-4.

[4] DE DONCKER, R., D. W. J. PULLE and A. VELTMAN. *Advanced Electrical Drives: Analysis, Modeling, Control*. Dordrecht: Springer Netherlands, 2011. ISBN 978-94-007-0181-6.

[5] CHIASSON, J. *Modeling and High Performance Control of Electric Machines*. Hoboken: Wiley-IEEE Press, 2005. ISBN 978-0-471-68449-7.

[6] GLUMINEAU, A. and J. L. MORALES *Sensorless AC Electric Motor Control: Robust Advanced Design Techniques and Applications*. New York: Springer, 2015. ISBN 978-3-319-14586-0.

[7] LIPO, T. A. *Analysis of Synchronous Machines*. Boca Raton: CRC Press, 2012. ISBN 978-1-4398-8067-8.

[8] HOLMES, D. G. and T. A. LIPO. *Pulse Width Modulation for Power Converters: Principles and Practice*. Hoboken: Wiley, 2003. ISBN 978-0-471-20814-3.

[9] VUKOSAVIC, S. N. *Electrical Machines*. New York: Springer, 2013. ISBN 978-1-4614-0400-2.

About Authors

Pavel Grigorievich KOLPAKHCHYAN was born in Novocherkassk, Russian Federation. He received his Dr.Sc. from South-Russian State Technical University in 2006. His research interests include power electronic, electrical engineering, design of electrical machines, mathematical modelling of electrical machines.

Alexey Rifkatovich SHAIKHIEV was born in Taganrog, Russian Federation. He received his Ph.D. in Rostov State Transport University in 2004. His research interests include power electronic, electrical engineering, design of electrical machines, switched-reluctance electrical machines.

Alexander Evgenievich KOCHIN was born in Torez, Ukraine. He received his Ph.D. in National Mining University in 2005. His research interests include power electronic, electrical engineering, design of electrical machines, mathematical modelling of electrical machines.

Konstantin Stepanovich PERFILIEV was born in Kolomna, Russian Federation. He received his Ph.D. from St. Petersburg State Transport University in 2005. His research interests include power electronic, electric traction drive, electrical engineering.

Jan OTYPKA was born in Celadna, Czech Republic. He is student of Ph.D. program at Department of Electrical Engineering at Faculty of Electrical Engineering and Computer Science VSB–Technical University of Ostrava. His research interests include analysis problems of induction motor supplied by non-harmonics sources.

Andrey Valerievich SUKHANOV was born in Zernograd, Russian Federation. He received his M.Sc. from Rostov State Transport University in 2013. His research interests include data mining and anomaly detection in time series.

13

Reliability Characteristics of Power Plants

Zbynek MARTINEK[1], *Ales HROMADKA*[1], *Jiri HAMMERBAUER*[2]

[1]Department of Electric Power Engineering and Ecology, Faculty of Electrical Engineering,
University of West Bohemia, Univerzitni 2732/8, 306 14, Pilsen, Czech Republic,
[2]Department of Applied Electronics and Telecommunications, Faculty of Electrical Engineering,
University of West Bohemia, Univerzitni 2732/8, 306 14, Pilsen, Czech Republic

martinek@kee.zcu.cz, aleshrom@kee.zcu.cz, hammer@kae.zcu.cz

Abstract. *This paper describes the phenomenon of reliability of power plants. It gives an explanation of the terms connected with this topic as their proper understanding is important for understanding the relations and equations which model the possible real situations. The reliability phenomenon is analysed using both the exponential distribution and the Weibull distribution. The results of our analysis are specific equations giving information about the characteristics of the power plants, the mean time of operations and the probability of failure-free operation. Equations solved for the Weibull distribution respect the failures as well as the actual operating hours. Thanks to our results, we are able to create a model of dynamic reliability for prediction of future states. It can be useful for improving the current situation of the unit as well as for creating the optimal plan of maintenance and thus have an impact on the overall economics of the operation of these power plants.*

Keywords

Modelling of dynamic behaviour, power plant unit, reliability, Weibull distribution.

1. Introduction

The phenomenon of reliability is a very important aspect of technical research at present and will certainly remain so in the future. We are able to achieve permanent sustainability thanks to the determination of the reliability of the whole system. This reliability has also a significant impact on electric power engineering, because all the components in power engineering systems have certain parameters of reliability, such as probability of failure-free operation and the mean time

of operation. If these parameters are determined correctly, we can obtain relatively exact models and also achieve optimization of the key components of these systems. This paper describes the most important parameters of reliability in power plants. These parameters are derived for two types of distribution – the exponential and the Weibull. Parameters for the Weibull distribution are more realistic because this distribution respects the dynamics of the cycle. The exponential distribution does not respect these dynamic features and is used only for verification of the hypothesis. Equations obtained for the reliability parameters can be used as optimizing tools. The optimization process reduces the maintaining costs as well as the additional costs connected with blackouts of the redundant key components. This procedure is also very important for maximizing the efficiency of the whole electric power engineering system.

2. Overview of Current State of Reliability

The reliability may be explained as ability of a unit to successfully operate in required time of operation. Another explanation of the reliability can be this statement: Reliability is equal to probability that a unit will be operating without failures in certain time of operation [1].

Reliability function $p(t)$ describes probability, that the unit works without failures in the time (t) which is longer than the time of operation (T). In the time of operation (T) there is certain reliability guaranteed by the producer:

$$p(t) = P\{T < t\}. \qquad (1)$$

Failure probability function $q(t)$ describes probability, that the unit has one or more failures in the

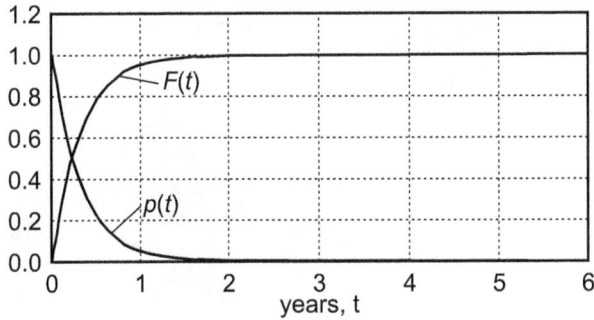

Fig. 1: Reliability function $p(t)$ and distribution function $F(t)$ of a power unit $\lambda = 3$ [1].

time (t):

$$q(t) = 1 - p(t). \quad (2)$$

Distribution function of time without failure:

$$F(t) = P\{T < t\} = q(t). \quad (3)$$

Density function of time without failures:

$$f(t) = \frac{\mathrm{d}F(t)}{\mathrm{d}(t)}. \quad (4)$$

If we can consider intensity of failures as a constant, the exponential distribution can be used:

$$p(t) = e^{-\lambda t}, \quad (5)$$

and

$$f(t) = \lambda e^{-\lambda t}. \quad (6)$$

On the basis of density function it is possible to determine the expression for the mean time of operation [1]:

$$m_s = \int_0^\infty t \cdot f(t)\mathrm{d}t. \quad (7)$$

3. Materials and Methods

3.1. Reliability Characteristics of a Power Plant Units

This part describes equations derived for determining the important reliability parameters of a given power plant. The probability of failure-free operation is further referred to as the first parameter and the mean time of operation as the second parameter of the reliability characteristics of the unit. These characteristics are solved below using the exponential distribution for the unrepairable unit and the Weibull distribution for the repairable unit. The Weibull distribution can describe the realistic operation better than the exponential distribution because it includes failures of all the components during the operation as well as the realistic situations which may occur.

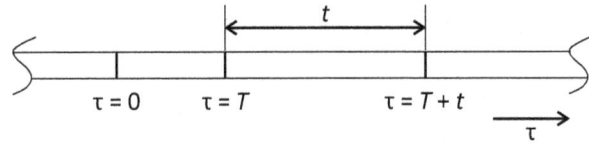

Fig. 2: Situation of the realistic unit shown on time axis [3].

3.2. Characteristics of Time of Failure-Free Operations Using the Exponential Distribution

Firstly, it is necessary to obtain solutions for the unrepairable unit by using the exponential distribution, because this situation is much easier to explain than the situation for the repairable unit. Therefore, it is very important to explain first the situation of the unrepairable unit. For this situation, it is crucial to determine the intensity of repair (μ) and mean-time to repair (r) as follows:

$$\mu = 0 \rightarrow r = \infty. \quad (8)$$

This statement is given by unrepairability of the unit. During operation of the unrepairable unit the intensity of failures (λ) is given as a constant value when using the exponential distribution. It means that the probability of failure-free operation does not depend on the time of operation (T) of the unit [3].

Explanation (Fig. 2):

- Interval from 0 to $\tau = 0$ represents repair.
- Interval from $\tau = 0$ to $\tau = T$ represents real-time operation.
- Interval from $\tau = T$ to $\tau = T + t$ represents prediction of the future state.

Verification can be performed by equation given for conditional probability:

$$P_{B(A)} = \frac{P_{(B \cap A)}}{P_{(B)}}. \quad (9)$$

This equation determines the probability of phenomenon A in the case of success of phenomenon B, [4].

Explanation (Fig. 2):

- Phenomenon A represents the unit in which failure occurs in the interval between T and $T + t$.
- Phenomenon B represents the unit in which failure does not occur in the interval between 0 and T.
- Intersection of these phenomena $A \cap B$ represents both these phenomena simultaneously.

The probability of failure-free operation between time 0 and T is determined as follows:

$$P_{(B)} = P_{1(t)} = e^{-\lambda \cdot T}, \qquad (10)$$

and the probability of intersection of phenomena $A \cap B$ is determined as follows:

$$P_{(B \cap A)} = \int_{T}^{T+t} f_t \mathrm{d}t = \int_{T}^{T+t} \lambda * e^{-\lambda t} \mathrm{d}t. \qquad (11)$$

Using the equation for conditional probability the following equation is obtained:

$$P_{B(A)} = \frac{e^{-\lambda \cdot T} - e^{-\lambda \cdot T} * e^{-\lambda \cdot T}}{e^{-\lambda \cdot T}} = 1 - e^{-\lambda \cdot T}. \qquad (12)$$

Independence of conditional probability $P_{B(A)}$ on the time of operation of unit T is a feature of the exponential distribution [2].

3.3. Characteristics of Time of Failure-Free Operations Using Weibull Distribution

Secondly, we need to find solutions for the unit of the power plant considered repairable. This state is described using the Weibull distribution below. Again, it is necessary to determine the value of $P_{B(A)}$ for the Weibull distribution. The definition of $P_{B(A)}$ is the same for both distributions. However, contrary to the exponential distribution, this state depends on the time of operation T of the unit [3].

Again, it is necessary to determine the probability of failure-free operation between time 0 and T for the Weibull distribution:

$$P_{(B)} = P_{0(t)} = 1 - \int_0^T f(t)\mathrm{d}t = \cdots$$
$$= 1 - \int_0^T k \cdot t^m e^{-(\frac{k}{m+1}*t^{m+1})}. \qquad (13)$$

As before, we can obtain the equation for $P_{B(A)}$ by using the equation for conditional probability:

$$P_{B(A)} = \frac{\int_{T}^{T+t} f(t)\mathrm{d}t}{1 - \int_0^T f(t)\mathrm{d}t}. \qquad (14)$$

Equation (14) is solved by the method of progressive integration in the following steps [3].

3.4. Solution of Conditional Probability

In the Weibull distribution there are two possible expressions of $\int f(t)\mathrm{d}t$:

$$f(t) = k \cdot t^m e^{-(\frac{k}{m+1}t^{m+1})}, \qquad (15)$$

$$f(t) = \frac{b}{d} * \left(\frac{t}{d}\right)^{b-1} e^{-(\frac{t}{d})^b}. \qquad (16)$$

This paper focuses on both expressions of the equation. Equation (15) is solved first and Eq. (16) is solved next [3].

1) Calculation for Eq. (15)

The first step is addition of the following integral to Eg. (8):

$$\int f(t)\mathrm{d}t = \int k \cdot t^m e^{-(\frac{k}{m+1}t^{m+1})}\mathrm{d}t. \qquad (17)$$

The second step is substitution for t^{m+1}:

$$t^{m+1} = z, (m+1)t^m \mathrm{d}t = \mathrm{d}z, t^m \mathrm{d}t = \frac{\mathrm{d}z}{m+1}. \qquad (18)$$

The third step is insertion of the substitution into Eq. (10):

$$\int f(t)\mathrm{d}t = \int \frac{k}{m+1} e^{-(\frac{k}{m+1}z)}\mathrm{d}z =$$
$$= -e^{-(\frac{k}{m+1}t^{m+1})}. \qquad (19)$$

2) Calculation for Eq. (16)

Same procedure:

$$\int f(t)\mathrm{d}t = \int \frac{b}{a} * \left(\frac{t}{d}\right)^{b-1} e^{-(\frac{t}{d})^b}\mathrm{d}t. \qquad (20)$$

Substitution for $\frac{t}{d}$:

$$\frac{t}{d} = x \rightarrow \frac{\mathrm{d}t}{d} = \mathrm{d}x \qquad (21)$$

And next substitution for x^b:

$$x^b = y \rightarrow bx^{b-1}\mathrm{d}x = \mathrm{d}y \qquad (22)$$

Equation 19 can be expressed as:

$$\int e^{-y}\mathrm{d}y = -e^{-y} = -e^{-x^b} = -e^{-(\frac{t}{d})^b}. \qquad (23)$$

3) Determination of Probability of Failure-Free Operation $P_{B(A)}$ for Eq. (15) and Eq. (16)

If we insert Eq. (15) into Eq. (14), we get the following result of $P_{B(A)}$:

$$P_{B(A)} = \frac{\left[-e^{-\left(\frac{k}{m+1}t^{m+1}\right)}\right]_T^{T+1}}{1-\left[-e^{-\left(\frac{k}{m+1}t^{m+1}\right)}\right]_0^T} =$$

$$= \frac{e^{-\left(\frac{k}{m+1}T^{m+1}\right)}-e^{-\left(\frac{k}{m+1}(T+t)^{m+1}\right)}}{1-\left[1-e^{-\left(\frac{k}{m+1}T^{m+1}\right)}\right]} = \qquad (24)$$

$$= \frac{e^{-\left(\frac{k}{m+1}T^{m+1}\right)}-e^{-\left(\frac{k}{m+1}(T+t)^{m+1}\right)}}{e^{-\left(\frac{k}{m+1}T^{m+1}\right)}}.$$

After next simplifications we obtain the following result of $P_{B(A)}$:

$$P_{B(A)} = 1 - e^{\left[\frac{k}{m+1}T^{m+1}-\frac{k}{m+1}(T+t)^{m+1}\right]}. \qquad (25)$$

The next step is insertion of Eq. (16) into into Eq. (14) which looks as follows:

$$P_{B(A)} = \frac{\left[e^{-\left(\frac{t}{d}\right)^b}\right]_T^{T+t}}{1-\left[e^{-\left(\frac{t}{d}\right)^b}\right]_0^T} = 1 - e^{\left[\left(\frac{T}{d}\right)^b-\left(\frac{T+t}{d}\right)^b\right]}. \qquad (26)$$

To verify our results, we consider the value of parameter m from Eq. (15), $m = 0$. This situation is typical of the exponential distribution. The resulting expression is obtained independently of parameter T:

$$P_{B(A)} = 1 - e^{[-k(T+t-T)]} = 1 - e^{-kt}. \qquad (27)$$

We perform similar verification also for Eq. (26), $b = 1$:

$$P_{B(A)} = 1 - e^{-\left(\frac{T+t-T}{d}\right)} = 1 - e^{-\left(\frac{t}{d}\right)}. \qquad (28)$$

Correctness of those results Eq. (15) and Eq. (16) can also be verified by insertion of the parameter $T = 0$.

We obtain the resultant expression for the probability of failure of the unit until time t; however, we have to know that the unit was in operation until time T, which is for Eq. (15) as follows:

$$P_{B(A)} = P_{1(t)} = 1 - e^{\left[\frac{k}{m+1}T^{m+1}-\frac{k}{m+1}(T+t)^{m+1}\right]}. \qquad (29)$$

If we determine the probability of failure of the unit by insertion of Eq. (16), we obtain the following equation:

$$P_{B(A)} = P_{1(t)} = 1 - e^{\left[\left(\frac{T}{d}\right)^b-\left(\frac{T+t}{d}\right)^b\right]}. \qquad (30)$$

3.5. Derivation of the Mean Time of Operation without the Time of Operation

Calculation of the mean time of operation m_s for the exponential distribution of time of failure-free operations is as follows [2]:

$$m_s = \int_0^\infty t \cdot f(t)\mathrm{d}t = \int_0^\infty t \cdot \lambda^{-\lambda \cdot t}\mathrm{d}t = \frac{1}{\lambda}. \qquad (31)$$

For the Weibull distribution of time of failure-free operations it is as follows:

$$m_s = \int_0^\infty t \cdot k \cdot t^m e^{-\left(\frac{k}{m+1}t^{m+1}\right)}\mathrm{d}t. \qquad (32)$$

Calculation of the mean time of operation requires more difficult operations for the Weibull distribution as described below. Firstly, substitution is necessary for kt^{m+1}:

$$t^{m+1} = x \rightarrow t = \sqrt[m+1]{x},$$
$$(m+1)t^m\mathrm{d}t = \mathrm{d}x \rightarrow t^m\mathrm{d}t = \frac{\mathrm{d}x}{(m+1)}. \qquad (33)$$

Secondly, the boundaries have to be recapitulated:

$$t \rightarrow 0, \infty,$$
$$x \rightarrow 0, \infty. \qquad (34)$$

The third step is the calculation for Eq. (15):

$$m_s = k \int_0^\infty t^{m+1}e^{-\left(\frac{k}{m+1}t^{m+1}\right)}\mathrm{d}t =$$

$$= k \int_0^\infty \sqrt[m+1]{x} \cdot e^{-\left(\frac{k\cdot x}{m+1}\right)}\frac{\mathrm{d}x}{(m+1)} = \qquad (35)$$

$$= \frac{k}{(m+1)} \int_0^\infty \sqrt[m+1]{x} \cdot e^{-\left(\frac{k\cdot x}{m+1}\right)}\mathrm{d}x.$$

Similar calculation is performed for Eq. (16):

$$x = t^b \rightarrow t = \sqrt[b]{x},$$
$$b \cdot t^{b-1}\mathrm{d}t = \mathrm{d}x,$$
$$t \rightarrow 0, \infty, \qquad (36)$$
$$x \rightarrow 0, \infty.$$

$$m_s = k \int_0^\infty t \left(\frac{b}{d}\right)\left(\frac{t}{d}\right)^{b-1}e^{-\left(\frac{t}{d}\right)^b}\mathrm{d}t =$$

$$= \frac{1}{d^b} \int_0^\infty \sqrt[b]{x} \cdot e^{-\left(\frac{x}{d}\right)}\mathrm{d}x. \qquad (37)$$

For verification of the resulting expressions, expression for the exponential distribution can be obtained by determination of parameters $m = 0$ and $b = 1$. For

Eq. (25) we obtain, by insertion of parameter $m = 0$, this expression:

$$m_s = k \int_0^\infty x e^{-x} \mathrm{d}x = \int_0^\infty k \cdot t \cdot e^{-k \cdot t} \mathrm{d}t = \frac{1}{k}, \quad (38)$$

and the same result must be obtained if we use Eq. (26) with constant parameter $b = 1$:

$$m_s = k \int_0^\infty x \cdot e^{-\frac{x}{d}} \mathrm{d}x = \int_0^\infty \frac{t}{d} e^{-\frac{t}{d}} \mathrm{d}t = d. \quad (39)$$

After verification we know that our equations connected with the two types of expressions of the Weibull distribution for the mean time of operation are both correct. In the next step it is necessary to perform numerical calculation of these integrals, because the given integrals are unsolvable by standard exact methods [3].

3.6. Derivation for the Unit which was in Operation for T Hours

For exponential distribution, independence on time T was shown by Eq. (12). The probability of failure of the unit in time t working during time T is given by Eq. (25) for the Weibull distribution and for the second expression by Eq. (30). Density of failure $f(t)$ is given by differentiating these equations [3].

3.7. Calculation of the Mean Time of Operation of the Unit which was in Operation for T Hours

The first important parameter for the calculation is the probability of failure. The probability of failure in time t is given by Eq. (29) and Eq. (30). The next step is determining the density of failure of these equations, which is given by their differentiating [3]. The density of failure for Eq. (29) is given by the following expression:

$$f(t) = -k(T+t)^m e^{\left[\frac{k}{m+1}T^{m+1} - \frac{k}{m+1}(T+t)^{m+1}\right]} \quad (40)$$

and the mean time of operation for Eq. (29) is determined by the following expression:

$$m_s = \int_0^\infty t \cdot f(t) \mathrm{d}t =$$
$$= \int_0^\infty -k \cdot t(T+t)^m e^{\left[\frac{k}{m+1}T^{m+1} - \frac{k}{m+1}(T+t)^{m+1}\right]}. \quad (41)$$

Firstly, substitution for $k(T+t)^{m+1}$ is necessary:

$$k(T+t)^{m+1} = z \to t = \sqrt[m+1]{\frac{z}{k}} - T$$
$$k(m+1)(T-t)^m \mathrm{d}t = \mathrm{d}z \to k(T+t)^m \mathrm{d}t = \frac{\mathrm{d}z}{m+1}. \quad (42)$$

Secondly, boundaries have to be recalculated:

$$t \to 0, \infty$$
$$z \to k(T)^{m+1}, \infty. \quad (43)$$

The third step is insertion of substitution into Eq. (20):

$$m_s =$$
$$= \frac{1}{(m+1)^{m+1}\sqrt{k}} \int_{kT^{m+1}}^\infty \sqrt[m+1]{z} e^{\left[\frac{k \cdot T^{m+1} - z}{m+1}\right]} \mathrm{d}z -$$
$$- \frac{T}{(m+1)} \int_{k \cdot T^{m+1}}^\infty e^{\left[\frac{k \cdot T^{m+1} - z}{m+1}\right]} \mathrm{d}z. \quad (44)$$

We also need to create substitution for $\dfrac{-kT^{m+1} + z}{m+1}$:

$$\frac{-k \cdot T^{m+1} + z}{m+1} = x \to z = k \cdot T^{m+1}x,$$
$$\frac{\mathrm{d}z}{m+1} = \mathrm{d}x \to \mathrm{d}z = (m+1)\mathrm{d}x. \quad (45)$$

Recalculation of boundaries:

$$z \to k(T)^{m+1}, \infty \quad x \to 0, \infty. \quad (46)$$

Insertion of substitution into Eq. (29):

$$m_s = \frac{1}{(m+1)\sqrt[m+1]{k}} \cdot$$
$$\cdot \int_0^\infty \sqrt[m+1]{kT^{m+1} + (m+1)x} \cdot e^{-x}(m+$$
$$+1)\mathrm{d}x - \frac{T}{(m+1)} \int_0^\infty e^{-x}(m+1)\mathrm{d}x = \quad (47)$$
$$= \frac{1}{\sqrt[m+1]{k}} \int_0^\infty e^{-x} \mathrm{d}x - T.$$

To verify this Eq. (47), $m = 0$ is inserted into it. Again we transform the equation for the Weibull distribution into the equation for the exponential distribution:

$$m_s = \int_0^\infty \frac{(k \cdot T + x)}{k} e^{-x} \mathrm{d}x - T =$$
$$= \frac{1}{k} \int_0^\infty x \cdot e^{-x} \mathrm{d}x + \frac{1}{k} \int_0^\infty k \cdot T \cdot e^{-x} \mathrm{d}x - T = \quad (48)$$
$$= \frac{1}{k} \int_0^\infty x \cdot e^{-x} \mathrm{d}x = \frac{1}{k}.$$

For the second expression of the Weibull distribution the mean time of operation is calculated as follows. Firstly, we need to perform conversion for the specific constants:

$$m + 1 = b \to \frac{m+1}{k} = d^b. \quad (49)$$

Secondly, we insert these converted constants into Eq. (41):

$$m_s = d \int_0^\infty \sqrt[b]{\frac{T^b}{d^b} + x} \cdot e^{-x}\mathrm{d}x - T. \quad (50)$$

Finally, for verification we choose the constant $b = 1$ and transform it into the exponential distribution. Due to this operation, we obtain the following expression:

$$m_s = d \int_0^\infty \left(x + \frac{T}{d}\right) e^{-x}\mathrm{d}x - T =$$
$$= d \int_0^\infty (x) e^{-x}\mathrm{d}x + T \int_0^\infty e^{-x}\mathrm{d}x - T = d. \quad (51)$$

The resulting Eq. (48) and Eq. (50) for the mean time of operation are correctly determined.

4. Achieved Results

This part is focused on the main results and their explanation. Firstly, the probability of failure-free operation for the exponential distribution was determined by the following equation [2]:

$$P_{B(A)} = 1 - e^{-\lambda t}. \quad (52)$$

After that it was derived from the same expression of probability also for the Weibull distribution. We obtained these two expressions:

$$P_{B(A)} = P_{1(t)} = 1 - e^{\left[\frac{k}{m+1}T^{m+1} - \frac{k}{m+1}(T+t)^{m+1}\right]}, \quad (53)$$

$$P_{B(A)} = P_{1(t)} = 1 - e^{\left[\left(\frac{T}{d}\right)^b - \left(\frac{T+t}{d}\right)^b\right]}. \quad (54)$$

Secondly, it was necessary to determine the expression of the mean time of operation for the exponential distribution. In this case the time of operation T was not respected:

$$m_s = \frac{k}{(m+1)} \int_0^\infty \sqrt[m+1]{x} e^{-\frac{kx}{m+1}}\mathrm{d}x, \quad t^{m+1}=x \quad (55)$$

$$m_s = \frac{1}{d^b} \int_0^\infty \sqrt[b]{x} e^{-\frac{x}{d}}\mathrm{d}x. \quad t^b=x \quad (56)$$

Finally, calculation of the mean time of operation was performed with respect to the time of operation. Here two expressions representing the most realistic mean time of operation were obtained.

For analytical solution, the substitutions and recalculation of boundaries must be performed twice:

$$k(T+t)^{m+1} = z, \quad \frac{-k \cdot T^{m+1} + z}{m+1} = x, \quad (57)$$

$$z \to k(T)^{m+1}, \infty \quad x \to 0, \infty, \quad (58)$$

$$m_s = \frac{1}{\sqrt[m+1]{k}} \int \sqrt[m+1]{\frac{k \cdot T^{m+1}}{m+1} + x} \cdot e^{-x}\mathrm{d}x - T. \quad (59)$$

For the second expression, the solution looks like this:

$$m+1 = b \to \frac{m+1}{k} = d^b, \quad (60)$$

$$m_s = d \int_0^\infty \sqrt[b]{\frac{T^b}{d^b} + x} \cdot e^{-x}\mathrm{d}x - T. \quad (61)$$

Integration of Eq. (35) and Eq. (38) is impossible by exact methods. This is the reason for using an approximate numerical method, such as Simpson's rule or Quadrature rule.

4.1. Practical Application of the Final Equations for a Real System

We used real parameters of a feed pump from the Czech nuclear power plant Dukovany. The nuclear power plant Dukovany has four units. Each unit of Dukovany contains the same components, i.e. a turbine with the performance of 250 MWe, condenser, steam generator and feed pump. In this case we have focused only on feed pumps [5].

Parameters used in the following script have been determined from a database of operation values in the software Access by statistical methods. The times of operation have been chosen on the basis of the data of the feed pump used in the power plant. Characteristics of wear $b = 0.65$ and complex characteristics of lifetime $d = 3150$ were also determined by these statistical methods. The script for calculation characteristics of reliability by the Quadrature rule is created and it is shown below:

```
T1=0;
T2=72;
T3=120;
T4=168;
T5=300;
T6=500;
T7=1000;
T8=2000;
T9=3500;
T10=5000;
T11=6000; %Required times
b=0.65; %Characteristics of wear
d=3150; %Characteristics of lifetime
try
sum_old = sum; %Saving previous value
of numerical integral
catch
sum_old = 0;
```

```
end
sum = 0;
h=0.01; %Determination of the step
x_in = 0 :  h :  2e1;
yout = zeros(size(x_in));
s_out = yout;
i = 1;
for i = 1 :  length(x_in) %Calculation
of the numerical integral by the
Quadrature rule
x = x_in(i);
y = d*(((T1./d).^b)+x).^(1./b)*exp(-x)-T1;
if y < 0 %Elimination of negative
values of the integral
break;
end
sum = sum + y * h;
yout(i) = y;
s_out(i) = sum;
end
subplot(2,1,1); %Figure of the function
before integration
plot(x_in,yout);
subplot(2,1,2);
plot(x_in,s_out); %Figure of the
function after integration
```

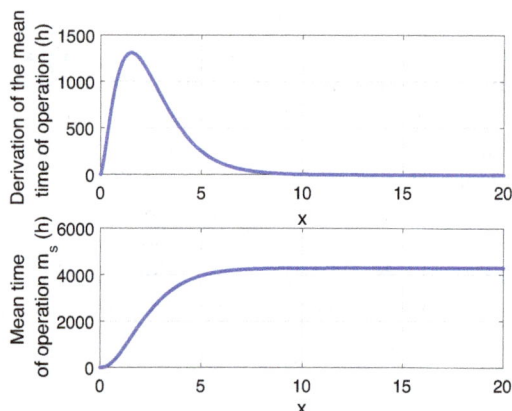

Fig. 4: Graphic expression of interdependence between time of operation and mean-time of operation.

5. Discussion

Dynamic reliability is a very promising field and offers a lot of possibilities for future research. The solution using the Weibull distribution has a significant impact in the field of dynamic reliability because the Weibull distribution is convenient to be used for repairable systems. It is the best tool of dynamic modelling of real systems.

The general procedure described in the experimental section may be used in most types of power plant units. All the necessary steps for calculation of the key parameters of dynamic reliability are described in that section. The final expressions are verified using the exponential distribution.

The results interpreted in the previous section concern the probability of failure-free operation and the mean time of operation. These parameters are representative and give a realistic view of the reliability of our nuclear power unit. Thanks to our results, we are able to create a model of dynamic reliability for prediction of future states. Prediction of future states may be used for optimal planning of maintenance and also for supporting the sustainability of these units. This section is concluded by Tab. 1, which presents the calculated values of the mean time of operation depending on the length of the time of operation, and Fig. 4 which shows a graphic view of the values.

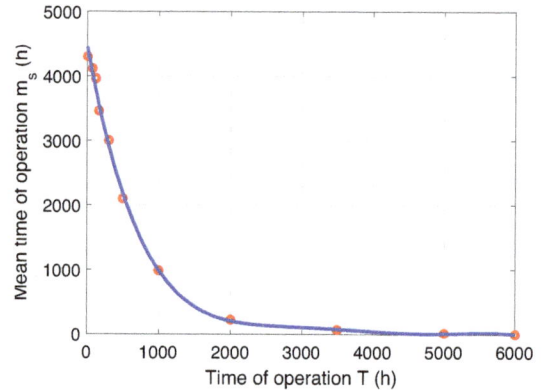

Fig. 3: Graphic expression of calculation before and after integration for $T1$ by the Quadrature rule.

The numerical results in Tab. 1 for Eq. (57) were calculated by the script in Matlab. The graphical expression of Tab. 1 is shown in Fig. 4 below.

In Tab. 1, an example of calculation $m_s = f_{((m,k,T))}$ of a nuclear power plant unit is shown with realistic parameters of reliability using the Weibull distribution. The trend of the mean time of operation is given in Fig. 4. This shows that the mean-time of operation falls with the rising time of operation.

6. Conclusion

In this paper the methodology of reliability characteristics, such as failure-free operation and the mean time of operation, is described. The methodology contains calculations and verifications of unit parameters. The final results were obtained by calculation and verified by the exponential distribution. This methodology is

Tab. 1: Interdependence between mean time of operation (m_s) and time of operation (T).

T (h)	0	72	120	168	300	500	1000	2000	3500	5000	6000
m_s (h)	4304	4113	3964	3821	3465	2999	2102	987	229	17	0.02

applicable to any system. However, this system must be repairable and periodically checked. Of course, we must know the key parameters of the system.

Obtained results of the mean-time of operation for the Weibull distribution is shown in Fig. 2. The results of the mean-time of operation for the Weibull distribution are shown in Fig. 2. The trend in Fig. 2 respects the theoretical assumptions of the theory of reliability. The rising time of operation causes the decreasing hyperbolic trend of the mean time of operation.

These results can be useful for improving the current situation of the unit and also for creating the optimal plan of maintenance.

Acknowledgment

This research was partially supported by the SGS grant from VSB–Technical University of Ostrava (No. SP2016/95) and by the project TUCENET (No. LO1404).

References

[1] VALDMA, M., M. KEEL, H. TAMMOHA and K. KILK. Reliability of electric power generation in power systems with thermal and wind plants. *Oil Shale*. 2007, vol. 24, no. 2, pp. 197–208. ISSN 0208-189X.

[2] MENTLIK, V., P. TRNKA, M. TRNKOVA and L. SASEK. *Spolehlivostni aspekty elektrotechnolgie*. Praha: BEN, 2011. ISBN 978-80-7300-412-5.

[3] TUMA, J., Z. MARTINEK, M. TESAROVA and I. CHEMISINEC. *Security, quality and reliability of electrical energy*. Praha: Conte, 2007. ISBN 978-80-239-9056-0.

[4] SKVARKA, P. *Spolahlivosl v Jadrovej Energetike*. Bratislava: Alfa, 1990. ISBN 80-05-00095-2.

[5] NOHACOVA, L. and J. MERTLOVA. *Electrical substations and lines*. Praha: BEN, 2010. ISBN 978-80-7300-517-7.

About Authors

Zbynek MARTINEK was born in 1955 in Pilsen, Czech Republic. In 1983 he graduated (M.Sc.) with distinction from the department of Power Engineering of the Faculty of Electrical Engineering of the University of West Bohemia in Pilsen. He defended his Ph.D. in the field of Reliability of Power grid in 1990, the title of his habilitation thesis was „Synthesis of reliability of power plant unit in the Czech Republic". Since 1990 he has worked as a tutor at the Department of Electric Power Engineering and Ecology, detachment production and distribution of electricity. His scientific research is focused on reliability of power grid and devices in power engineering, heating industry and electrical installation design. Since 1990 he has regularly been named a member of the final examination committee at the Department of Electric Power Engineering and Ecology. He has also been appointed the chairman of the final examination committee of the bachelor programmes since 2005.

Ales HROMADKA was born in 1990 in Pilsen, Czech Republic. He received his M.Sc. from electrical engineering at the University of West bohemia in 2015. In present he is a Ph.D. student and a junior researcher at the University of West Bohemia. His research interests include modelling of dynamic reliability of electricity systems and also modelling of thermal cycles in thermal power plants.

Jiri HAMMERBAUER born in 1964, Education: 1989 Ing.(M.Sc.), Institute of Technology, Pilsen, 1999 Ph.D. Applied Electronics, Univ. of West Bohemia, Pilsen, 2003 - Assoc. Professor, University of West Bohemia, Pilsen. Position at the University: Dean of the Faculty of Electrical Engineering at the University of West Bohemia, Assoc. Professor of the department Applied Electronics and Telecommunications. A member of the Board for Science of the Faculty of Electrical Engineering. Member of the Board for Science of University of West Bohemia. Research and expertise: Development of DC/DC converters with high efficiency and small dimensions, and controlled switched current converters for high luminance LEDs. Design and development of mains driven power supplies with emphasis on high efficiency, small size, high switching frequency, wide range of input voltage, reliability, and EMC. Development of chargers for all types of accumulators, especially of modern Li-ion and Li-pol types. Development of methods and systems for the measurement and diagnostics of standby power supplies used in consumer and industrial electronics (UPS systems). Research and design of electronic control systems for modern methods of accumulation of energy generated by photovoltaic cells, e.g. by means of super- and ultracapacitors. AC and DC current regulators.

Appendix A
Abbreviations

The following abbreviations are used in this manuscript:

- P $\cdots$ Probability of something

- μ $\cdots$ Failure probability function

- $q(t)$ $\cdots$ Intensity of repair

- r $\cdots$ Mean time to repair

- λ $\cdots$ Intensity of failure

- b $\cdots$ Characteristics of wear

- d $\cdots$ Characteristics of lifetime

- m $\cdots$ Mean time of failure-free operation

- k $\cdots$ Constant inclusive characteristics of d and b

- T $\cdots$ Real time of operation for the unit

- t $\cdots$ Required time for prediction of probability of failure-free operation

- $P_{(B)}$ $\cdots$ Probability of phenomenon B

- $P_{(B \cap A)}$ $\cdots$ Probability of intersection of phenomenon A and phenomenon B simultaneously

- $P_{B(A)}$ $\cdots$ Conditional probability of phenomenon A, if success of phenomenon B

- $P_{1(t)}$ $\cdots$ Probability of failure-free operation

- $f_{(t)}$ $\cdots$ Density of probability of failure-free operation

- $m_{(s)}$ $\cdots$ Mean time of operation

MATERIAL TRACKING WITH DYNAMIC TORQUE ADAPTATION FOR TENSION CONTROL IN WIRE ROD MILL

Tomas BOROVSKY[1], Karol KYSLAN[2], Frantisek DUROVSKY[2]

[1]Slovakia Steel Mills, a.s., Priemyselna 720, 072 22 Strazske, Slovak Republic
[2]Department of Electrical Engineering and Mechatronics, Faculty of Electrical Engineering,
Technical University of Kosice, Letna 9, 042 00 Kosice, Slovak Republic

tomas.borov@gmail.com, karol.kyslan@tuke.sk, frantisek.durovsky@tuke.sk

Abstract. *Material tracking is an important part of the automation control system which has a major impact on the product quality. This paper addresses a stand load identification in wire rod mill as a new algorithm added to existing control system. Tension control approaches are described and a modification of existing tracking system is proposed in order to eliminate tracking faults. Proposed method is based on dynamic torque calculation and its performance was experimentally verified on the industrial wire rod mill. Experimental results show significant reduction of the errors.*

Keywords

Dynamic torque, material tracking, tension control, wire rod.

1. Introduction

Wire Rod Mill (WRM) usually consists of roughing, intermediate and finishing mill. Each mill consists of several Stands (STD) [1] and [2] and each stand is driven individually by electrical motors supplied from power converters. Typically, one quadrant or four quadrant controlled DC machines are used. To maintain a stable and high quality rolling process, it is required to control the speed of the stands according to the tension conditions between the stands. Speed ratios between these stands must remain constant to maintain stable material flow. If there is a change in a gap between rolls, it causes a deviation in rolling parameters, what can increase or decrease interstand tension [3]. The tension between stands has a great influence on the properties of the workpiece produced in the mill.

Wire rod rolling is a periodical process. The workpieces pass through roll stands sequentially and one workpiece follows another. For that reason, material tracking system is one of the most important parts of the WRM control structure. It manages the tension control and it is used also for cobble identification. Material tracking function provides accurate information of the workpiece head and tail end positions in the mill. This is a fundamental requirement for automatic control sequence, data collecting systems, main and auxiliary drives and services. Even more, the speed reference distribution, automatic loop control, minimum tension control and automatic cutting of the flying shears are based on precise material tracking.

Following sources of feedback signals are usually used in WRM [10] and [11]:

- *Hot Metal Detector (HMD)* - the most commonly used sensor in the mill lines, it uses infra-red radiation emitted by hot materials which is received by an optical system in the sensor,

- *Loop scanner* - used in the automatic loop control, optically scans the field to be controlled and does not need any optical adjustments,

- *Stand threading signal* - generated from the peak torque detector in the drive or PLC unit.

Obviously, installation of the first two sensor systems results in additional costs.

While the end of the workpiece passes through the roll, the tension of the bar suddenly turns to zero and size of the bar changes [4]. The changes in cross-sectional area along the workpiece can lead to the cobble in the intermediate and finishing mill. Massive tension can cause total unloading of the stand that consequently leads to the cobble due to the failure of interstand tension control. On the other side, no tension

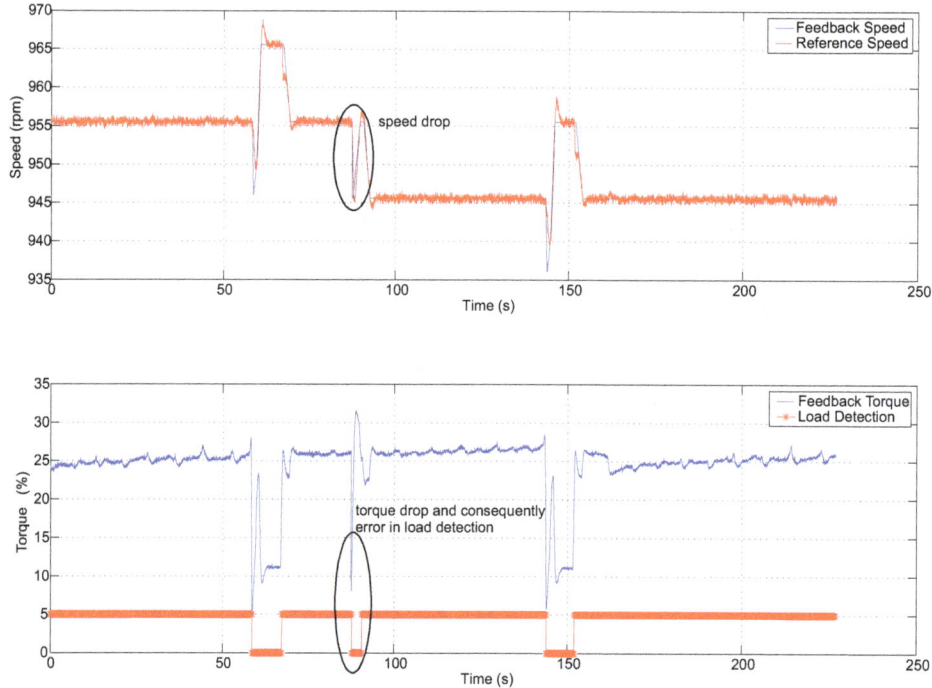

Fig. 1: Material tracking fault due to the raped machine speed decrease.

means loopering of the material between the stands, that also leads to cobble in the mill [6].

All these deviations have a big impact on product quality. Even more, these deviations increase amount of the time spent by engineers for the mill maintenance.

2.   Problem Formulation

In case there is no sensor for tracking of the workpiece, motor torque and speed are used to track the material. Load Detection (LD) signal is a control signal indicating that the stand is under load i.e. the workpiece has been rolled by stand. If the actual torque value exceeds LD signal level, head position of the workpiece is traced. When workpiece leaves the stand, motor torque decreases and workpiece tail end is detected. However, WRM is highly dynamic system consisting of drives and processing material coupled together. While the mill speed is decreased rapidly, motor torque can decrease close to the zero, even though it is fully loaded. On the other side, if the mill speed is increased rapidly, motor torque increases and, under certain conditions, it can exceed LD signal level. This situation is shown in Fig. 1 and, obviously, it causes significant errors. It can be observed in Fig. 1 that 3 workpieces have been rolled. During the rolling of the 2nd workpiece (since $t = 60$ s to $t = 140$ s), LD signal is active, but the speed change in $t = 85$ s coming from superior control results in torque drop. Torque value had dropped under LD level, therefore, LD signal became inactive, although

the stand was fully loaded. This phenomenon leads to faulty tension control with consequent cobble. In order to eliminate this problem, a modification of the classical tension control has been proposed, as shown in Section 5.

Total motor torque T of each stand can be expressed as [6], [15]:

$$T = T_0 + T_d + T_r + T_p + T_{ft} + T_{bt}, \qquad (1)$$

where:

- T_0 is friction torque,

- T_d is dynamic torque,

- T_r is rolling torque,

- T_p is torque from pins friction during the rolling,

- T_{ft} is forward tension torque,

- T_{bt} is backward tension torque.

For a period of a free rolling, when there is no billet in the stand, only T_0 and T_d act on motor shaft. T_0 depends on actual motor speed and friction conditions and T_d is given as:

$$T_d = J\frac{\mathrm{d}\omega}{\mathrm{d}t}, \qquad (2)$$

where J is a total stand inertia and ω is motor speed of given stand [10]. Inertia of each stand depends on invariable factors, such as gearboxes or motor shafts, and variable factors, which are working roll parameters. Mass and size of the roll varies according to pass schedule and roll wear. However these changes, compared to the total inertia of the stand, are small enough to be neglected.

During the rolling, additive torque components T_r, T_p, T_{ft} and T_{bt} act on the motor shaft. These components are closely related and depend on many rolling parameters and rolling and tension forces, which influence the stand. For example, rolling torque T_r is given by [1]:

$$T_r = 4 \int_0^{L_r} b\sigma_y(L_r - x)\mathrm{d}x, \qquad (3)$$

where b is a width of approximated workpiece-roll surface area, σ_y is a stress component in vertical direction and L_r is a roll-bite length. Torque from pins' friction during the rolling T_p is:

$$T_p = \frac{1}{2}F_r f_p D_p, \qquad (4)$$

where F_r is a rolling force [1], f_p is a friction coefficient of the pins and D_p is a diameter of the pins. Forward and backward tension torque, T_{ft} and T_{bt}, are given by [1]:

$$T_{ft} = \frac{1}{2}D_{ef}A_f\sigma_f, \qquad (5)$$

$$T_{bt} = \frac{1}{2}D_{ef}A_b\sigma_b, \qquad (6)$$

where D_{ef} is an effective diameter of the rolls, A_b and A_f is a workpiece cross-sectional area in front and back of the roll-bite lenght, respectively; and σ_f and σ_b is a front and back axial stress, respectively [12].

There is no tension force measurement in the mill and so the components of the motor torque cannot be identified separately. Estimation of all these components requires knowledge of many rolling parameters, which are varying during the rolling process. Therefore, components T_r, T_p, T_{ft} and T_{bt} will be used only as one lumped variable for further calculations. It is important to note, that the torque acting on the motor shaft consists of these parts:

- rolling part T_r, T_p, T_{ft}, T_{bt},
- free running part T_0, T_d.

Tracking system has to securely identify the time interval, in which the stand is under rolling load. When compared to the torque of rolling part, T_0 is quite small, so it is neglected in Eq. (7). Hence, by calculation of dynamic torque T_d in each time of rolling sequence and consequently by subtracting it from the motor torque feedback, free running part is excluded

from the total motor torque and only rolling part remains. This is the way how to securely indicate total rolling load. Load torque T_{load} is then given by:

$$T_{load} = T_{fbk} - J\frac{\Delta\omega}{\Delta t}, \qquad (7)$$

where T_{fbk} is actual motor torque feedback and $\Delta\omega$ is the change of angular speed during defined time interval Δt. Note, that the value of T_0 in Eq. (7) is neglected, but it has to be identified for the calculation of inertia.

3. Tension Control

In high-speed WRM with fixed speed ratios between the stands, interstand tensions are used to obtain stable rolling conditions [5]. Traditional tension control systems are based on the *minimum tension control* for roughing mill, the *loop control* for intermediate mill and the *tension rolling* for finishing mill.

3.1. Roughing Mill

Since it is not possible to control the loop between the stands due to big dimensions of the workpiece in roughing mill, a looperless control scheme has to be used. So called *minimum tension control* assumes that the workpiece dimension and material temperature profile remain constant along the whole workpiece. Motor torque feedback during the time interval with and without interaction of the workpiece with the stand is used as a tension indicator [6], [8] and [9]. Recent approach is to use *Interstand Dimension Control* (IDC) with the complex system of workpiece cross-sectional area measurements in the interstand area. U-Gauge sensors measure dimensions and IDC control system is used to automatically adjust the gap set–up and speed [7]. WRM described in this paper has a conventional sensorless control system with the minimum tension control. Tension free rolling conditions of i-th stand STD_i are represented by the time interval, when the workpiece is rolled by STD_i but does not reach following stand yet. Once the workpiece enters stand STD_{i+1}, motor torque of STD_i is affected by the tension between STD_i and STD_{i+1} and tension control turns on. To control the intervals, the knowledge of actual front and tail end position of workpiece in the mill is required.

3.2. Intermediate Mill

To keep the constant material flow in the intermediate mill, the loopers are installed between the stands. Looper consists of a roller mounted on a pivoted arm

that is initially lowered below the mill pass line. The
pivoted arm is raised after the head of the workpiece
enters the stand. This causes a deflection of the workpiece and a small loop is formed. Loop scanner measures loop height. The speed of a given stand is then
adjusted to maintain the loop height on desired value
[1]. When workpiece leaves the stand, pivot arm is lowered and loop height is decreased. Recent approach is
to use IDC, where, as in the roughing mill, U-gauge
sensors measure dimensions and IDC system is used
to automatically adjust roll gap and speed [7]. WRM
under experiment has a conventional looper control
scheme without IDC.

3.3. Finishing Mill

Configuration of the finishing mill is different from the
previous ones. To meet the demand for increased production rates, the rolling speed must be drastically
increased. Exit material speed can reach more than
100 m·s^{-1}; therefore the conventional loop control cannot be used. Stands are equipped with changeable roll
rings and there is no space for a measuring sensor or
device, so measurement of any quantity in the finishing
block is hardly possible. Rolling speed of each stand is
determined by gear ratio and working diameter of the
roll ring.

Since there is no possibility to change the speed ratios between the stands, roll rings of the finishing block
usually creates so-called roll rings families, which have
to be used for each individual pass design. Pass schedule is designed to maintain a tension between each two
consequent stands to prevent the loopering [1] and [2].
WRM under experiment has a conventional block with
the ten stands coupled by gearboxes and driven by one
DC motor. Loop control is used only between the last
stand of the intermediate mill and first stand of this
block. Recent approach is to equip each stand with
a maller motor instead of a complex gearing with one
large motor.

4. Parameter Identification

4.1. Identification of Friction

The value of load torque friction component T_0 has
to be identified at first in order to calculate inertia
of each stand. T_0 is determined as a speed-depended
polynomial function. The function was obtained using
the MATLAB Curve Fitting Toolbox using the data
recorded during motor acceleration with slow speed reference ramp in order to suppress the effect of dynamic
torque. This measurement is shown in Fig. 2, where actual torque and speed feedback during the slow speed

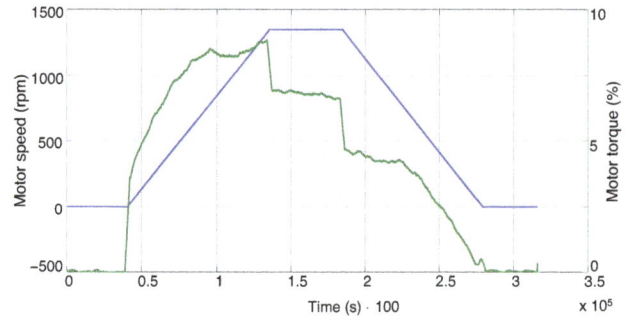

Fig. 2: Measurement for the identification of friction torque.

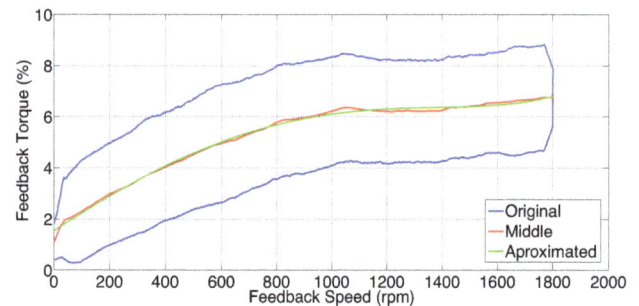

Fig. 3: Friction torque as a function of actual speed.

ramp up (approx. 100 s), free running and slow speed
ramp down is shown for stand *STD16*. The relation
between speed and torque and the approximation by
the 4th degree polynomial equation is shown in Fig. 3,
where *original* line shows measured motor torque during acceleration, free running and deceleration, *middle*
line shows the difference between acceleration and deceleration torque and *approximated* line shows the result of the approximation of the *middle* line with MAT-
LAB.

4.2. Identification of Inertia

Identification of the total stand inertia is shown in
Fig. 4. Motor was accelerated to the maximum speed
and actual speed and torque were recorded. Green line
shows the interval used for further calculations. Partial calculation of J for each measured time interval
Δt was performed according to:

$$J = (T - T_0)\frac{\Delta t}{\Delta \omega}, \qquad (8)$$

where $\Delta \omega$ is the speed difference during each time
interval, T is actual motor torque and T_0 is friction
torque given by polynomial function. Total stand inertia is then estimated as the average value of partial
inertia calculations. The results of these calculations
of inertia are shown in Fig. 5. It shows the difference
between the inertia calculated exactly as in Eq. (8) and
the inertia calculated as in Eq. (8) but for $T_0 = 0$. It
can be observed, that the difference is considerable and

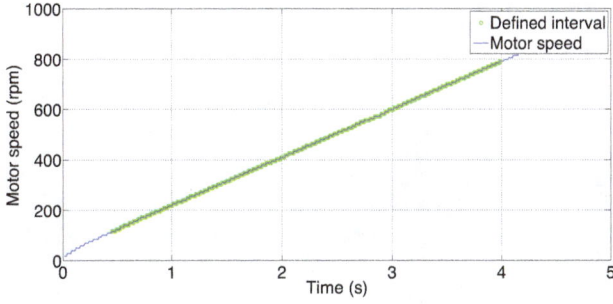

(a)

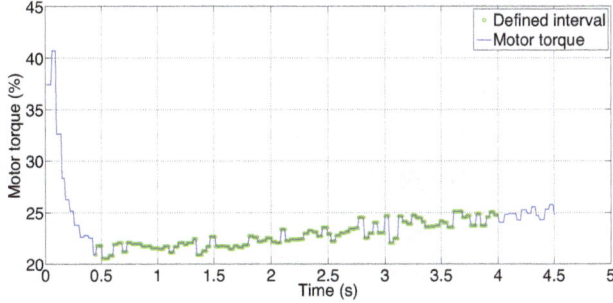

(b)

Fig. 4: Identification of total inertia - the 1st method.

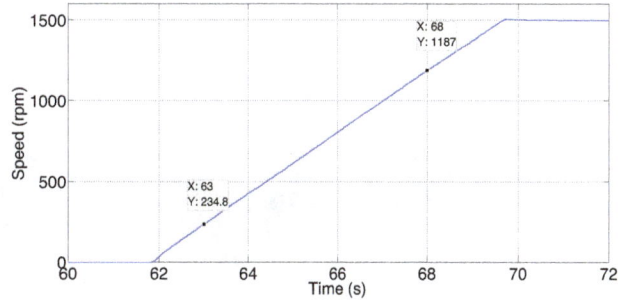

(a)

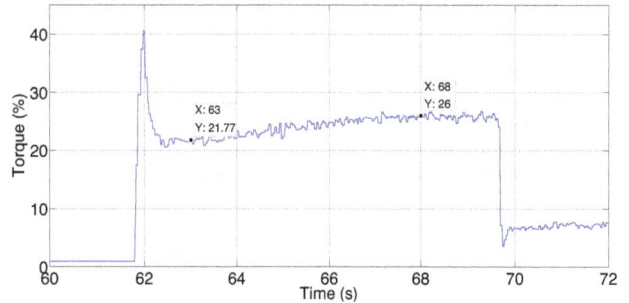

(b)

Fig. 6: Identification of J - the 2nd method.

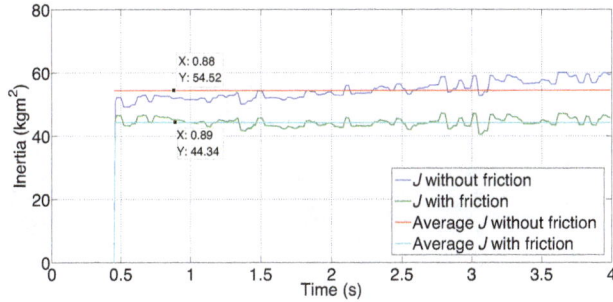

Fig. 5: Partial calculations of J in *STD16*.

to be calculated in order to implement the dynamic torque adaptation algorithm described in the next section.

Tab. 1: Inertia values of the individual WRM stands.

Stand	STD01	STD02	STD03	STD04
J (kg·m²)	8.97	21.87	10.41	22.67
Stand	STD05	STD06	STD09	STD10
J (kg·m²)	18.49	29.97	19.47	26.27
Stand	STD11	STD12	STD13	STD14
J (kg·m²)	20.81	28.74	19.97	35.16
Stand	STD15	STD16	FBLCK	
J (kg·m²)	23.39	41.13	1960.66	

so friction torque always has to be identified for inertia calculations.

Much faster estimation of inertia with less computational burden can be obtained during the acceleration of the motor without any additional measurement. If the average value of friction torque in steady state is used (e.g. T_0 is approximately 7 % for 1500 rpm in Fig. 6, J can be calculated again with Eq. (8). The measurement points are shown in Fig. 6. $\Delta\omega$ and Δt are given:

$$\Delta\omega = \omega_2 - \omega_1, \qquad (9)$$

$$\Delta t = t_2 - t_1. \qquad (10)$$

In this case, T is an average value of motor torque during acceleration within time interval $t_1 - t_0$. The difference between values of inertia calculated by the 1st and by the 2nd approach is less than 1 %. Hence, the 2nd method was used to calculate the inertias for all stands. These values are shown in Tab. 1. They need

5. DTA Algorithm

The algorithm of Dynamic Torque Adaptation (DTA) was implemented into existing control system based on Siemens S7 400 Programmable Logic Controller (PLC). Cycle interrupt time for proposed algorithm is 10 ms. There is no need for modification of the existing hardware, the solution is purely based on the software modification. DTA algorithm is shown in Fig. 7. The aim of the algorithm is to to calculate the value of T_{load}, which is used for stand load identification, where:

- *cycle time* is PLC cyclic interrupt time interval,

- *speed FBK* $(t - 1)$ is motor speed feedback from the previous cycle,

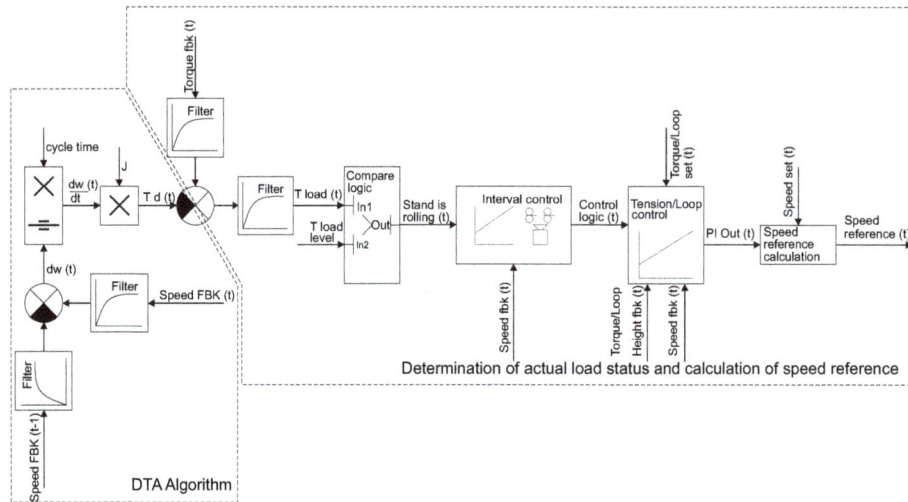

Fig. 7: Speed reference control chain for one stand with determination of actual load status and DTA algorithm implementation.

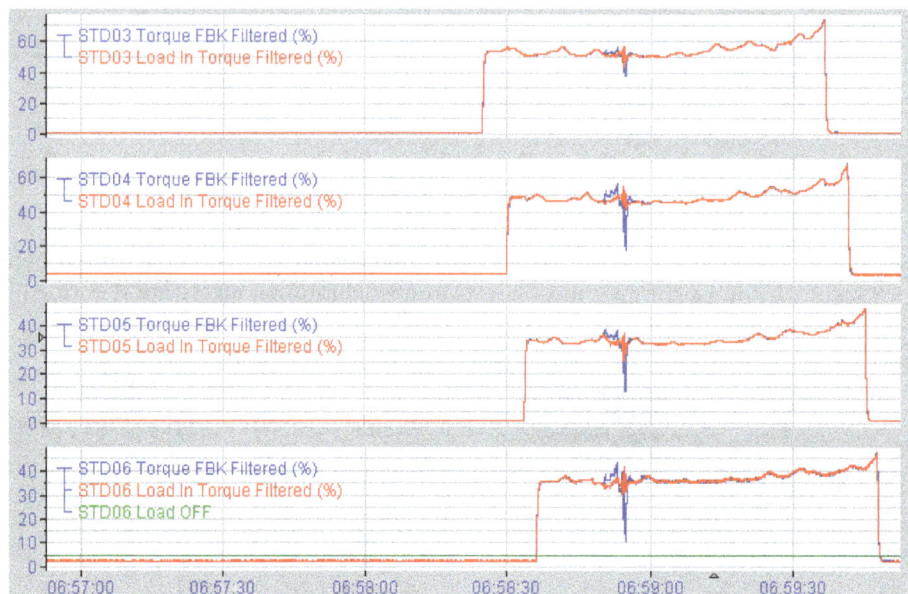

Fig. 8: Experimental results of the WRM control after the implementation of DTA algorithm.

- *speed FBK* (t) is motor speed feedback of the actual cycle,

- J is total stand inertia, as given by Tab. 1,

- $T(t)$ is motor torque feedback,

- T_{load_level} is a torque decision-making level for identification of rolling (LD signal).

Actual motor torque and actual motor speed are obtained from the drive converter via Profibus network. Calculated dynamic torque is subtracted from filtered actual torque, what gives T_{load} value. This value is compared with LD signal level in order to obtain logical signal showing that stand is under load. The logic for auxiliary devices (shears, loopers, cooling) is calculated in *Interval control* block. One of its outputs

presents the control logic signal for *Tension/Loop control* block. Here, the tension control or the loop control is executed: the tension control for roughing mill and the loop control for intermediate and finishing mill. Output of this controller is led into *Speed reference calculation* block. Reference speed calculation, together with the speed components from other drives and basic speed from the pass schedule, is executed here. Drive speed is controlled in closed loop and encoder is used for motor speed measurement. If the speed feedback is noised, it is necessary to use filtered speed signal, otherwise it would result in the ripple of actual torque, which then cannot be used for the torque level comparison. The ripple effect was significant especially for the Finishing Block (FBLCK) torque calculation, since its inertia is much higher when compared to other stands.

Detailed description of the speed control chain is beyond the scope of this paper. If necessary, it can be found in [13] or [16].

Experimental results of DTA algorithm implementation are shown in Fig. 8. The blue lines are the actual torque values T_{fbk} of individual stands (STD03 - STD04 - STD05 - STD06) and the red lines are the values of total filtered load torque T_{load} obtained by DTA algorithm.

The time responses show values during the rolling of one workpiece. Rolling started in $t = 06$ h 58 m 26 s. Rapid speed changes occurred in $t = 06$ h 58 m 50 s due to the control signal from intermediate mill. These speed changes cause torque droop, which is clearly visible on the value of T_{fbk} of STD06. This torque droop is very close to the LD level value (green line). However, the tension control uses values of T_{load} instead of T_{fbk}. The value calculated by DTA is invariant to the speed changes and so it is preferable for LD. All these measurements were obtained by *ibaAnalyzer* [14] during full rolling operation with high dynamic speed changes.

6.   Conclusion

The stand rolling load identification in wire rod rolling was analysed in this paper. Load identification was improved by dynamic torque adaptation. With the calculation of the dynamic part of the motor torque, which is subsequently excluded from the motor torque feedback, the effect of rapid speed change to the stand load identification is suppressed. Functional description of the algorithm, the procedure of rolling parameters identification, the measurements and the final implementation to the control system was described in detail. Experimental results measured in real rolling mill demonstrate applicability and correctness of the proposed technique.

Moreover, the results of the dynamic torque adaptation algorithm have some other benefits. In conventional tension controllers, the motor torque actual value obtained from the converter is used for the tension controller. With DTA algorithm, the load torque value is available as a new process value for the tension controller independently, without the dynamic and the friction components, and so it can be used for further tension control quality improvement. However, this issue has not been investigated in the paper and it may be a subject of further research.

DTA algorithm was successfully implemented in Slovakia Steel Mills company in Slovakia. There were circa 3 tracking faults per 30 000 rolled billets (on average per month) before DTA implementation, sometimes with consequent cobble in the mill. After the DTA implementation, the rolling load was determined securely. At present, there are no tracking faults due to the rapid speed changes in the mill.

Acknowledgment

This work was supported by the Slovak Research and Development Agency (APVV), under the Project code APVV-15-0750. This work was supported by the Scientific Grant Agency of the Ministry of Education of the Slovak Republic and Slovak Academy of Sciences (VEGA), under the Project code 1/0464/15.

References

[1] LEE, Y. *Rod and Bar Rolling: Theory and Applications*. 1st ed. New York: CRC Press, 2004. ISBN 978-0-824-75649-9.

[2] SMS Group - Wire Rod Mills. Sophisticated solutions for economic success: Information brochure. *SMS Group* [online]. 2017. Available at: http://meer.sms-group.com.

[3] LI, G., F. JANABI-SHARIFI and L. WITNISKY. Decoupling of multiple stand interactions in looperless rolling control process. In: *IEEE International Conference on Industrial Technology (ICIT'02)*. Bangkok: IEEE, 2002, pp. 821–826. ISBN 0-7803-7657-9. DOI: 10.1109/ICIT.2002.1189273.

[4] OGAI, H., A. FUJII, K. BABA, Y. NOGUCHI, H. ISHII and M. BABA. Free tension set-up of rod bar rolling by fuzzy inference method. In: *International Conference on Industrial Electronics, Control and Instrumentation (IECON'91)*. Kobe: IEEE, 1991, pp. 60–64. ISBN 0-87942-688-8. DOI: 10.1109/IECON.1991.239328.

[5] ERIKSSON, C. *Roll Pass Design for Improved Flexibility and Quality in Wire Rod Rolling*. Stockholm, 2004. Doctoral thesis. Royal Institute of Technology. Supervisor prof. Ulf Stahlberg. Available at: http://kth.diva-portal.org/smash/get/diva2:14308/FULLTEXT01.pdf.

[6] BOROVSKY, T. Rolling Sped and Torque Prediction in Wire Rod Mill. In: *13th Scientific Conference of Young Researchers (SCYR 2013)*. Herlany: Technical University of Kosice, 2013, pp. 395–397. ISBN 978-80-553-1422-8. Available at: http://hron.fei.tuke.sk/scyr/data/templates/Proceedings_2013.pdf.

[7] SOLLANDER, D. and S. LINDER. Interstand Dimension Control Implementing U-gauge Technology in Rod and Bar Mills. *ABB review*. 2000, vol. 1, iss. 1, pp. 47–53. ISSN 1013-3119.

[8] LIU, J. *Design and Analysis of Intelligent Fuzzy Tension Controllers for Rolling Mills*. Waterloo, 2002. Master thesis. University of Waterloo. Supervisor prof. Farrokh Janabi-Sharifi. Available at: http://hdl.handle.net/10012/848.

[9] LI, G. and F. JANABI-SHARIFI. Fuzzy Tension Control Scheme for Roughing and Intermediate Rolling Mills. In: *Artificial Neural Networks in Engineering Conference (ANNIE 2002)*. St. Louis: University of Missouri-Rolla, 2002, pp. 347–352. ISBN 0-79180-191-8.

[10] SUN, X. and H.-F. SUN. Speed cascade control system for bar and wire rod mills. *ABB China Ltd* [online]. 2015. Available at: https://library.e.abb.com/public/4cb4384eb21 fc2738525761f004fc27b/1618%20Speed%20Cas cade%20VP.pdf.

[11] Hot Metal Detector and Loop Scanners for Rolling Mills and Casters. *Pacific International Pvt* [online]. 2017. Available at: http://www.pacificinternational.co.in.

[12] BAYOUMI, L. S. and Y. LEE. Effect of Interstand Tension on Roll Load, Torque and Workpiece Deformation in the Rod Rolling Process. *Journal of Materials Processing Technology*, 2004, vol. 145, iss. 1, pp. 7–13. ISSN 0924-0136. DOI: 10.1016/S0924-0136(03)00581-8.

[13] KHRAMSHIN, V. R., A. S. EVDOKIMOV, G. P. KORNILOV, A. A. RADIONOV and A. S. KARANDAEV. System for speed mode control of the electric drives of the continuous train of the hot-rolling mill. In: *International Siberian Conference on Control and Communications (SIBCON)*. Omsk: IEEE, 2015, pp. 1–6. ISBN 978-1-4799-7103-9. DOI: 10.1109/SIBCON.2015.7147264.

[14] ibaAnalyzer. *iba AG* [online]. 2017. Available at: http://www.iba-ag.com/en/iba-system/iba-system/analyze/.

[15] JANDURA, P., J. CERNOHORSKY and A. RICHTER. Electric Drive and Energy Storage System for Industry Modular Mobile Container Platform, Feasibility Study. *IFAC-PapersOnLine*. 2016, vol. 49. iss. 25, pp. 448–453. ISSN 2405-8963. DOI: 10.1016/j.ifacol.2016.12.056.

[16] MAMATOV, A. and V. DROZDOV. Cross-coupled Synchronous Control of Telescopes Complex. In: *International Conference on Power Drives Systems (ICPDS)*. Perm: IEEE, 2016, pp. 1–5. ISBN 0-79180-191-8. DOI: 10.1109/ICPDS.2016.7756678.

About Authors

Tomas BOROVSKY was born in Trebisov, Slovak Republic. He received his M.Sc. at the Faculty of Electrical Engineering and Informatics, Technical University of Kosice in 2012. Since then he has been an engineer at Slovakia Steel Mills, a.s. in Strazske. He is currently working towards his Ph.D. degree at Technical University of Kosice in electrical engineering. His research interests include control of steel-mills and multi-motor drives.

Karol KYSLAN was born in Humenne, Slovak Republic. He received his M.Sc. and Ph.D. at the Faculty of Electrical Engineering and Informatics, Technical University of Kosice in 2009 and 2012, respectively. He is currently an Assistant Professor at Department of Electrical Engineering and Mechatronics TUKE. His research interests include control of electrical drives, hardware-in-the-loop simulation and motion control.

Frantisek DUROVSKY was born in Kremnica, Slovak Republic. He received his M.Sc. and Ph.D. degree at Faculty of Electrical Engineering and Informatics, Technical University of Kosice in 1983 and 1993, respectively. He is currently an Associated Professor and Deputy Head of Department of Electrical Engineering and Mechatronics TUKE. His field of research interests are motion control, electrical drives in industrial and automotive applications, control and simulation of mechatronic systems.

15

Wireless Health Monitoring System for Rotor Eccentricity Faults Detection in Induction Machine

Ilias OUACHTOUK, Soumia EL HANI, Khalid DAHI

Electrical Laboratory Researche, Ecole Normale Superieure de Enseignement Technique, Mohammed V University, Avenue des Nations Unies, Agdal, Rabat, Morocco

ilias_ouachtouk@um5.ac.ma, s.elhani@um5s.net.ma, khalid.dahi@um5s.net.ma

Abstract. *Condition monitoring and fault detection of induction machines become an important area of research. Many techniques have been applied in this field including vibration, thermal, chemical and acoustic emission monitoring, but Motor Current Signature Analysis monitoring techniques usually are applied to detect the various classes of induction machine faults such as rotor, short winding, eccentricity and bearing fault etc. This paper presents a wireless system detection for rotor eccentricity faults in induction machine based on LabVIEW platform. Moreover, it is demonstrated that eccentricity fault generates a series of low frequency components in the form of sidebands around the fundamental frequency and its harmonic. In addition, the amplitudes of those components increase in proportion to the load and fault severity. The Power Spectral Density techniques and Short Time Frequency Transform spectrogram of current signals are used to detect the presence of those fault signatures.*

Keywords

Eccentricity, condition monitoring, induction machine, stator current signature, wireless system.

Nomenclature

Fs	Stator supply frequency.
fr	Rotational frequency.
N_r	Number of rotor bars.
s	Slip.
p	Pole pair number.
MCSA	Motor Current Signature Analysis.
PSD	Power spectral density.
STFT	Short time Fourier transforms.
n_d	Eccentricity order.
h	Stator time harmonic (1, 3, 5, ...).
k	Any integer (0, 1, 2, ...).

1. Introduction

Condition Monitoring (CM) is an effective strategy to maintain the performance of modern industrial equipment. In developed countries, much of the energy in industry is consumed by motor systems. Among these, three-phase induction motors are dominant because of their robustness and easy maintenance [1] and [2]. Moreover Induction Machines (IM) are the most commonly type of electrical rotating drives used in industry. However, due to the combination of poor working environment and installation, internal faults frequently occur on rotor, such as broken rotor bars, end ring connectors, and rotor eccentricity [3], [4] and [5]. Most of the faults in IM have a relationship with air-gap eccentricity which is the condition of the unequal air-gap between the stator and the rotor. This fault can result from variety of sources such in a stator rotor rub, thereby causing severe damage to the motor as well as acoustic noise and vibration. The creation of an unequal air gap may involve many different factors including unbalanced load, bearing wear, bent rotor shaft, and mechanical resonance at the critical load [6] and [7].

For the purpose of detecting such fault-related signals, many diagnostic methods have been developed so far, such as electromagnetic field monitoring, temperature measurements, Radio-Frequency (RF) emissions monitoring, vibration monitoring, chemical analysis, acoustic noise measurements and Motor-Current Signature Analysis (MCSA). Both, Vibration [8] and

[9] and MCSA [4], [10] and [11] are well proven methods for electrical machines fault diagnostics, but can be classified as the most promising fault-detection method for IM faults and can be done either online or offline. A great number of researchers have reported success in using MCSA as fault detection method which is based on current monitoring of induction motor; therefore it is not very expensive. The MCSA uses the current spectrum of the machine for locating characteristic fault frequencies. To diagnose the rotor eccentricity fault, MCSA based fault detection techniques such as Fast Fourier Transform (FFT), Power Spectral Density (PSD), Short Time Fourier Transform (STFT) and Wavelet Transform (WT) are used to diagnose the faults of IM. In this paper, the PSD and STFT spectrogram of current signals are used to detect the presence of frequency fault of rotor eccentricity around the fundamental and its harmonics [12] and [13].

Eccentricity fault in induction machine is usually detected by analyzing the stator line current spectrum. To detect and diagnose the static eccentricity, dynamic eccentricity and mixed eccentricity using frequency spectrum of stator current, two fundamental parameters need to be calculated:

- frequency of sideband components due to eccentricity fault around fundamental and

- frequency of sideband components due to eccentricity fault around the 3rd harmonic [14].

It is notable that low frequency components are more sensitive to torque variations. In addition, eccentricity fault and variation of the load does also affect considerably the above mentioned parameters. In this case, eccentricity fault detection is more challenging, because more sophisticated signal processing is required. For this purpose, time-frequency domain signal processing instantaneous frequency of fault components has been employed to detect and diagnose of mixed eccentricity fault [15].

The paper is organized as follows: Section 2. describes the Theoretical background of origins and effects of eccentricity fault. Section 3. describes the Condition Monitoring Techniques. Section 4. presents the detailed hardware design for the Laboratory Experimental Setup. Section 5. deals with the measurement results. Finally, conclusions are given in Sec. 6.

2. Theoretical Background

Most of the faults in IM have a relationship with air-gap eccentricity which is a condition in which there is an uneven air gap between the stator and the rotor. Normally, the rotor that is centered at the stator

bore of healthy motor results in identical air-gap among the stator and rotor. However, when the eccentricity emerges, it may lead to severe damage to the stator and rotor core. Thus, it is critical to detect the air gap eccentricity at an early stage to protect the motor system. In general, there are three forms of air-gap eccentricity: static, dynamic, and mixed. Static eccentricity (where the rotor is displaced from the stator bore center but is still turning upon its own axis), or dynamic eccentricity (where the rotor is still turning upon the stator bore center but not on its own center), if both static and dynamic eccentricities exist together then mixed eccentricity must be considered. In case of mixed eccentricity, the center of rotor, the center of stator, and the center of rotation are displaced with respect to each other. Figure 1 shows the cross sections of the induction motor with different types of eccentricities [14] and [16].

With MCSA, eccentricity can be detected by observing the rotor rotational speed frequency f_r and/or the rotor slot passing frequency components. The frequency equation for determining air-gap characteristics [13], [17] and [18] is as follows:

$$f_{ecc} = Fs \left[(k \cdot N_r \pm n_d) \frac{(1-s)}{p} \pm h \right], \quad (1)$$

where $n_d = 0$ in case of static eccentricity, and $n_d = 1, 2, 3, \ldots$; in case of dynamic eccentricity. Furthermore, if mixed eccentricity exist, the case in most air-gap related failures, there will be low-frequency components near the fundamental frequency and its harmonics which can be expressed by the following equation:

$$f_{Lecc} = h \cdot Fs \pm k \cdot fr. \quad (2)$$

However, in the line current spectrum of the faulty motor, the most important components are obtained around the fundamental frequency by substituting $k = 1$ as follows:

$$f_{Lecc} = Fs \pm fr. \quad (3)$$

Vibration signals can also be monitored to detect eccentricity faults. In case of mixed eccentricity, the low-frequency stator vibration components are given by [7] and [19]:

$$f_{Sv} = 2Fs \pm fr. \quad (4)$$

However, vibration sensors are delicate and expensive. They also have special installation requirements to avoid damage due to shock and vibration.

3. Condition Monitoring Techniques

Condition Monitoring (CM) techniques are effective to maintain the performance of modern industrial equip-

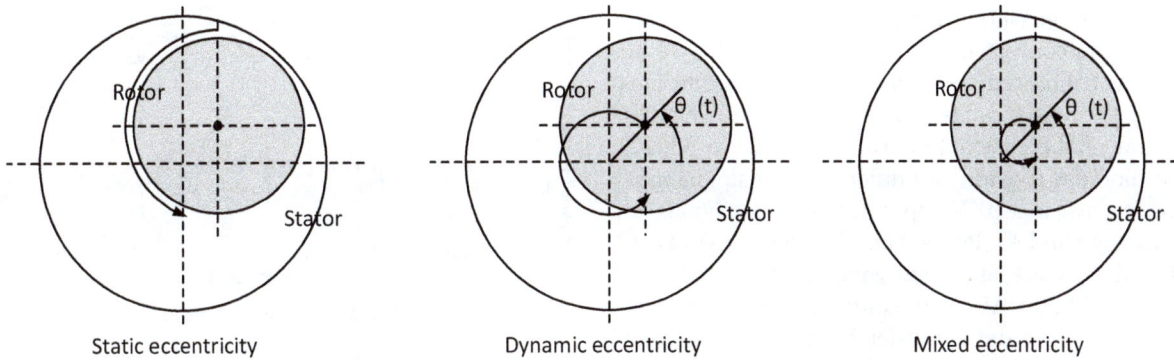

Fig. 1: Different types of eccentricity.

ment and have been successfully used by many industries to diagnose and identify the causes of machinery fault. Technically, IM CM methods are performed on-line or off-line. Off-line tests require the interruption of the motor operation or shutdown, while on-line methods provide motor diagnostics during the operation. However, on-line CM methods have attracted a great attention, as they offer adequate warning of the imminent failures, diagnosing present maintenance needs, schedule future preventive maintenance, and repair work [13].

Currently, Wireless System is gaining popularity in online CM fields and has been successfully employed in industry because of its inherent advantages, such as ease of installation, low cost, low latency, self-organization, and high reliability [2]. A block diagram

can be analyzed by different digital signal processing techniques to extract features, which are sensitive to the presence of faults as mentioned in block (C). In next step, the characteristic fault frequencies of rotor eccentricity fault are detected from the processed signals as shown in block (D). Finally, in block (E), the fault frequency characteristics that relate to different fault conditions are analyzed. This helps to detect the eccentricity fault and tracks the evolution of fault components over time.

This paper presents the design of wireless CM system shown in Fig. 3. The wireless data acquisition system is used to record the current signals in digital form. These signals can be analyzed by different digital signal processing techniques (PSD and STFT) to extract features, which are sensitive to the presence of rotor eccentricity fault.

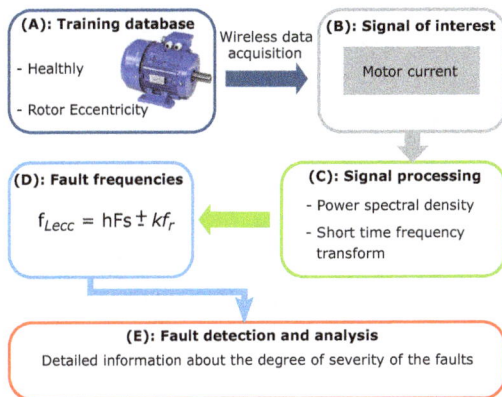

Fig. 2: Schematic diagram of the proposed fault detection procedure.

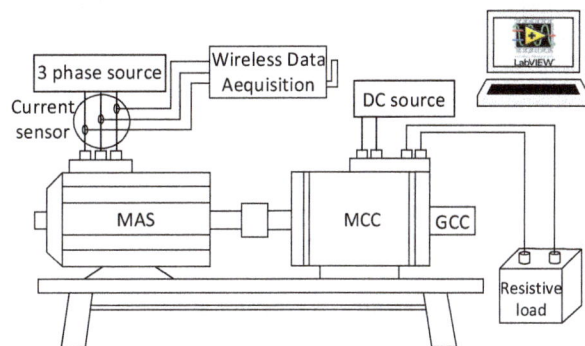

Fig. 3: Line diagram of the wireless monitoring system.

of a wireless CM system is shown in Fig. 2. In this stage, all the related terms used in CM during this study are described briefly. The experimental tests were developed under a variety of load conditions as mentioned in block (A). The eccentricity faults affect the symmetry of the machine and as a result produce characteristic fault frequencies. A wireless data acquisition system is used to acquire the motor current signals in digital form as shown in block (B). The signals

4. Laboratory Experimental Setup

The experimental system was conducted using the test rig and data acquisition system as shown in Fig. 4. The experimental tests were developed on a 3 kW, 50 Hz, 220 V = 380 V, 4-poles Induction Machine. The motor was directly coupled to a DC machine, which was loaded using a variable-resistance bank, the current of the motor is measured using LEM modules (LT 100-S/SP30) and converted to voltage signals. The voltage range was calibrated such that start-up currents measured by the sensor can fall within the voltage ranges on the input of the national instrument modules NI9222. These modules provide four differential analogue input channels with a ±10 V input range. All signals are sampled by wireless data acquisition device NI cDAQ-9191 and they are analyzed using the LabVIEW [18] and [20]. The cDAQ-9191 controls the timing, synchronization, and data transfer between C Series I/O modules and an external host. You can send data from the cDAQ-9191 to a host PC over Ethernet or IEEE 802.11 Wi-Fi.

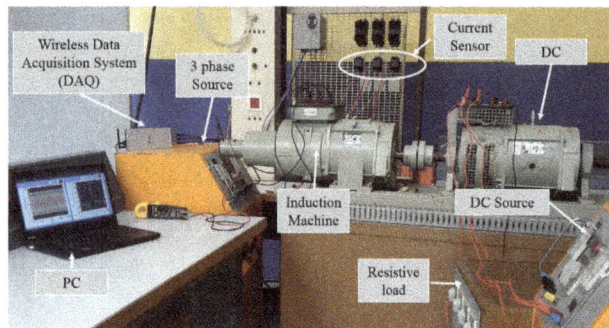

Fig. 4: Experimental setup.

5. Measurement Results

The experiments test have been performed to detect mixed eccentricity faults under no load and loaded condition to study its effect on current components. The results obtained from these experiments are given below.

First, the IM was tested at no load condition so the results can be compared with the loaded test. Figure 5 shows the motor current power spectrum of faulty machine with Mixed Eccentricity under no load and loaded condition around fundamental and the 3ʳᵈ harmonic. This power spectrum shows the fault frequencies at 25 Hz, 74 Hz, 125 Hz, and 174 Hz. It is observed from the figures that magnitude of those components increases in proportion to the load and severity of a fault. Table 1 shows the analysis of power spectrums of IM with mixed eccentricity.

Figure 6 shows the stator vibration spectrum of faulty machine with Mixed Eccentricity under loaded condition.

The Diagnosis in the time-frequency analysis can be performed through different continuous transforms such as STFT, CWT, WVD, etc. The standard result of these transforms is a two-dimensional graph usually plotted as a colored map. This map provides an infor-

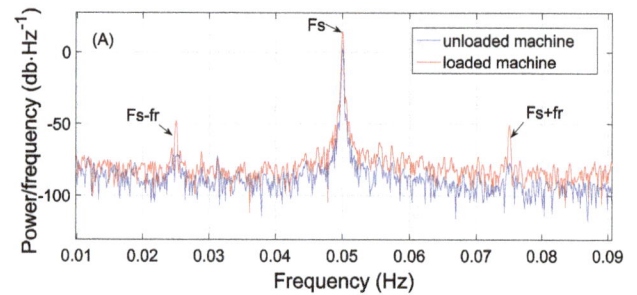

(a) Around fundamental.

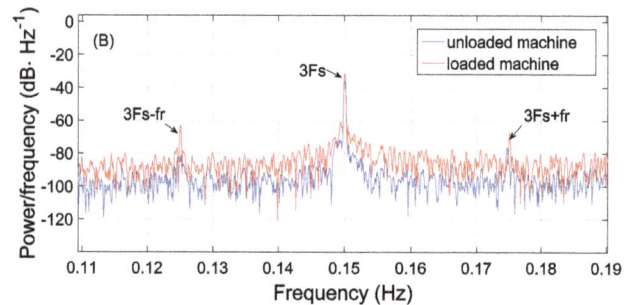

(b) Around 3ʳᵈ harmonic.

Fig. 5: Power spectrum of faulty machine with Mixed Eccentricity under no load and loaded condition.

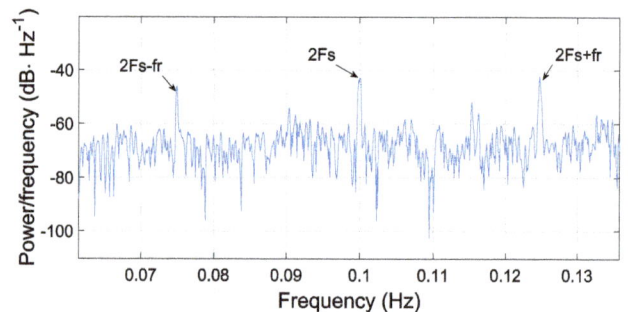

Fig. 6: Stator vibration spectrum of faulty machine with Mixed Eccentricity.

mation about the distribution of signal energy among different frequencies. This enables evolution tracking of fault components during transients and nonstationary operating conditions. Figure 7 shows the STFT spectrogram of faulty machine with Mixed Eccentricity under loaded condition around fundamental and the 3ʳᵈ harmonic.

6. Conclusion

This paper presents a wireless system to perform real-time condition monitoring of induction machine for early detection and monitoring of rotor eccentricity fault. Moreover, in condition monitoring algorithms, measurements are taken for a loaded and unloaded induction machine at the time of commissioning by wireless data acquisition device. The fault algorithm moni-

Tab. 1: Measurement results.

	Load Condition	Slip	Fault Frequency (Hz)				Magnitude Frequency (db)				Observations
			Fs-fr	Fs+fr	$3Fs$-fr	$3Fs$+fr	Fs-fr	Fs+fr	$3Fs$-fr	$3Fs$+fr	
Fig. 4	No load	0.0055	25.13	74.86	125.14	174.86	−76.66	−75.74	−92.06	−75.7	Magnitude increases with increase of load
	Loaded	0.0289	25.72	74.27	125.72	174.27	−47.77	−50.6	−62.96	−67.06	

tors the amplitude of fault frequencies and tracks the evolution of fault components over time. The wireless communications are based on the IEEE 802.11 Wi-Fi. This ensures the data transmission and a synchronous acquisition, which are critical elements in a condition monitoring system based on a wireless system network. Experimental results confirm the theatrical background of rotor eccentricity fault, and its usefulness for the preventive maintenance for induction machine.

References

[1] MEDINA-GARCIA, J., T. SANCHEZ-RODRIGUEZ, J. A. GALAN, A. DELGADO, F. GOMEZ-BRAVO and R. JIMENEZ. A wireless sensor system for real-time monitoring and fault detection of motor arrays. *Sensors*. 2017, vol. 17, no. 3, pp. 469. ISSN 1424-8220. DOI: 10.3390/s17030469.

[2] FENG, G.-J., J. GU, D. ZHEN, M. ALIWAN, F.-S. GU and A. D. BALL. Implementation of envelope analysis on a wireless condition monitoring system for bearing fault diagnosis. *International Journal of Automation and Computing*. 2015, vol. 12, no. 1, pp. 14–24. ISSN 1751-8520. DOI: 10.1007/s11633-014-0862-x.

[3] OUACHTOUK, I., S. EL HANI, S. GUEDIRA, L. SADIKI and K. DALI. Modeling of squirrel cage induction motor a view to detecting broken rotor bars faults. In: *International Conference on Electrical and Information Technologies*. Marrakech: IEEE, 2015, pp 347–352. ISBN 978-1-4799-7479-5. DOI: 10.1109/EITech.2015.7163001.

[4] OUACHTOUK, I., S. EL HANI, S. GUEDIRA, K. DALI and L. SADIKI. Advanced Model of Squirrel Cage Induction Machine for Broken Rotor Bars Fault Using Multi Indicators. *Advances in Electrical and Electronic Engineering*. 2016, vol. 14, no. 5, pp. 512–521. ISSN 1804-3119. DOI: 10.15598/aeee.v14i5.1705.

[5] FAIZ, J. and S. M. M. MOOSAVI. Eccentricity fault detection – From induction machines to DFIG—A review. *Renewable and Sustainable Energy Reviews*. 2016, vol. 55, iss. 1, pp. 169–179. ISSN 1364-0321. DOI: 10.1016/j.rser.2015.10.113

[6] KIM, D.-J., H.-J. KIM, J.-P. HONG and C.-J. PARK. Estimation of Acoustic Noise and Vibration in an Induction Machine Considering Ro-

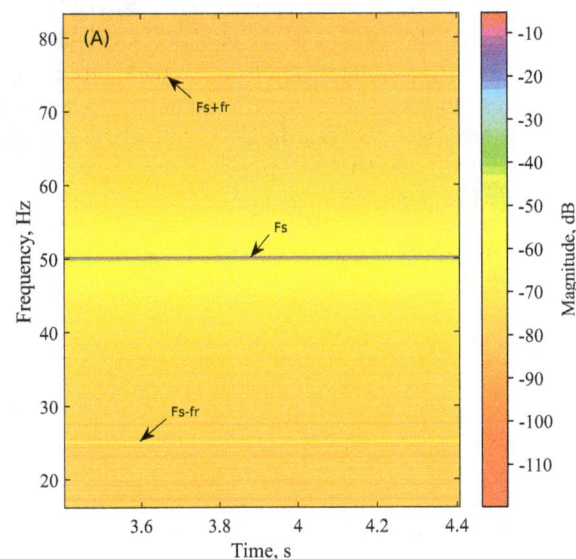

(a) Around fundamental.

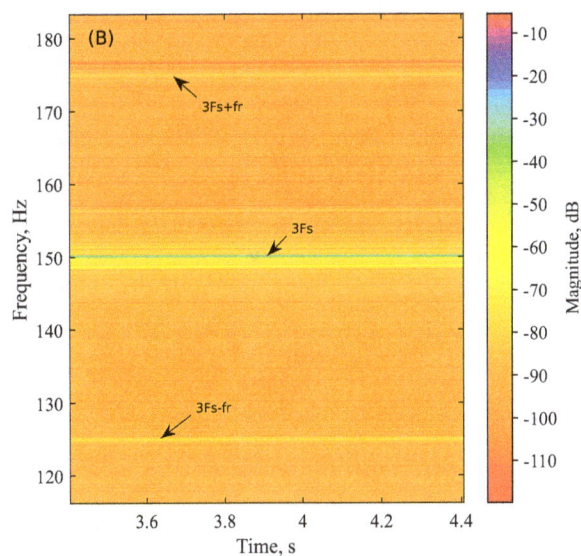

(b) Around 3rd harmonic.

Fig. 7: STFT spectrogram of faulty machine with Mixed Eccentricity under loaded condition.

tor Eccentricity. *IEEE Transactions on Magnetics*. 2016, vol. 50, no. 2, pp. 857–860. ISSN 1941-0069. DOI: 10.1109/TMAG.2013.2285391.

[7] HYUN, D., J. HONG, S. B. LEE, K. KIM, E. J. WIEDENBRUG, M. TESKA, S. NANDI and I. T. CHELVAN. Automated Monitoring of Airgap Eccentricity for Inverter-Fed Induction Motors Under Standstill Conditions. *IEEE Transactions on Industry Applications*. 2011, vol. 47, no. 3, pp 1257–1266. ISSN 0093-9994. DOI: 10.1109/TIA.2011.2126010.

[8] SOBRA, J., A. BELAHCEN and T. VAIMANN. Vibration and stator current spectral analysis of induction machine operating under dynamic eccentricity. In: *International Conference on Electrical Drives and Power Electronics*. Tatranska Lomnica: IEEE, 2015, pp. 285–290. ISBN 978-1-4673-7376-0. DOI: 10.1109/EDPE.2015.7325307.

[9] NARWADE, S., P. KULKARNI and C. Y. PARTIL. Fault Detection of Induction Motor Using Current and Vibration Monitoring. *International Journal of Advanced Computer Research*. 2013, vol. 3, no. 4, pp. 272. ISSN 2249-7277.

[10] ZAGIRNYAK, M., D. MAMCHUR and A. KALINOV. Comparison of induction motor diagnostic methods based on spectra analysis of current and instantaneous power signals. *Przeglad Elektrotechniczny*. 2012, vol. 88, no. 12b, pp. 221–224. ISSN 0033-2097.

[11] EL BOUCHIKHI, E. H., V. CHOQUEUSE and M. BENBOUZID. Induction machine faults detection using stator current parametric spectral estimation. *Mechanical Systems and Signal Processing*. 2015, vol. 52, iss. 1, pp. 447–464. ISSN 1096-1216. DOI: 10.1016/j.ymssp.2014.06.015.

[12] HENAO, H., G.-A. CAPOLINO, M. FERNANDEZ-CABANAS, F. FILIPPETTI, C. BRUZZESE, E. STRANGAS, R. PUSCA, J. ESTIMA, M. RIERA-GUASP and S. HEDAYATI-KIA. Trends in fault diagnosis for electrical machines: A review of diagnostic techniques. *IEEE Industrial Electronics Magazine*. 2014, vol. 8, no. 2, pp. 31–42. ISSN 1932-4529. DOI: 10.1109/MIE.2013.2287651.

[13] MEHRJOU, M. R., N. MARIUN, M. H. MARHABAN and N. MISRON. Rotor fault condition monitoring techniques for squirrel-cage induction machine—A review. *Mechanical Systems and Signal Processing*. 2011, vol. 25, no. 8, pp. 2827–2848. ISSN 1096-1216. DOI: 10.1016/j.ymssp.2011.05.007.

[14] FAIZ, J., B. M. EBRAHIMI, B. AKIN and H. A. TOLIYAT. Finite-Element Transient Analysis of Induction Motors Under Mixed Eccentricity Fault. *IEEE Transactions on Magnetics*. 2008, vol. 44, no. 1, pp. 66–74. ISSN 1941-0069. DOI: 10.1109/TMAG.2007.908479.

[15] FAIZ, J. and S. M. M. MOOSAVI. Review of eccentricity fault detection techniques in IMs focusing on DFIG. In: *IEEE 5th International Conference on Power Engineering, Energy and Electrical Drives*. Riga: IEEE, 2015, pp. 513–520. ISBN 978-1-4799-9978-1. DOI: 10.1109/PowerEng.2015.7266370.

[16] AHMED, I., M. AHMED, K. IMRAN, M. SHUJA KHAN and S. JUNAID AKHTAR. Detection of Eccentricity Faults in Machine Using Frequency Spectrum Technique. *International Journal of Computer and Electrical Engineering*. 2011, vol. 3, no. 1, pp. 111. ISSN 1793-8163. DOI: 10.7763/IJCEE.2011.V3.300.

[17] ESFAHANI, E. T., S. WANG and V. SUNDARARAJAN. Multisensor Wireless System for Eccentricity and Bearing Fault Detection in Induction Motors. *IEEE/ASME Transactions on Mechatronics*. 2014, vol. 19, no. 3, pp. 818–826. ISSN 1083-4435. DOI: 10.1109/TMECH.2013.2260865.

[18] HAMMADI, K. J., D. ISHAK and M. KAMAROL. On-Line Monitoring and Diagnosis Broken Rotor Bars in Squirrel-Cage Induction Motor By Using Labview. *Australian Journal of Basic and Applied Sciences*. 2011, vol. 5, no. 9, pp. 1525–1528. ISSN 1991-8178.

[19] NANDI, S., H. A. TOLIYAT and X. LI. Condition Monitoring and Fault Diagnosis of Electrical Motors—A Review. *IEEE Transactions on Energy Conversion*. 2005, vol. 20, no. 4, pp. 719–729. ISSN 1558-0059. DOI: 10.1109/TEC.2005.847955.

[20] KULKARNI, V. V., M. M. NADAKATTI and A. A. DESHPANDE. LabView based Bearing Failure Prediction Using Data Acquisition System. *Indian Journal of Advances in Chemical Science S1*. 2016, vol. 142, pp. 142–145. ISSN 2320-0928.

About Authors

Ilias OUACHTOUK was born in Foum Zguid-Tata, Morocco, in 1991. He received his M.Sc. degree in electrical engineering from the Mohammed V University in Rabat, Morocco, in 2014. Where he is currently working toward the Ph.D. degree, in the department of electrical engineering. Since 2015, his research interests include modeling and diagnosis of electrical drives, in particular, synchronous and asynchronous motors.

Soumia EL HANI has been Professor at the ENSET (Ecole Normale Superieure de l'Enseignement Technqique, Rabat, Morroco) since October 1992. She is a Research Engineer at the Mohammed V University in Rabat, Morocco, in charge of the research team electromechanical, control and diagnosis, IEEE member, member of the research laboratory in electrical engineering at ENSET, Rabat. Author of several publications in the field of electrical engineering, including robust control systems, diagnosis and control systems of wind electric conversion. She has been general co-Chair the two editions of "the International Conference on Electrical and Information Technologies", held in Marrakech, March 2015 and Tangier, May 2016 respectively.

Khalid DAHI was born in Errachidia, Morocco, in 1988. He received the M.Sc. degree in electrical engineering in 2012 from the Mohammed V University in Rabat, Morocco. Where he is currently working toward the Ph.D. degree in the department of electrical engineering. Since 2012, his research interests are related to electrical machines and drives, diagnostics of induction motors. His current activities include monitoring and diagnosis of induction machines in wind motor.

16

Non-Adaptive Methods of Fetal ECG Signal Processing

Radana KAHANKOVA[1], Rene JAROS[1], Radek MARTINEK[1], Janusz JEZEWSKI[2], He WEN[3], Michal JEZEWSKI[4], Aleksandra KAWALA-JANIK[5,6]

[1]Department of Cybernetics and Biomedical Engineering, Faculty of Electrical Engineering and Computer Science, VSB–Technical University of Ostrava, 17. listopadu 15, 708 33 Ostrava, Czech Republic
[2]Institute of Medical Technology and Equipment ITAM, Roosevelt Street 118 , 41-800 Zabrze, Poland
[3]College of Electrical and Information Engineering, Hunan University, Lushan Road, Yuelu District, 410082 Changsha, Hunan Province, China
[4]Institute of Electronics, Faculty of Automatic Control, Electronics and Computer Science, Silesian University of Technology, Akademicka 2A, 44-100 Gliwice, Poland
[5]Automatic Control and Informatics, Faculty of Electrical Engineering, Opole University of Technology, Proszkowska 76/1, 45-758 Opole, Poland
[6]Department of Biomedical Engineering, College of Engineering, University of Kentucky, 43 Graham Avenue, 40508 Lexington, United States of America

radana.kahankova@vsb.cz, rene.jaros@vsb.cz, radek.martinek@vsb.cz, jezewski@itam.zabrze.pl, hewen@hnu.edu.cn, michal.jezewski@polsl.pl, kawala84@gmail.com

Abstract. *Abdominal fetal ElectroCardioGrams (fECGs) carry a wealth of information about the fetus including fetal Heart Rate (fHR) and signal morphology during different stages of pregnancy. Here we report our results on the implementation and evaluation of two non-adaptive signal processing methods suitable for fECG signal extraction, namely: the Independent Component Analysis (ICA) and the Principal Component Analysis (PCA) Methods. We used the fetal heart rate extracted from fECG signals (in Beats Per Minute - BPM) and Signal-to-Noise Ratio (SNR) as effective performance evaluation metrics for our applied methods. Our findings demonstrated that given adequate SNR, these methods produced excellent results in accurate determination of fHR. Furthermore, we found out that compared to the PCA Method, the ICA Method produces a lower variance in the detection of the fHR.*

Keywords

Blind source separation, ECG extraction, fetal ElectroCardioGram (ECG), independent component analysis, non-adaptive filtration, non-invasive fetal monitoring, principal component analysis.

1. Introduction

ElectroCardioGraphy (ECG) is a diagnostic method which detects the electrical activity of the cardiac muscle. In clinical practice, ECG is utilised to diagnose heart arrhythmia, ischemia, and to assess the efficiency of the treatment with drugs. For fetal monitoring, fetal ElectroCardioGraphy (fECG) can be used. From the fECG, it is possible to determine fetal Heart Rate (fHR), which can provide information about fetal hypoxia [1]. Fetal ECG contains potentially valuable information that could not be acquired by conventional ultrasound-based methods [33], [34] and [35]. The methods for fECG measuring can be invasive or non-invasive. Invasive method is the most accurate method for measuring fHR and is performed by direct transvaginal Fetal Scalp Electrode (FSE) attached directly to the fetus. Nevertheless, it is dangerous and inconvenient for both mother and fetus due to its invasive nature [2]. For these reasons, invasive method is being replaced by non-invasive method, which is measured by means of electrodes placed on maternal abdomen. This signal (abdominal ECG, aECG) contains both maternal and fetal component and also some noise caused by maternal and fetal muscle activity, potentials generated by respiration and stomach, noise generated from electrode-skin contact, etc. Equation (1) illustrated the above mentioned relations, where x_{aECG} is aECG,

x_{mECG} is materal ECG (mECG), x_{fECG} is fECG, and n is noise [9]:

$$x_{\text{aECG}}(n) = x_{\text{mECG}}(n) + x_{\text{fECG}}(n) + n(n). \quad (1)$$

The value of Signal to Noise Ratio (SNR) depends on the abdominal electrodes placement [4] and [32] gestational age, and fetal position [2]. The placement of the electrodes is not standardized making it difficult to automate the fHR measurement [4]. Normal fetal Heart Rate (fHR) usually ranges from 120 to 160 Beats Per Minute (BPM) compared to maternal heart rate, which ranges from 70 to 80 BPM [3]. In addition, maternal signal amplitude significantly differs from the fetal signal amplitude, which is 10 to 30 times weaker. Although there is no direct neural connection between mother and the fetus, hormones and placenta can affect fHR and fetal blood pressure. Figure 1 shows that the blood circulation in the fetus varies from the circulation of a newborn and adult person.

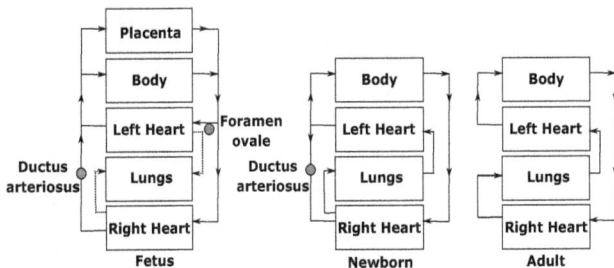

Fig. 1: Circulation of fetus, newborn and adult person.

Non-invasive measurement of fECG is performed by means of a single-channel or multichannel source signals [2]. These signals are processed by using the adaptive and non-adaptive methods. Although several techniques and fECG extraction algorithms have been tested, an optimal solution has not been found yet.

1.1. Adaptive Methods

Adaptive methods are characterized by an ability to automatically set its coefficients according to varying circumstances. Adaptive algorithms use aECG as the primary output, whereas the signal recorded on the maternal thorax (mECG) is used as the reference input due to the fact that it is considered to contain only the maternal component. Non-linear adaptive techniques include Artificial Neural Networks (ANN), methods using a Hybrid Neural Network (HNN), and apply the techniques of Adaptive Neuro-fuzzy Inference System (ANFIS) [2] and [5]. Linear adaptive methods include the methods based on the theory of Kalman filtering (KF), Least Mean Squares algorithm (LMS) [6], Recursive Least Squares algorithm (RLS) [7], and methods based on Adaptive Linear Neuron (ADALINE) [2].

1.2. Non-Adaptive Methods

This paper is mainly focused on the non-adaptive methods, which can be used for the elimination of the unwanted signal and for fECG signal extraction without any adaptation of the system. Figure 2 shows different non-adaptive methods using multichannel or single channel signal sources.

1) Single Channel Signal Source

Many non-adaptive methods use the Single channel signal source, e.g. methods based on Wavelet Transform (WT), Complex Wavelet Transform (CWT) [8], Pitch Synchronous Wavelet Transform (PSWT) [11] or Discrete Wavelet Transform (DWT) [9] and [10]. Hassanpour et al., 2006 [9] and Bhoker et al. 2013 [10], tested the DWT for fECG extraction. The results showed that this method is able to correctly detect R-R interval.

Karvounis et al., 2004 [8], tested CWT, which is used for automatic fECG extraction from aECG. They found out that this algorithm is very fast and accurate and could be used for simultaneous monitoring of fECG and mECG in order to obtain mHR. Kumar et al., 2016 [11], evaluated PSWT and they reached better SNR and correct estimation of fHR. Another non-adaptive method, Correlation Technique (CT), was introduced by Bemmel et al., 1968 [12]. However, this method is not suitable for estimating fECG. Levkov et al., 2005 [9], introduced Subtraction Technique (ST) and suggested that method does not defect spectrum of the fECG during elimination of network disturbance when compared to the other methods. Hon et al., 1964 [14], improved SNR ranging from 10 to 20 dB during fECG estimation by using Averaging Technique (AT).

Su et al., 2016 [15], dealt with nonlinear time-frequency analysis called De-shape Short-time Fourier Transform (STFT) and non-local median method and concluded that these methods have better performance than adaptive methods. These methods can estimate fECG even if aECG contains more noise and provide more information included in single aECG such as non-linear relationships between consecutive cardiac activities.

Lee et al., 2016 [16], investigated the method of Sequential Total Variation Denoising (STVD) and demonstrated that fECG can be obtained with lower errors and it is feasible for real time fHR monitoring in the future. Tan et al., 2015 [17], introduced Fuzzy C-means Clustering Method (FCM) and their results showed that the method is extremely effective and safe in the monitoring during the pregnancy and it is very simple and suitable method for monitoring multiple fetuses in the womb.

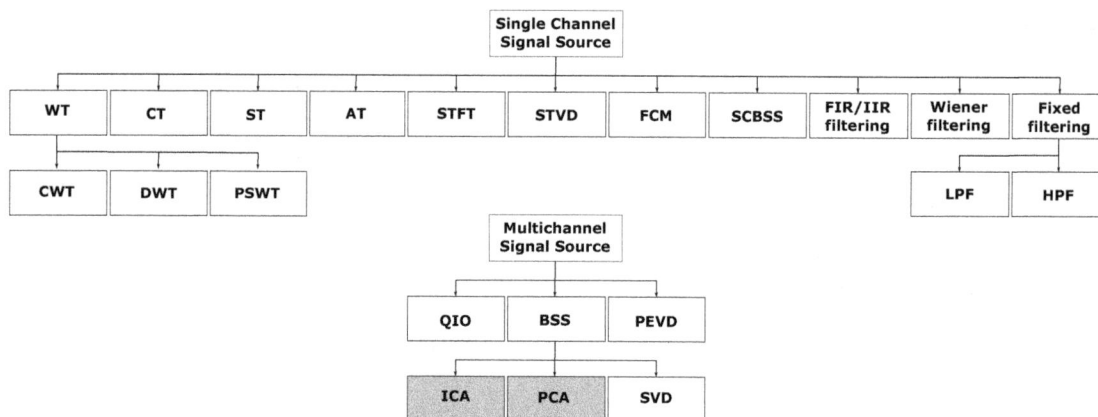

Fig. 2: Non-adaptive methods.

Peng Ju He et al., 2016 [18], focused on a Single Channel Blind Source Separation (SCBSS) and proved that this method is able to detect fHR in case of multiple pregnancy. Additionally, this method is able to extract fECG from aECG. Non-adaptive methods also include regression techniques, frequency selective filters with Finite Impulse Response (FIR) and Infinite Impulse Response (IIR), methods based on Wiener filtering theory and fix filtering, which includes Low-Pass Filter (LPF) and High-Pass Filter (HPF) [2].

2) Multichannel Signal Sources

Multichannel signal sources are used mainly for the implementation of the methods based on Blind Source Separation (BSS). These methods include Independent Component Analysis (ICA), Principal Component Analysis (PCA), and Singular Value Decomposition (SVD). Raj et al., 2015 [19], proposed Fast ICA algorithm for fECG extraction and the results showed that this method has very good performance. It is the most commonly used method, for more information, please see [20], [21] and [28]. Bacharakis et al. [22], focused on the use of PCA and proved that this method has good results but ICA method shows better performance. ICA and PCA, the main methods tested in this paper, will be explained in more detail in Sec. 2. For more information about PCA, see [29].

Leach et al. [23], discussed about SVD method and concluded that this method is very effective for fECG extraction and noise filtering. Unfortunately, the algorithm is computationally demanding. Varanini et al., 2016 [24], introduced the method of Quality Index Optimization (QIO) and concluded that this method can be used even if the fECG has a low amplitude. Kumar et al., 2016 [11], used the combination of SVD method and polynomial classifiers. The results showed that this combination improves SNR than when using SVD method alone.

Gao et al., 2003 [21], tested the combination of SVD and ICA methods and found out that this combination can be used, when mECG and fECG are overlapped. Liu et al., 2015 [25], proposed a novel integrated algorithm based on ICA, Ensemble Empirical Mode Decomposition (EEMD) and Wavelet Shrinkage (WS). They concluded that the tested combination improves SNR, correlation coefficient (R), and Mean Squared Error (MSE).

Ayat et al., 2015 [26], introduced the combination of polynomial networks and Savitzky-Golay smoothing filters. The results proved that this combination provides better performance and can be use in real-time fECG monitoring. Redif et al., 2016 [27], discussed the method using Polynomial Matrix Eigenvalue Decomposition (PEVD). According to the results, this method is not accurate in detecting P and T waves. On the other hand, in the detection of the R waves the method has proven itself.

3) Steps of This Work

Based on the extensive research of the literature discussed above, we chose ICA and PCA methods. Moreover, according to our initial testing, they provided the best results. In this paper, Sec. 2. deals with the algorithms of ICA and PCA methods, describes generator of synthetic data, and the parameters used to evaluate the quality of the experiments. In Sec. 3. we introduce the results which are then discussed in Sec. 4.

2. Methods

2.1. ICA

Independent Component Analysis is the most performed method of non-adaptive methods using multichannel signal sources. It is a method for finding hidden vectors in the data file [21]. ICA estimates the fECG signal from the signal mixture. ICA intends to find non-Gaussian data with independent components, which are statistically independent, or at least almost statistically independent. Statistical independency means that information contained in one variable does not provide information about another one. One limitation of this method is that the signals in abdominal mixture overlap. In addition, this method requires at least two abdominal electrodes to record the input signals. The Algorithm is very quick and effective in extracting fECG.

The principle of ICA can be described simply as a room with two persons that are communicating. In this room, there are also two microphones located at different places providing two signals $x_1(t)$ and $x_2(t)$, where x_1 and x_2 are amplitudes and t denotes the time. Each signal is a sum of speech signals and marked as $s_1(t)$ and $s_2(t)$. This problem, when two or more people are talking, is a so-called cocktail-party problem [28]. In case of fetal monitoring, the maternal and fetal components in the abdominal signals are considered as the two voices in the previous example. Thus, ICA is an ideal method for extracting fECG. The principle is described by Eq. (2) and Eq. (3), where a_{11}, a_{12}, a_{21} and a_{22} are parameters depending on distance of a speaker from a microphone:

$$x_1(t) = a_{11}s_1 + a_{12}s_2, \qquad (2)$$

$$x_2(t) = a_{21}s_1 + a_{22}s_2. \qquad (3)$$

A problem is that the parameters a_{ij} are unknown. The solution is to assume that $s_1(t)$ and $s_2(t)$ are statistically independent (it is true in many cases). That allows to separate the original signals from the abdominal mixture [28]. For ICA, linear signals x_1 to x_n from n independent components are defined by Eq. (4):

$$x_j = a_{j1}s_1 + a_{j2}s_2 + \cdots + a_{jn}s_n. \qquad (4)$$

Time index t is obtained and then every mixture of signals and every independent component s_k are random variables. In addition, it is assumed that mixture of signals and independent components have a zero mean value. If not, observed variables x_i can be always centered by subtracting mean of the samples, thus creating a zero mean model. It is very beneficial to use vector-matrix notation instead of the sum. Then matrix $\mathbf{A}_{mix}$ is used with elements a_{ij} as it is shown in Eq. (5), which has rows with transposed vectors $\vec{x}^T$ [28]:

$$\vec{x} = \mathbf{A}_{mix} \cdot \vec{s}. \qquad (5)$$

Sometimes, the columns of matrix $\mathbf{A}_{mix}$ are needed and for this reason, Eq. (5) is modified by model a_j and

then we obtain Eq. (6). If we assume that components are statistically independent and have non-Gaussian distribution, it is possible to assume that mixture matrix is square and can be calculated with its inverse matrix $\mathbf{W}$ for estimation of matrix $\mathbf{A}_{mix}$, [28]. Then independent components are obtained from this matrix as in Eq. (7):

$$\vec{x} = \sum_{i=1}^{n} a_i \cdot s_i, \qquad (6)$$

$$\vec{s} = \mathbf{W} \cdot \vec{x}. \qquad (7)$$

Fast ICA algorithm is divided into 6 steps [20]. First, given mixed signals are converted into other signals such that covariance matrix $\mathbf{B}$ computed using the converted signals is the identity matrix. Then initialize values for the matrix $\mathbf{B}$ to achieve $\mathbf{B}^T\mathbf{B} = 1$. Third step is updating elements of the matrix $\mathbf{B}$ using iteration formula (update all elements of this matrix). In step four, columns of matrix $\mathbf{B}$ are orthonormalized. Fifth step is repeating step three and four for each iteration. Finally, the component is obtained by multiplying $\mathbf{B}^T$.

It is necessary to do pre-processing of signal by centering and whitening before applying ICA algorithm. Centering creates a vector with zero mean value and then whitening creates a vector which is white, its components are uncorrelated and their variances equal unity. Figure 3 shows block diagram of ICA. For more information about ICA and FastICA, please see [20], [21] and [28].

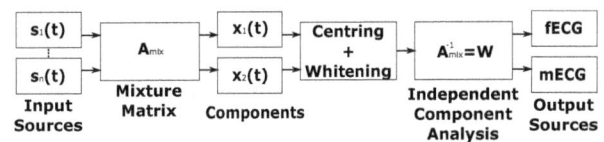

Fig. 3: Block scheme of independent component analysis.

2.2. PCA

Principal component analysis replaces original variables, which are correlated, with principal components that are uncorrelated and in the most cases are linear combination of original variables. Input of PCA is the matrix $\mathbf{X}$, which contains n samples for p original variables. Output of PCA is the matrix $\mathbf{Z}$, which contains n samples, but for p principal variables [29]. When assuming that matrix $\mathbf{X}$ is centered by columns, which indicates that means of columns of matrix $\mathbf{X}$ equals to zero, then matrix $\mathbf{Z}$ contains columns of principal components created by linear combination of columns of matrix $\mathbf{X}$. This applies for Eq. (8), where $\mathbf{A}$ the orthogonal (uncorrelated) matrix and its inverse transformation is defined by Eq. (9):

$$\mathbf{Z} = \mathbf{X} \cdot \mathbf{A}, \qquad (8)$$

$$\mathbf{X} = \mathbf{Z} \cdot \mathbf{A}^{\mathrm{T}}. \qquad (9)$$

From Eq. (8) and Eq. (9) following equality $\mathbf{X} \cdot \mathbf{X}^{\mathrm{T}} = \mathbf{Z} \cdot \mathbf{Z}^{\mathrm{T}}$ can be defined. It indicates that both coordinate systems have the same Euclidean distance between the points and have the same angle between the vectors connecting points and coordinate origin. Matrix $\mathbf{G}$ is created because matrix $\mathbf{A}$ causes rotation around coordinate origin [29]. This new matrix $\mathbf{G}$ causes rotation around coordinate origin and for this matrix, principal components are orthogonal as in Eq. (10):

$$\mathbf{Z} = \mathbf{X} \cdot \mathbf{G}. \qquad (10)$$

Statistically, PCA is identified as multivariate method, which is based on the decomposition of the co-variance matrix. For analysis, PCA ussually uses two or three components. These components are graphically displayed in the space, which provides easy detection of structures, such as a group of points. To estimate structures, different two or three principal components can be used and take PCA as the projection of 2D or 3D data. Usually, chart of columns from columns of matrix $\mathbf{Z}$ is created. It is influenced by the transformation of the data.

There are several limitations of PCA procedure. Some components, the variablility of which is low, are important for analysis of multivariate data. It is difficult to assess which part of variability of data is unimportant [29].

Four steps are given in PCA data analysis: transformation of the data, distribution of covariance or correlation matrix, determination of the number of relevant principal components, and graphical representation of multivariate data [29].

Sometimes it is difficult to determine number of relevant principal components. For purposes of ECG signal processing, we will use two components to separate mECG and fECG. Graphical representation is performed for the specific pairs of principal components. It usually adds vectors of projections as rows of matrix $\mathbf{P} = \mathbf{G} \cdot \mathbf{L}$ that create combination chart.

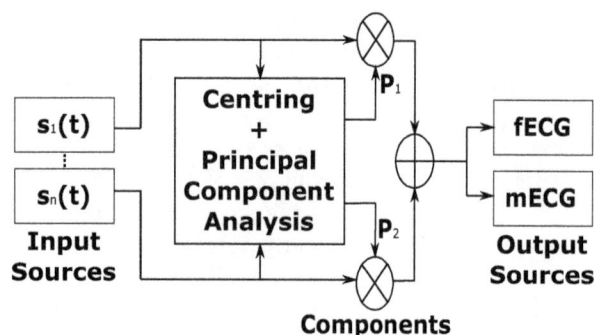

Fig. 4: Block scheme of principal component analysis.

Equation (11) shows that basis of PCA method is the spectral decomposition of covariance matrix on eigenvalues and vectors. This method uses SVD method directly as in Eq. (11) [29]. Mostly, shortened form of the SVD method, which has variables $\mathbf{U}$ and $\mathbf{S}$ with changed dimensions, is used and PCA method is calculated by Eq. (12):

$$\mathbf{Y} = \mathbf{U} \cdot \mathbf{S} \cdot \mathbf{V}^{\mathrm{T}}, \qquad (11)$$

$$\mathbf{Z} = \mathbf{U} \cdot \mathbf{S}. \qquad (12)$$

It is necessary to do pre-processing of the signal only by centering. Centering creates a vector with zero mean value, similarly as in case of ICA algorithm, but whitening is not usually necessary. In Fig. 4, we can see block diagram of PCA. More information about PCA can be found in [22] and [29].

2.3. Dataset

For the experiments, synthetic data were used. The data were created by the signal generator introduced by Martinek et al., 2016 [30]. It is a multi-channel generator which allows for the creation of synthetic signals nearly identical to the real signals. The biggest advantage of this generator is that it provides a reference fECG and mECG for the selected electrodes (abdominal or thoracic). Reference fECG is used to check the accuracy of proposed methods. This generator can determine fHR, mHR, interference, gestational age, or simulate the hypoxic conditions during 20th to 42nd week of pregnancy. Another advantage of this generator is the possibility to generate the signal by setting properties for six leads, four of them are abdominal and two of them are thoracic.

Figure 5 shows 5 abdominal electrodes, which were chosen for estimation in this work because they provide ideal position for evaluation. Non-adaptive methods require at least two abdominal electrodes and do not use thoracic electrodes. Using the generator, we set fHR on value 130, mHR on value 75, and recording time of data on 30 seconds. Records of aECG data from these 5 electrodes are generated for different levels of input signals in range from -5 dB to -50 dB. From these 5 abdominal electrodes, we get 10 combinations by using two of them, 10 combinations by using three of them, 5 combinations by using four of them and 1 combination by using all of them. That is in total 26 combinations for evaluation of proposed methods as we can see in the first columns of all tables in Sec 3.

2.4. Evaluation Parameters

Evaluation of extracted fECG by proposed methods can be performed subjectively or objectively. Subjec-

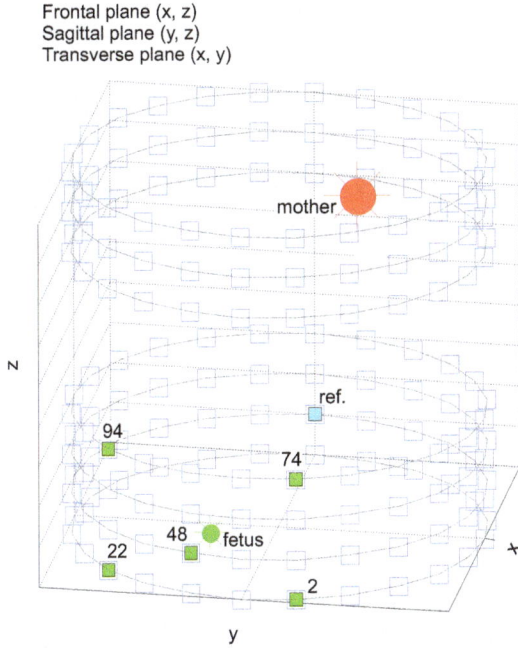

Fig. 5: Chosen abdominal signals from generator.

tively, we can evaluate the graph of extracted fECG and evaluate if this fECG is similar to ideal form fECG visually. For this work, objective evaluation, using parameters such as BPM and SNR, is more relevant.

Evaluation by using SNR is used to define the relationship between the useful signal and the noise. The resulting SNR is calculated by subtracting input SNR (SNR_{in}) from output SNR (SNR_{out}). If SNR_{in} and SNR_{out} is known, we can calculate resulting SNR and use it for the evaluation of the filtering by using proposed non-adaptive method. In Eq. (13), we can see calculation of SNR_{in} and in Eq. (14) we can see calculation of SNR_{out}, where $fECG_{ideal}$ is generated fECG by generator, $aECG_{input}$ is aECG which contains maternal and fetal component, and $fECG_{extract}$ is the extracted fECG by proposed non-adaptive methods. We need to note that $aECG_{input}$ contains mECG and fECG, so in Eq. (13) it is necessary to subtract $fECG_{ideal}$ from $aECG_{input}$ in the denominator. Similarly, it is necessary to subtract $fECG_{ideal}$ from $fECG_{extract}$ in Eq. (14):

$$SNR_{in} = 10\log_{10}\frac{\sum\limits_{n=1}^{N-1}(fECG_{ideal})^2}{\sum\limits_{n=1}^{N-1}(fECG_{input}-fECG_{ideal})^2}, \quad (13)$$

$$SNR_{out} = 10\log_{10}\frac{\sum\limits_{n=1}^{N-1}(fECG_{ideal})^2}{\sum\limits_{n=1}^{N-1}(fECG_{extract}-fECG_{ideal})^2}. \quad (14)$$

Heart rate is a very important evaluation parameter. To detect more accurate fHR, the algorithm does

not use fix amplitude level. In this work, the number of BPM in a recording is solved by using Detector of R waves. We used full implementation of the Pan-Tompkins filter [31].

3. Results

This section will be mainly focused on evaluation ICA and PCA by BPM, fHR, and mHR, respectively. Next, evaluation by using SNR is only implemented for PCA, since ICA change the amplitude of obtained fECG as we can see in Fig. 9 and Fig. 10. Therefore, it is impossible to calculate SNR.

3.1. Heart Rate (HR)

As it was mentioned before, this paper is mainly focused on fHR determination. In Tab. 2 and Tab. 3, we can see results of fHR determination for ICA and PCA. First columns of these tables show 26 combinations of electrodes. All these combinations use signals with different input quality levels. Input quality levels of signals are marked by Roman numerals from I to X and the values of these signals for each electrode on a certain level are included in the Tab. 1.

1) Determination of fHR by Using ICA

The left part of Tab. 2 shows results of determination of fHR by ICA from extracted fECG. In this part, we can see that ICA is good in detection fHR for the most cases in range of quality level of input signals from I to VI. In quality level of input signals VII, this method is not that effective. In last three quality levels, there is the obtained HR of maternal component (mHR).

In the right part of Tab. 2, we can see that determination of mHR from extracted component mECG is good for the most quality levels of input signals.

Results from Tab. 2 are also shown in Fig. 6. We can see the most of combinations by using ICA have approximately same value of fHR as ideal (reference) fECG. Figure 6(a) shows detection of fHR in extracted fECG. Blue, pink, and black circles represent quality levels from VIII to X, have mainly different fHR than ideal form of fECG. Figure 6(b) shows detection of mHR in extracted mECG. Lines in both parts of Fig. 6 represent generated HR value which was 130 for fECG and 75 for mECG.

2) Determination of BPM by Using PCA

Again, in left part of Tab. 3 we can see determination of fHR from extracted fECG but by using PCA. In

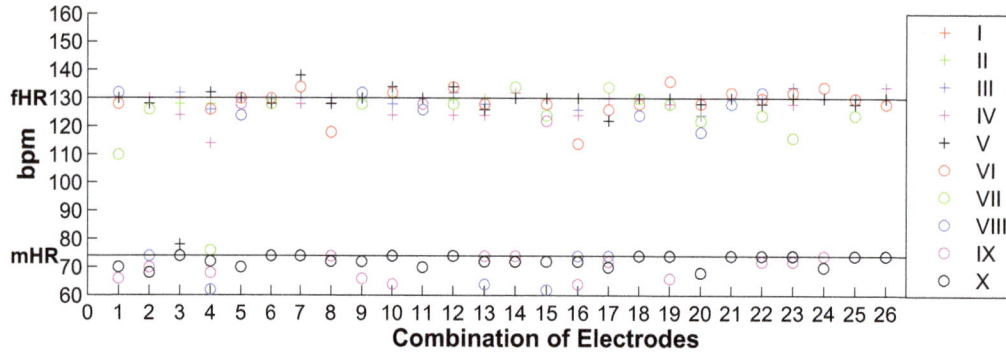

(a) Determination of BPM in extracted fECG.

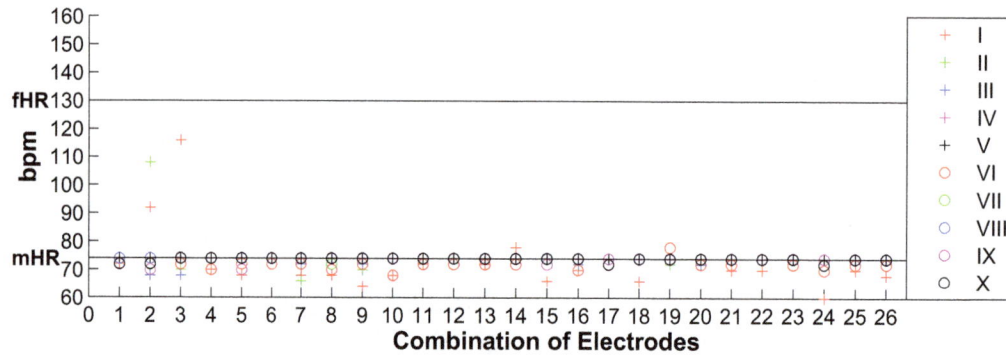

(b) Determination of BPM in extracted mECG.

Fig. 6: Recorded detection accuracy of fHR and mHR by using ICA.

Tab. 1: Table of SNR_{in} for different quality levels.

Electrode	SNR_{in}									
	I	**II**	**III**	**IV**	**V**	**VI**	**VII**	**VIII**	**IX**	**X**
2	4.1	-3.1	-6.9	-13.2	-16.8	-21.4	-30.0	-32.6	-37.8	-41.5
22	6.8	-0.2	-4.0	-10.2	-14.0	-18.6	-26.2	-29.9	-35.0	-38.7
48	10.1	2.6	-1.1	-7.2	-10.7	-15.2	-22.9	-26.9	-31.4	-35.6
74	0.7	-6.7	-10.4	-16.7	-20.1	-24.6	-32.2	-36.0	-41.0	-44.9
94	-0.2	-7.0	-11.0	-17.1	-20.9	-25.7	-33.1	-36.7	-42.1	-45.6

this left part, a good detection of fHR also prevails. So again, method of blind source separation proves to be effective in fHR determination for the most quality levels of input signals from I to VI. In quality level VII, this method is not so effective and in last thee quality levels, the value of obtained HR is equal to maternal component (mHR) instead of fetal one (fHR).

In the right part of Tab. 3, same as in Tab. 2, there is determination of mHR from second extracted maternal component (mECG) by PCA. PCA is suitable for most of the quality levels of the input signals. Determination of mHR is not sufficient only in the first two quality levels due to high SNR_{in} at these levels.

Figure 8 illustrates the results from Tab. 3. We can see that most of the combinations using PCA have fHR approximately same as the ideal (reference) fECG. Figure 8 shows detection of fHR in extracted fECG. Blue,

pink, and black circles again represent quality levels from VIII to X, have mainly different fHR than the ideal fECG. Figure 7(b) shows detection of mHR in extracted mECG. Lines in both parts of Fig. 7 represent the HR set in the generator, i.e. 130 for fECG and 75 for mECG.

3) Summary of fHR and mHR Detection

As we assumed, both of the proposed methods are very accurate in detection of fHR from extracted fetal component and mHR from extracted maternal component. Both methods stop working in quality index of input signals VIII, i.e. approximately for the values of SNR_{in} in the range from −30 to −35 dB. From upper figures in Fig. 6 and Fig. 7, we can see that in case of determination of fHR, ICA shows slightly better results than PCA.

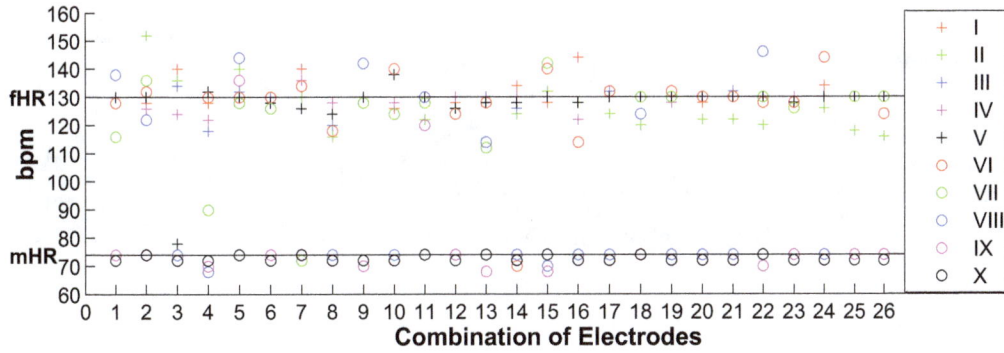

(a) Determination of BPM in extracted fECG.

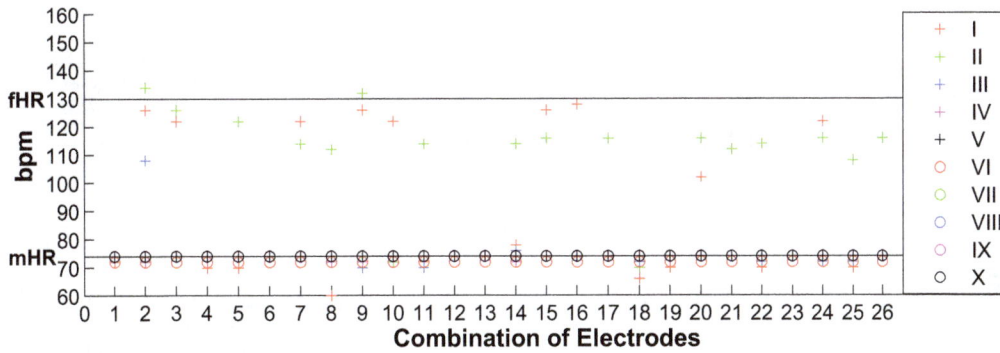

(b) Determination of BPM in extracted mECG.

Fig. 7: Recorded detection accuracy of fHR and mHR by using PCA.

Tab. 2: Table of BPM detected from the extracted components by using ICA.

Combination	Determination of fHR by ICA									
of electrodes	I	II	III	IV	V	VI	VII	VIII	IX	X
2, 22	130	130	130	130	130	128	110	132	66	70
2, 48	130	130	130	130	128	126	126	74	70	68
2, 74	130	128	132	124	78	74	74	74	74	74
2, 94	130	128	126	114	132	126	76	62	68	72
22, 48	130	130	130	130	130	130	128	124	128	70
22, 74	130	130	130	128	128	130	128	74	74	74
22, 94	130	130	130	128	138	134	74	74	74	74
48, 74	130	130	130	128	128	118	74	74	74	72
48, 94	130	130	130	130	130	128	128	132	66	72
74, 94	130	130	128	124	134	132	64	74	64	74
2, 22, 48	130	130	130	130	130	128	128	126	128	70
2, 22, 74	130	130	132	124	134	134	128	74	74	74
2, 22, 94	130	130	128	124	126	128	72	64	74	72
2, 48, 74	130	130	130	132	130	72	134	74	74	72
2, 48, 94	130	130	130	130	130	128	124	62	122	72
2, 74, 94	130	130	126	124	130	114	72	74	64	72
22, 48, 74	130	130	130	130	122	126	134	74	72	70
22, 48, 94	130	130	130	130	130	128	130	124	74	74
22, 74, 94	130	130	130	128	130	136	128	74	66	74
48, 74, 94	130	130	124	130	128	128	122	118	68	68
2, 22, 48, 74	130	130	130	130	130	132	74	128	74	74
2, 22, 48, 94	130	130	130	130	128	130	124	132	72	74
2, 22, 74, 94	130	130	134	128	130	132	116	74	72	74
2, 48, 74, 94	130	130	130	130	130	134	74	74	74	70
22, 48, 74, 94	130	130	130	128	128	130	124	74	74	74
2, 22, 48, 74, 94	130	130	130	134	130	128	74	74	74	74

(a)

(b)

(c)

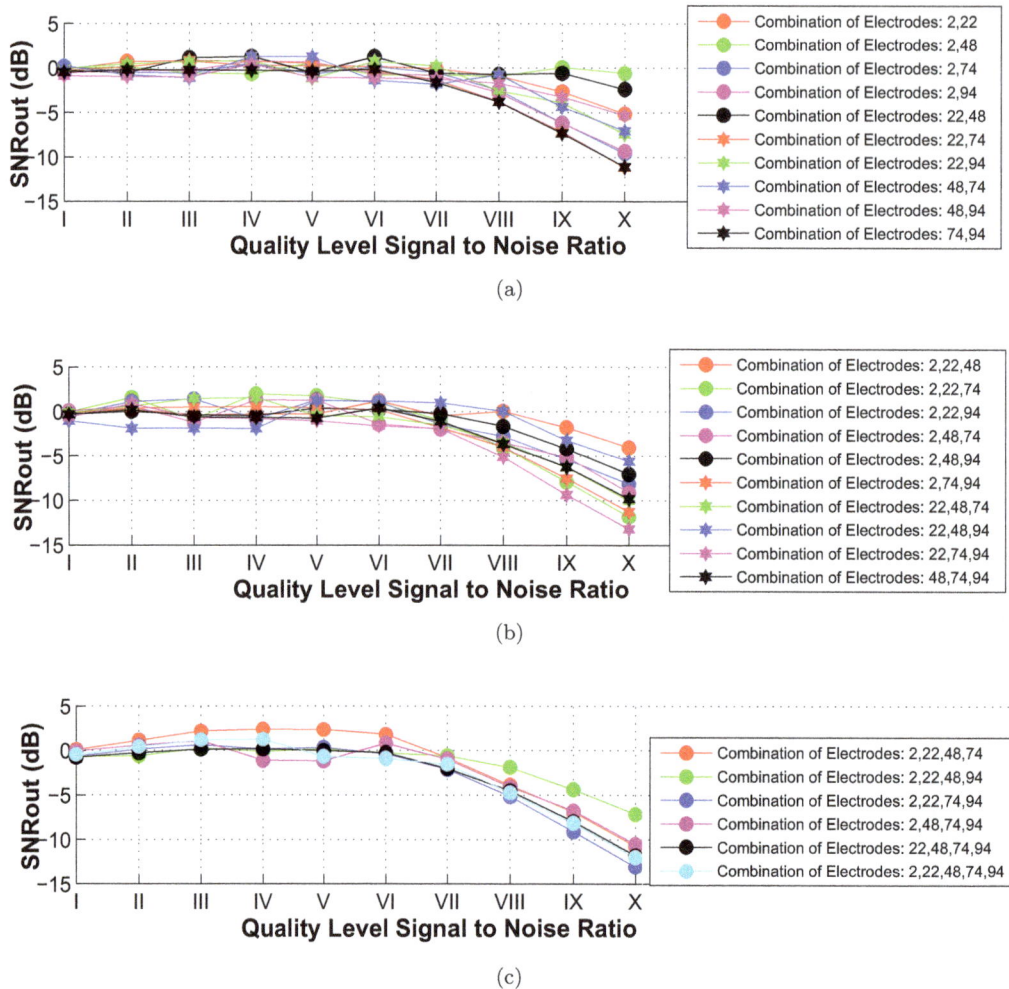

Fig. 8: Comparison of results by PCA for different combinations of electrodes. Charts SNR_{out} dependence on different quality levels of input signals.

3.2.   SNR

Second part of the evaluation is focused only on PCA since ICA changes the amplitudes of both components (see in Fig. 9 and Fig. 10) and changes order of the estimated components. Similarly, as in evaluation of HR, input signals with different quality levels are used (see Tab. 1). Table 4 shows averaged values of computed SNR_{out}. It shows if method on a certain quality level of input signals still works or not. Table 4 shows average values of SNR_{out} and resulting SNR for all combinations of different quality levels of corresponding input signals after using PCA.

In this paper, only averaged values are used because the ideal fECG signals, used in dominator in Eq. (14), differ for a certain combination of electrodes. For example in case of electrodes 2 and 22, we must compute the ideal form of fECG to determine the final SNR_{out}. We get one table of SNR_{out} values just for one quality level of input signals. For these 10 quality levels of input signal, we get 10 tables for SNR_{out} and

10 tables for SNR. Then we average the values computed for one combination in certain quality level. Figure 8 shows process of all 26 combinations for all quality levels of input signals. Figure 8(a) shows the combinations of two electrodes, Fig. 8(b) shows the combinations of three electrodes and Fig. 8(c) shows the combinations of four and five electrodes. According to Fig. 8, most of the electrodes combinations stop working in the quality level of input signals ranging from VI to VII. So similarly as in previous evaluation of HR, this evaluation shows that PCA stops working with input signals in range from −30 to −35 dB and in this range, PCA improves SNR approximately up to 25 dB.

3.3.   Subjective Observations

Subjective evaluation is not suitable approach, some observations are interesting, though. One of them was already mentioned and concerns ICA. This method changes amplitudes of the components as we can see

(a)

(b)

(c)

(d)

(e)

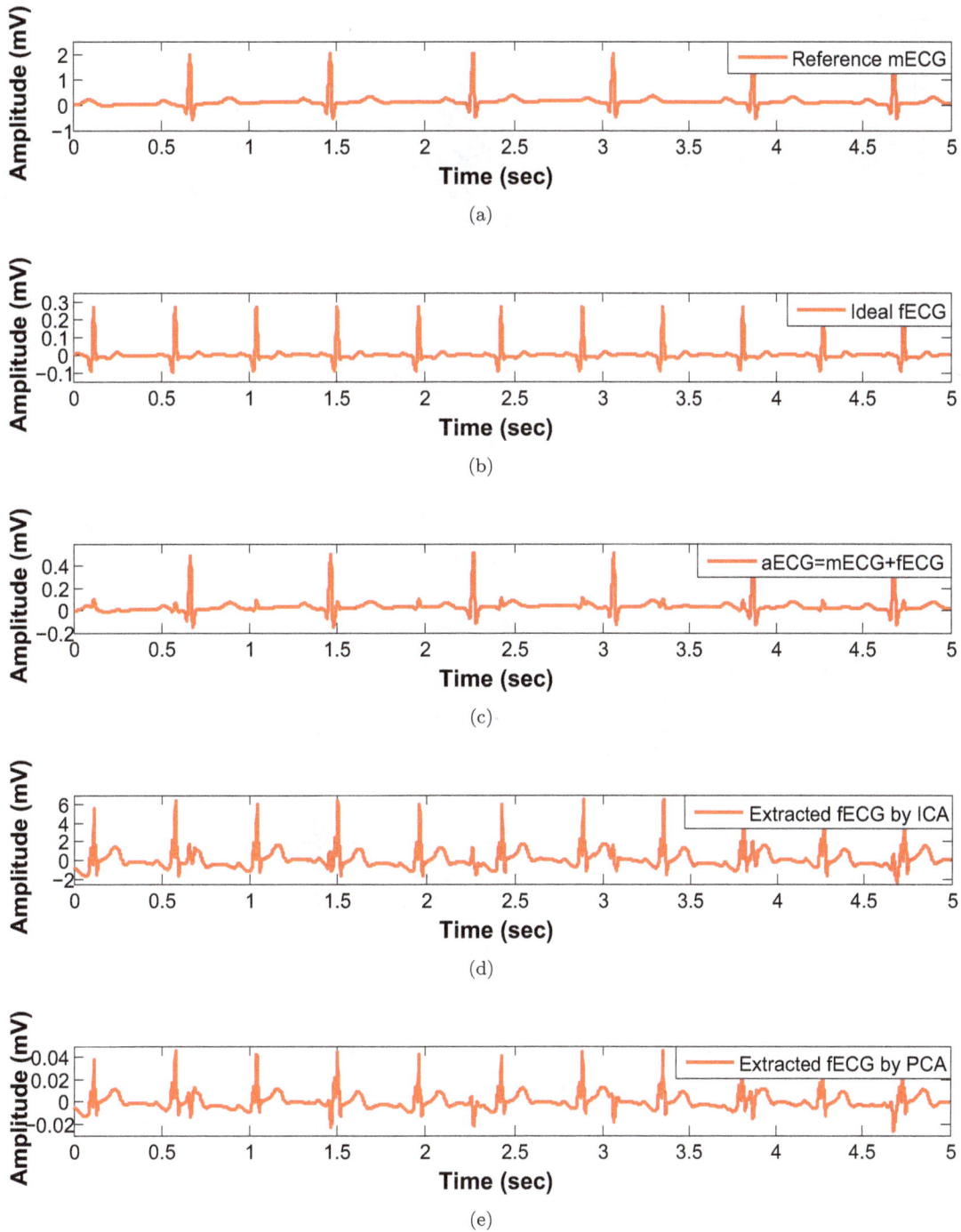

Fig. 9: ICA and PCA output for fECG estimation.

in Fig. 9 and Fig. 10, where fECG is extracted using ICA and PCA, respectively. This method also changes order of the estimated components. That is important in case of creating a program to display the extracted components. Figure 9 shows an example of fECG extraction by ICA and PCA. For the extraction, input signals with quality level V were used, which ensures the ideal accuracy for extraction of fECG. The signal in Fig. 9(a) is the generated mECG from electrode number 2. Figure 9(b) is the generated fECG from electrode number 2, Fig. 9(c) is aECG from electrode number 2. Figure 9(d) is the extracted fECG by using ICA for combination of electrodes 2 and 48. Figure 9(e) is extracted fECG by using PCA for combination of electrodes 2 and 48. We can see that mECG signal is suppressed and only a small random noise remains. In Fig. 10 we can see the deformations caused by maternal QRS complexes on both extracted fECGs by ICA and PCA. These deformations are marked by in Fig. 10.

(a)

(b)

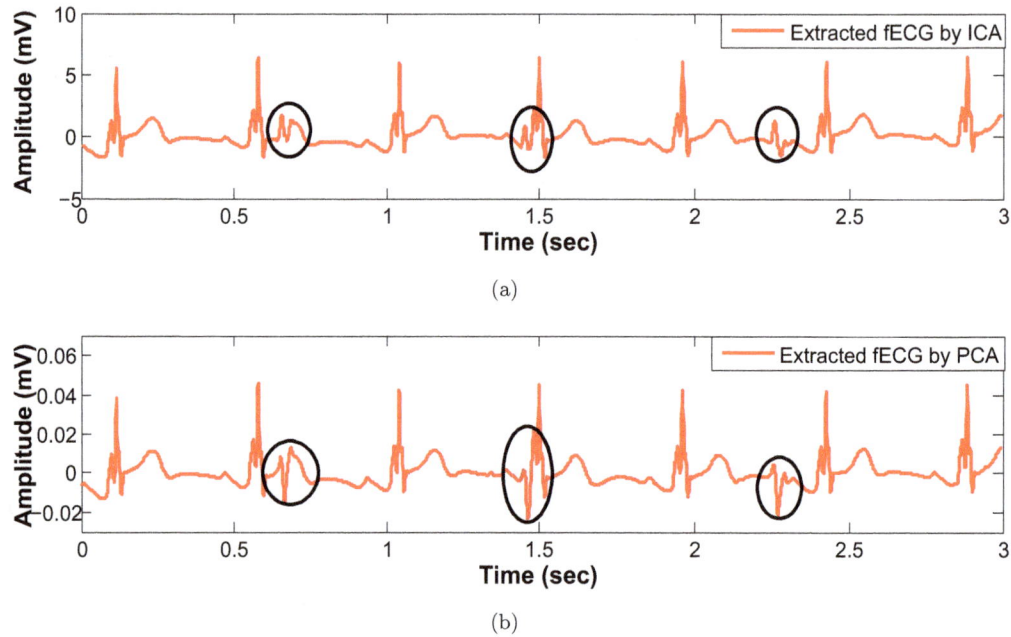

Fig. 10: Deformations of extraction fECG signal due to mHR.

Tab. 3: Table of BPM detected from the extracted components by using PCA.

Combination	Determinationof fHR by PCA									
of electrodes	I	II	III	IV	V	VI	VII	VIII	IX	X
2, 22	130	130	130	130	130	128	116	138	74	72
2, 48	128	152	130	126	130	132	136	122	74	74
2, 74	140	136	134	124	78	74	74	74	72	72
2, 94	128	118	118	122	132	130	90	68	70	72
22, 48	130	140	132	130	130	130	128	144	136	74
22, 74	130	130	130	128	128	130	126	74	74	72
22, 94	140	130	136	136	126	134	72	74	74	74
48, 74	128	116	120	128	124	118	72	74	72	72
48, 94	130	130	130	130	130	128	128	142	70	72
74, 94	126	128	128	128	138	140	124	74	72	72
2, 22, 48	130	122	130	130	130	128	128	130	120	74
2, 22, 74	128	126	130	130	126	124	74	74	74	72
2, 22, 94	130	128	130	128	128	128	112	114	68	74
2, 48, 74	134	124	126	130	128	70	74	74	72	72
2, 48, 94	128	132	130	130	130	140	142	70	68	74
2, 74, 94	144	128	122	122	128	114	74	74	72	72
22, 48, 74	130	124	132	130	130	132	74	74	72	72
22, 48, 94	130	120	130	130	130	130	130	124	74	74
22, 74, 94	130	130	130	128	130	132	130	74	72	72
48, 74, 94	128	122	130	130	130	130	74	74	72	72
2, 22, 48, 74	130	122	132	130	130	130	74	74	72	72
2, 22, 48, 94	130	120	130	130	130	128	130	146	70	74
2, 22, 74, 94	128	130	130	130	128	128	126	74	74	72
2, 48, 74, 94	134	126	130	130	130	144	74	74	72	72
22, 48, 74, 94	130	118	130	130	130	130	130	74	74	72
2, 22, 48, 74, 94	130	116	130	130	130	124	130	74	74	72

4.   Conclusion

In this paper, we have tested ICA and PCA mainly for fHR detection. Both methods showed good results, but the fHR detection using ICA showed smaller variance of values. Methods fail to work when input SNR ranges from −30 to −35 dB. In another evaluation, we used

SNR as the main parameter. However, this evaluation is possible only for PCA since ICA changes amplitude of extracted components. This evaluation showed similar results-PCA had high performance besides the range from −30 to −35 dB. The extracted fECG signal was deformed in case of using both algorithms by the maternal residues. These algorithms show very high performance, therefore it is possible to use them in the

Tab. 4: Table of calculated values of SNR_{out} for different input quality levels.

Combination of electrodes	SNR$_{out}$									
	I	II	III	IV	V	VI	VII	VIII	IX	X
2, 22	-0.13	0.79	0.81	0.31	0.27	0.17	-0.07	-0.87	-2.63	-5.09
2, 48	-0.02	0.14	-0.53	-0.63	0.61	-0.73	-0.75	-0.88	0.10	-0.54
2, 74	0.21	-0.45	-0.55	0.53	-0.67	0.14	-0.50	-2.43	-6.12	-9.52
2, 94	-0.53	-0.29	-0.27	0.81	0.65	-0.58	-0.82	-2.74	-6.15	-9.27
22, 48	-0.29	-0.57	1.14	1.26	-0.54	1.24	-0.59	-0.74	-0.60	-2.39
22, 74	-0.27	0.14	0.88	0.78	0.54	-0.35	-1.31	-3.73	-7.13	-11.10
22, 94	0.01	0.46	0.71	0.85	-1.15	0.74	0.24	-2.50	-3.85	-7.37
48, 74	0.21	-0.75	-1.13	1.31	1.30	-1.37	-1.79	-0.72	-4.32	-7.00
48, 94	-0.85	-0.94	-0.98	0.46	-1.05	-1.12	-1.21	-1.68	-3.19	-5.29
74, 94	-0.48	-0.22	-0.25	-0.28	-0.41	-0.18	-1.65	-3.80	-7.34	-11.07
2, 22, 48	-0.06	0.32	-0.33	-0.25	-0.23	1.24	-0.46	0.08	-1.73	-3.99
2, 22, 74	0.03	1.62	-0.81	2.05	1.84	1.02	-0.92	-3.96	-7.84	-11.74
2, 22, 94	-0.29	1.18	1.46	-0.78	1.35	1.14	-1.40	-2.71	-5.33	-8.08
2, 48, 74	0.13	0.64	-1.17	1.33	1.31	-1.37	-1.92	-3.43	-5.07	-9.04
2, 48, 94	-0.32	0.03	-0.39	-0.44	0.38	0.23	-0.22	-1.66	-4.18	-7.05
2, 74, 94	-0.11	0.43	0.56	0.58	0.50	0.24	-1.74	-3.93	-7.49	-11.22
22, 48, 74	-0.08	0.65	1.51	1.61	-0.37	-0.54	-1.46	-3.40	-6.22	-9.95
22, 48, 94	-1.00	-1.84	-1.84	-1.86	1.28	1.24	1.00	0.08	-3.18	-5.55
22, 74, 94	-0.60	0.95	-0.64	-0.73	-0.98	-1.57	-1.88	-5.07	-9.32	-13.16
48, 74, 94	-0.30	0.18	-0.60	-0.67	-0.75	0.40	-1.11	-3.62	-6.26	-9.82
2, 22, 48, 74	0.13	1.17	2.21	2.45	2.39	1.84	-0.77	-3.86	-6.86	-10.69
2, 22, 48, 94	-0.65	-0.54	0.30	-0.04	0.23	-0.24	-0.56	-1.83	-4.33	-7.10
2, 22, 74, 94	-0.65	0.19	0.66	0.20	0.37	-0.33	-2.07	-5.11	-9.06	-13.01
2, 48, 74, 94	-0.02	0.64	1.12	-1.07	-1.15	0.86	-0.94	-4.06	-6.73	-10.47
22, 48, 74, 94	-0.78	-0.28	0.13	0.16	-0.03	-0.26	-2.02	-4.53	-7.98	-11.80
2, 22, 48, 74, 94	-0.45	0.44	1.21	1.28	-0.63	-0.86	-1.52	-4.71	-8.11	-11.96

clinical practice for determining fHR for diagnosing fetal hypoxia. This research may be improved by testing obtained fECG by determining so-called T/QRS ratio. However, the deformation of T wave in extracted fECG must be minimal.

Acknowledgment

This paper has been elaborate in the framework of the project SP2017/128 of Student Grant system, VSB–Technical University of Ostrava.

References

[1] SAMENI, R. and G. D. CLIFFORD. A Review of Fetal ECG Signal Processing Issues and Promising Directions. *The Open Pacing, Electrophysiology and Therapy Journal.* 2010, vol. 5, iss. 3, pp. 4–20. ISSN 1876-536X. DOI: 10.2174/1876536X01003010004.

[2] JAGANNATH, D. J. and A. I. SELVAKUMAR. Issues and research on foetal electrocardiogram signal elicitation. *Biomedical Signal Processing and Control.* 2014, vol. 10, iss. 1, pp. 224–244. ISSN 1746-8094. DOI: 10.1016/j.bspc.2013.11.001.

[3] KITTNAR, O. *Medical physiology.* 1st ed. Prague: Grada, 2011. ISBN 978-80-247-3068-4.

[4] MARTINEK, R., M. KELNAR, P. KOUDELKA, J. VANUS, P. BILIK, P. JANKU, H. NAZERAN and J. ZIDEK. Enhanced processing and analysis of multi-channel non-invasive abdominal foetal ecg signals during labor and delivery. *Electronics Letters.* 2015, vol. 51, no. 22, pp. 1744–1746. ISSN 0013-5194. DOI: 10.1049/el.2015.2222.

[5] MARTINEK, R., H. SKUTOVA, R. KAHANKOVA, P. KOUDELKA, P. BILIK and J. KOZIOREK. Fetal ECG Extraction Based on Adaptive Neuro-Fuzzy Interference System. In: *10th International Symposium on Communication Systems, Networks and Digital Signal Processing (CSNDSP16).* Prague: IEEE, 2016, pp. 1–6. ISBN 978-1-5090-2526-8. DOI: 10.1109/CSNDSP.2016.7573973.

[6] MARTINEK, R., R. KAHANKOVA, H. SKUTOVA, J. ZIDEK and J. KOZIOREK. Adaptive Signal Processing Techniques for Extracting Abdominal Fetal Electrocardiogram. In: *10th International Symposium on Communication Systems, Networks and Digital Signal Processing (CSNDSP16).* Prague: IEEE, 2016, pp. 1–6. ISBN 978-1-5090-2526-8. DOI: 10.1109/CSNDSP.2016.7573974.

[7] KAHANKOVA, R., R. MARTINEK and P. BILIK. Non-Invasive Fetal ECG Extraction from Maternal Ab-dominal ECG Using LMS and RLS Adaptive Algorithms. In: *Third International*

Afro-European Conference for Industrial Advancement (AECIA 2016). Marrakesh: Springer, 2016, pp. 258–271. ISBN 978-3-319-60834-1. DOI: 10.1007/978-3-319-60834-1_27.

[8] KARVOUNIS, E. C., C. PAPALOUKAS, D. I. FOTIADIS and L. K. MICHALIS. Fetal heart rate extraction from composite maternal ECG using complex continuous wavelet transform. In: *Computers in Cardiology*. Chicago: IEEE, 2004, pp. 737–740. ISBN 0-7803-8927-1. DOI: 10.1109/CIC.2004.1443044.

[9] HASSANPOUR, H. and A. PARSAEI. Fetal ECG Extraction Using Wavelet Transform. In: *International Conference on Computational Inteligence for Modelling Control and Automation and International Conference on Intelligent Agents Web Technologies and International Commerce 2006 (CIMCA'06)*. Sydney: IEEE, 2006, pp. 179–182. ISBN 0-7695-2731-0. DOI: 10.1109/CIMCA.2006.98.

[10] BHOKER, R. and J. P GAWANDE. Fetal ECG Extraction Using Wavelet Transform. *ITSI Transactions on Electrical and Electronics Engineering*. 2003, vol. 1, iss. 4, pp. 19–22. ISSN 2320–8945.

[11] KUMAR, P., S. K. SHARMA and S. PRASAD. CAD for Detection of Fetal Electrocardiogram by using Wavelets and Neuro-Fuzzy Systems. *Second International Conference on Computational Intelligence and Communication Technology 2016 (CICT)*. Ghazibad: IEEE, 2016, pp. 587–590. ISBN 978-1-5090-0211-5. DOI: 10.1109/CICT.2016.122.

[12] VAN BEMMEL, J. H. Detection of weak foetal electro-cardiograms by autocorrelation and crosscorrelation of envelopes. *IEEE Transactions on Biomedical Engineering*. 1968, vol. BME-15, iss. 1, pp. 17–23. ISSN 0018-9294. DOI: 10.1109/TBME.1968.4502528.

[13] LEVKOV, C., G. MIHOV, R. IVANOV, I. DASKALOV, I. CHRISTOV and I. DOTSINSKY. Removal of power-line interference from the ECG: a review of the subtraction procedure. *BioMedical Engineering OnLine*. 2005, vol. 4, iss. 50, pp. 1–18. ISSN 1475-925X. DOI: 10.1186/1475-925X-4-50.

[14] HON, E. H. and S. T. LEE. Averaging techniques in fetal electrocardiography. *Medical electronics and biological engineering*. 1964, vol. 2, iss. 1, pp. 71–76. ISSN 0140-0118. DOI: 10.1007/BF02474362.

[15] SU, L. and H.-T. WU. Exract fetal ECG from single-lead abdominal ECG by De-Shape short time Fourier transform and nonlocal median.

In: *Frontiers in Applied Mathematics and Statistics* [online]. 2016. Available at: https://pdfs.semanticscholar.org/9e0e/34f844f605562dffbac37a7010168f0e601f.pdf.

[16] LEE, K. and B. LEE. Sequential Total Variation Denoising for the Extraction of Fetal ECG from Single-Channel Maternal Abdominal ECG. *Sensors*. 2016, vol. 16, iss. 7, pp. 1–15. ISSN 1424-8220. DOI: 10.3390/s16071020.

[17] TAN, B., Q. PENG, J. LIN and M. LI. A novel method for estimating source number of fetal ECG. In: *International Conference on Wireless Communications and Signal Processing (WCSP)*. Nanjing: IEEE, 2015, pp. 1–6. ISBN 978-1-4673-7686-0. DOI: 10.1109/WCSP.2015.7341070.

[18] HE, P. J., X. M. CHEN, Y. LIANG, H. Z. ZENG, W. P. SUNG and J. C. M. KAO. Extraction for fetal ECG using single channel blind source separation algorithm based on multi-algorithm fusion. *MATEC Web of Conferences*. 2016, vol. 44, iss. 1, pp. 1–9. ISSN 2261-236X. DOI: 10.1051/matecconf/20164401026.

[19] RAJ, C. G., V. S. HARSHA, B. S. GOWTHAMI and R. SUNITHA. Virtual Instrumentation Based Fetal ECG Extraction. *Procedia Computer Science*. 2015, vol. 70, iss. 1, pp. 289–295. ISSN 1877-0509. DOI: 10.1016/j.procs.2015.10.093.

[20] AHUJA, E. and F. I. SHAIKH. A Novel Approach to FEG Extraction based on Fast ICA. *International Research Journal of Engineering and Technology (IRJET)*. 2016, vol. 3, iss. 4, pp. 2450–2454. ISSN 2395-0072.

[21] GAO, P., E. C. CHANG and L. WYSE. Blind separation of fetal ECG from single mixture using SVD and ICA. In: *Proceedings of the Joint Conference of the Fourth International Conference on Information, Communications and Signal Processing and Fourth Pacific Rim Conference on Multimedia*. Singapore: IEEE, 2003, pp. 1418–1422. ISBN 0-7803-8185-8. DOI: 10.1109/ICICS.2003.1292699.

[22] BACHARAKIS, E., A. K. NANDI and V. ZARZOSO. Foetal ECG extraction using bling source separation methods. In: *8th European Signal Processing Conference, EUSIPCO*. Trieste: IEEE, 1996, pp. 1–4. ISBN 978-888-6179-83-6.

[23] LEACH, S. Singular Value Decomposition - A Primer. *CSAIL* [online]. 1995. Available at: http://www.dis.uniroma1.it/~visiope/Esercitazioni/es2008/es12_2/svd.pdf.

[24] VARANINI, M., G. TARTARISCO, R. BALOC-CHI, A. MACERATA, G. PIOGGIA and L. BILLECI. A new method for QRS complex detection in multichannel ECG: Application to self-monitoring of fetal health. *Computers in Biology and Medicine*. 2017, vol. 81, iss. 1, pp. 125–134. ISSN 0010-4825. DOI: 10.1016/j.compbiomed.2016.04.008.

[25] LIU, G. and Y. LUAN. An adaptive integrated algorithm for noninvasive fetal ECG separation and noise reduction based on ICA-EEMD-WS. *Medical and Biological Engineering and Computing*. 2015, vol. 53, iss. 11, pp. 1113–1127. ISSN 0140-0118. DOI: 10.1007/s11517-015-1389-1.

[26] AYAT, M., K. ASSALEH and H. AL-NASHASH. Extracting fetal ECG from a single maternal abdominal record. In: *8th GCC Conference and Exhibition (GCCCE)*. Muscat: IEEE, 2015, pp. 1–4. ISBN 978-1-4799-8422-0. DOI: 10.1109/IEEEGCC.2015.7060027.

[27] REDIF, S. Fetal electrocardiogram estimation using polynomial eigenvalue decomposition. *Turkish Journal of Electrical Engineering and Computer Sciences*. 2014, vol. 24, iss. 4, pp. 2483–2497. ISSN 1300-0632. DOI: 10.3906/elk-1401-19.

[28] HYVARINEN, A. and E. OJA. Independent component analysis: A Tutorial. In: *UCLA College: Statistics* [online]. 1999. Available at: http://www.stat.ucla.edu/~yuille/courses/Stat161-261-Spring14/HyvO00-icatut.pdf.

[29] MILITKY, J. and M. MELOUN. Principal component method and exploratory analysis of multivariate data. In: *Milan Meloun* [online]. 2005. Available at: https://meloun.upce.cz/docs/publication/127a.pdf.

[30] MARTINEK, R., M. KELNAR, P. KOUDELKA, J. VANUS, P. BILIK, P. JANKU, H. NAZERAN and J. ZIDEK. A novel LabVIEW-based multi-channel non-invasive abdominal maternal-fetal electrocardiogram signal generator. *Physiological Measurement*. 2016, vol. 37, iss. 2, pp. 238–256. ISSN 0967-3334. DOI: 10.1088/0967-3334/37/2/238.

[31] Mathworks. In: *Mathworks* [online]. 2014. Available at: http://www.mathworks.com/matlabcentral/fileexchange/.

[32] MARTINEK, R., M. KELNAR, P. VOJCINAK, P. KOUDELKA, J. VANUS, P. BILIK, P. JANKU, H. NAZERAN and J. ZIDEK. Virtual simulator for the generation of patho-physiological foetal ECGs during the prenatal period. *Electronics Letters*. 2015, vol. 51, iss. 22, pp. 1738–1740. ISSN 0013-5194. DOI: 10.1049/el.2015.2291.

[33] JEZEWSKI, J., D. ROJ, J. WROBEL and K. HOROBAL. Instrumentation for fetal monitoring-improvement in Doppler ultrasound technology. *Journal of Medical Informatics and Technologies*. 2001, vol. 10, iss. 92, pp. 1–17. ISSN 0006-3398.

[34] JEZEWSKI, J., J. WROBEL, K. HOROBA, S. GRACZYK and A. GACEK. Coping with limitations of Doppler ultrasound fetal heart rate monitors. In: *Engineering in Medicine and Biology Society, 1995 and 14th Conference of the Biomedical Engineering Society of India. An International Meeting, Proceedings of the First Regional Conference*. New Delhi: IEEE, 1995, pp. PS9–P10. ISBN 0-7803-2711-X. DOI: 10.1109/RCEMBS.1995.511699.

[35] JEZEWSKI, J., J. WROBEL and K. HOROBA. Comparison of Doppler ultrasound and direct electrocardiography acquisition techniques for quantification of fetal heart rate variability. *IEEE Transactions on Biomedical Engineering*. 2006, vol. 53, iss. 5, pp. 855–864. ISBN 0-7803-2711-X. DOI: 10.1109/TBME.2005.863945.

About Authors

Radana KAHANKOVA was born in 1991 in Opava, Czech Republic. She received her Bachelor's degree at the VSB–Technical University of Ostrava, the Department of Cybernetics and Biomedical Engineering in 2014. Two years later at the same department, she received her Master's degree in the field of Biomedical Engineering. She is currently pursuing her Ph.D. in Technical Cybernetics. Her current research is focused on improving the quality of electronic fetal monitoring.

Rene JAROS was born in Ostrava, Czech Republic. He graduated from the Faculty of Electrical Engineering and Computer Science, VSB–Technical University of Ostrava in 2015. In 2015, he wrote a bachelor thesis, which dealt with the transformation methods of electrocardiography to vectorcardiography. He is currently M.Sc. student at the same university. His research interests include in fECG extraction by using non-adaptive methods.

Radek MARTINEK was born in 1984 in Czech Republic. In 2009 he received Master's degree in Information and Communication Technology from VSB–Technical University of Ostrava. Since 2012 he worked here as a Research Fellow. In 2014 he successfully defended his dissertation thesis titled "The Use of Complex Adaptive Methods of Signal Processing for Refining the Diagnostic Quality of the Abdominal Fetal Electrocardiogram". He became an Associate Professor in Technical Cybernetics in 2017 after defending the habilitation thesis titled" Design and Optimization of Adaptive Systems for Applications of Technical Cybernetics and Biomedical Engineering Based on Virtual Instrumentation". He works as an Associate Professor at VSB–Technical University of Ostrava since 2017. His current research interests include: Digital Signal Processing (Linear and Adaptive Filtering, Soft Computing - Artificial Intelligence and Adaptive Fuzzy Systems, Non-Adaptive Methods, Biological Signal Processing, Digital Processing of Speech Signals); Wireless Communications (Software-Defined Radio); Power Quality Improvement. He has more than 70 journal and conference articles in his research areas.

Janusz JEZEWSKI is an Assistant Professor at the Institute of Medical Technology and Equipment–ITAM in Zabrze Poland, where he is also a Deputy Director for Science. He is a member of the Committee of Biocybernetics and Biomedical Engineering of the Polish Academy of Sciences. In 1979 he received the M.Sc. degree in electronic engineering from the Institute of Electronics of the Silesian University of Technology in Gliwice, the Ph.D. degree in biological sciences (1997) from the University of Medical Sciences in Poznan, and in 2012 the D.Sc. degree in biocybernetics and biomedical engineering from the Nalecz Institute of Biocybernetics and Biomedical Engineering of the Polish Academy of Sciences in Warsaw. His research interests include: biomedical instrumentation and digital signal processing, especially the analysis of biomedical signals: the fetal heart rate variability acquired via pulsed Doppler ultrasound as well as the transabdominal fetal electocardiogram and the uterine electromyographic activity. Currently, his main area of interest is an application of computational intelligence for the extraction of clinically relevant information, and the automated classification of fetal and maternal signals within a computer-aided fetal surveillance system. He is an author or co-author of numerous scientific papers in the international journals, conference proceedings and books.

He WEN was born in Hunan, China, in 1982. He received the B.Sc., M.Sc., and Ph.D. degrees in electrical engineering from Hunan University, Hunan, China, in 2004, 2007 and 2009, respectively. He is currently an Associate Professor with the College of Electrical and Information Engineering, Hunan University. His research interests include power system harmonic measurement and analysis, power quality, and digital signal processing.

Michal JEZEWSKI was born in 1982 in Zabrze, Poland. He received the M.Sc. degree in computer science from the Silesian University of Technology, Gliwice Poland, in 2006. In 2011 he received the Ph.D. degree in electronic engineering also from the Silesian University of Technology in Gliwice. He is currently at the Division of Biomedical Electronics from the Institute of Electronics in the Silesian University of Technology. His research interests include biomedical signal processing, machine learning, and computational intelligence methods with emphasis on fuzzy and neuro-fuzzy modelling.

Aleksandra KAWALA-JANIK was born in 1984, in Olesnica–Poland. In 2007 she has obtained her M.Sc. Degree in Computer Engineering from the Opole University of Technology. In 2012 she successfully defended her Ph.D. thesis at the University of Greenwich in London. Till the end of October 2013 she has been working there as a Researcher. From 2013 she was also appointed Assistant and in 2014–assistant professor at the Opole University of Technology. She is currently working as a postdoctoral research scholar at the University of Kentucky in United States. Her current research interest focus on analysis of various biomedical data, in particular EEG and EMG. She has authored and coauthored over 60 papers.

17

A Two-Level Sensorless MPPT Strategy Using SRF-PLL on a PMSG Wind Energy Conversion System

Amina ECHCHAACHOUAI[1], Soumia EL HANI[1],
Ahmed HAMMOUCH[2], Imad ABOUDRAR[1]

[1]Department of Electrical Engineering, Ecole Normale Superieure de Enseignement Technique,
Mohammed V University, Avenue des Nations Unies, Agdal, 10102 Rabat, Morocco
[2]National Center for Scientific and Technical Research, Angle avenues des FAR et Allal El Fassi,
Hay Ryad, 10102 Rabat, Morocco

amina.echchaachouai@um5s.net.ma, s.elhani@um5s.net.ma,
hammouch_a@yahoo.com, imad.aboudrar@um5s.net.ma

Abstract. *In this paper, a two-level sensorless Maximum Power Point Tracking (MPPT) strategy is presented for a variable speed Wind Energy Conversion System (WECS). The proposed system is composed of a wind turbine, a direct-drive Permanent Magnet Synchronous Generator (PMSG) and a three phase controlled rectifier connected to a DC load. The realised generator output power maximization analysis justifies the use of the Field Oriented Control (FOC) giving the six Pulse Width Modulation (PWM) signals to the active rectifier. The generator rotor speed and position required by the FOC and the sensorless MPPT are estimated using a Synchronous Reference Frame Phase Locked Loop (SRF-PLL). The MPPT strategy used consists of two levels, the first level is a power regulation loop and the second level is an extremum seeking bloc generating the coefficient gathering the turbine characteristics. Experimental results validated on a hardware test setup using a DSP digital board (dSPACE 1104) are presented. Figures illustrating the estimated speed and angle confirm that the SRF-PLL is able to give an estimated speed and angle which closely follow the real ones. Also, the power at the DC load and the power at the generator output indicate that the MPPT gives optimum extracted power. Finally, other results show the effectiveness of the adopted approach in real time applications.*

Keywords

Direct-drive PMSG, extremum seeking, FOC, sensorless two-level MPPT, SRF-PLL, WECS.

1. Introduction

The Earth receives every day an infinite renewable energy that we can exploit to increase sustainable development. The extracted energy is integrated in the ecosystem that stimulates the growth of new expertise, creates jobs and ensures the energy future. Among all known forms of renewable energy, wind energy conversion has become a major producer of electric power [1].

Recent research in this field is focused on minimizing the overall cost of the Wind Energy Conversion System (WECS) while improving the quality of the produced power. In order to achieve this objective, several works have been carried out to avoid the use of the mechanical sensors which are expensive to buy and maintain. These sensors are usually implemented to measure the generator rotation speed as well as the angle of the rotor that are necessary for the general control of the system and the search for the maximum points of the extractable power.

With the same principle, we consider for investigation in this work a WECS consisting of a wind turbine, a direct-drive Permanent Magnet Synchronous Generator (PMSG), a three phase active rectifier connected to a DC load. The PMSG choice is very advantageous [2], it allows direct-drive systems that avoid gearbox [3] use and this leads to low maintenance constraints. This type of machines is characterized by a high-power density and high efficiency (as there are no copper losses). The use of permanent magnets for the excitation consumes no extra electrical power. The absence of mechanical commutator and brushes or slip rings implies low mechanical friction losses. Finally, the PMSG drives achieve very high torque at low speeds with less noise.

Based on a generator output power maximization analysis, the Field Oriented Control (FOC) is implemented [4] to control the active rectifier whose performance compared to a diode rectifier has been confirmed in a previous work [6].

Betz limit [5] indicates that the maximum power that can be extracted from the wind is limited to approximately 0.59 of the kinetic energy. The power coefficient C_p, included in the expression of the generated power, takes into consideration this limit and varies with the rotational speed of the turbine. In a different way, the maximum points of the power vary with the wind speed and this justifies the implementation of a Maximum Power Point Tracking (MPPT) strategy to keep our system available to generate optimal power and improve its performance.

To reduce the cost of a WECS, a profitable solution is the sensorless MPPT that allows avoiding the use of mechanical sensors and then improving the relative cost/quality of the complete system. For this reason, there are many estimators used to estimate the generator rotor speed and position required by the FOC and the MPPT. In our previous works, two types of estimators were implemented; The fisrt is the Angle Tracking Observer (ATO) [6] and the second is the Extended Kalman Filter (EKF) [7]. In this work, we adopt a mechanical sensorless MPPT strategy that uses a Synchronous Reference Frame Phase Locked Loop (SRF-PLL) as estimator. It should be noted that the advantage of the SFR-PLL is that it avoids the double-frequency error problem of single-phase standard PLL. It has a simple structure that offers ease of parameter tuning and robust features.

The MPPT strategy adopted is a two-level strategy. The first level is the power regulation loop that gives the reference value of current i_{qref} to the FOC. A method of extremum seeking [8] is involved as a second level of the MPPT algorithm to generate the optimum value of the coefficient including turbine parameters in the expression of the output power. This coefficient is needed in the first level. Estimating this coefficient allows realizing this approach even when the turbine parameters are not determined.

This work is organized as follows: The first part presents the system modeling by detailing the equation models of both the different components of the wind conversion system considered and the control applied to the active rectifier and by giving their detailed structure illustrated in Fig. 1. The second part concerns the sensorless two-level MPPT strategy adopted and explains the estimators used. The third part gives the results obtained on the experimental test bed setup that enables the evaluation of the approach discussed in this paper.

2. System Modeling

Figure 1 presents the system considered in this work for investigation. It consists of two main parts; the hardware setup gathering elements of the wind energy conversion system which are: a wind turbine, a direct-drive PMSG connected to a three phase rectifier supplying a DC load, and the analog board dSPACE 1104 containing analog to digital and digital to analog converters making connection between the Real-Time Interface (RTI) in the monitoring PC and the hardware possible. In the interface, the Simulink model of the global control is implemented giving-out six Pulse Width Modulation (PWM) signals to the active rectifier.

2.1. Wind Turbine

The turbine considered in this paper is a Horizontal Axis Wind Turbine (HAWT) with 3 blades whose length is R in (m) and which are fixed on a drive shaft rotating at the speed Ω_{turbine} in (rad·s^{-1}). Considering V_w the wind speed in (m·s^{-1}), ρ the air density in (kg·m^{-3}) (approximately 1.22 at atmospheric pressure), and the area opposed to the wind A in (m^2), the kinetic power of wind is expressed as follows:

$$P_w = \frac{\rho A V_w^3}{2}. \tag{1}$$

Then the power at the output of the turbine has this expression:

$$P_t = C_p(\lambda) P_w. \tag{2}$$

The power coefficient C_p is a function of the relative speed $\lambda = \dfrac{R\Omega_{\text{turbine}}}{V_w}$. This coefficient depends on the type of the turbine since it depends on the surface swept by the rotor whose size is different for different types of turbine.

2.2. Permanent Magnet Synchronous Generator

The PMSG model [9] is represented in the Park referential related to the rotating field, voltages expressions are:

$$v_d = -R_s i_d + L_q \omega_e i_q - L_d \frac{di_d}{dt}, \tag{3}$$

$$v_q = -R_s i_q - L_d \omega_e i_d - L_q \frac{di_q}{dt} + \sqrt{\frac{3}{2}} \Phi_{sf} \omega_e, \tag{4}$$

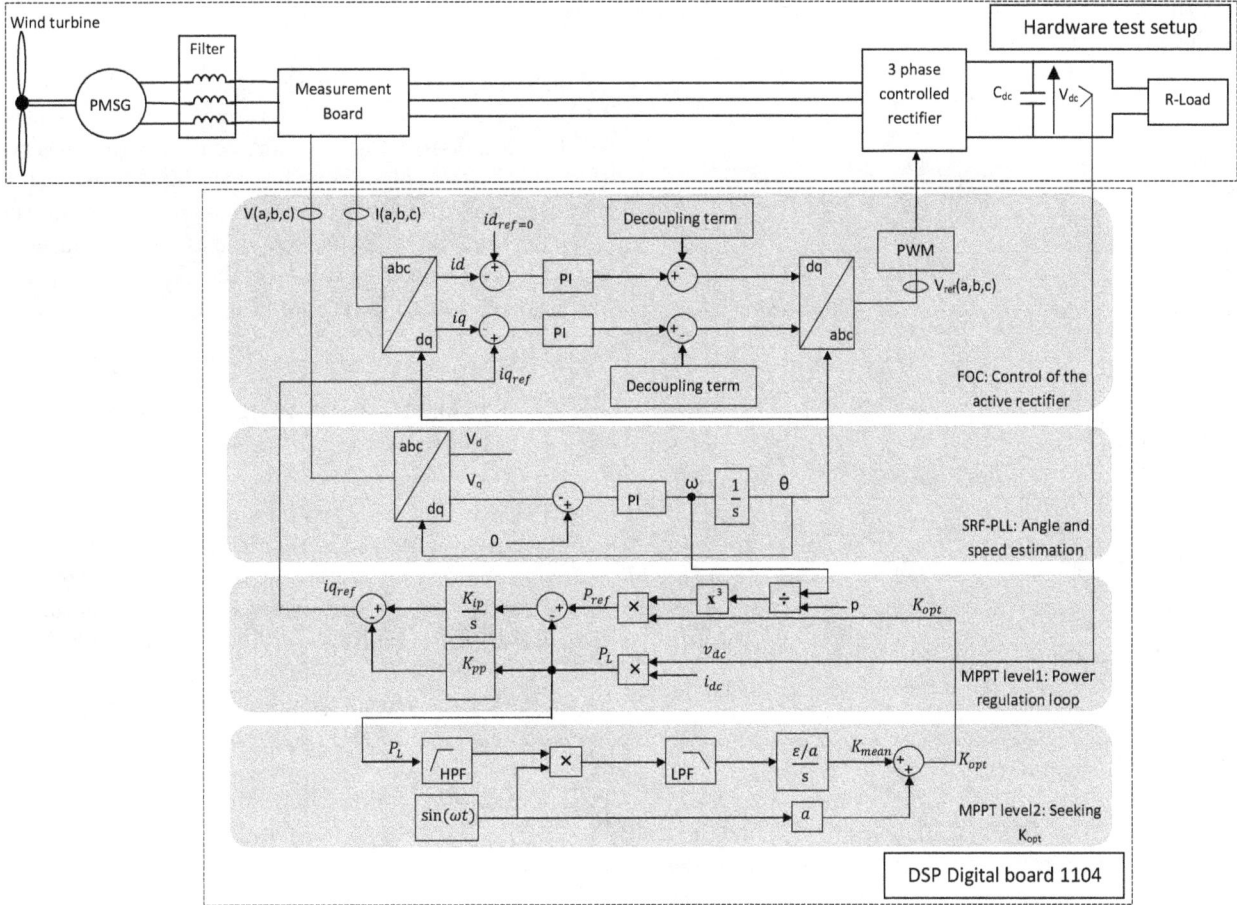

Fig. 1: Global structure.

where R_s is the stator resistance, L_d and L_q are the dq inductances, Φ_{sf} is the magnetic flux density, $\omega_e = p\Omega_{\text{turbine}}$ is the angular frequency and p is the number of pole pairs. Production of the electrical energy at the output of the generator causes a braking torque which has this expression:

$$T_e = -p\left(\sqrt{\frac{3}{2}}\Phi_{sf}i_q + (L_q - L_d)i_d i_q\right). \quad (5)$$

The mechanical equation is:

$$J\frac{d\Omega_{\text{turbine}}}{dt} = T_t + T_e - f\Omega_{\text{turbine}}, \quad (6)$$

where J is the total inertia, f is the viscous friction coefficient and $T_t = P_t/\Omega_{\text{turbine}}$ is the turbine torque.

2.3. FOC for the Active Rectifier

1) Study of Power Maximization

This study is realised to explain the motivation to use the FOC for the rectifier. The aim is to maximize the generator output power whose expression is:

$$P_g = v_a i_a + v_b i_b + v_c i_c = v_d i_d + v_q i_q. \quad (7)$$

Replacing elements of this equation by their expressions in Eq. (3) and Eq. (4) and considering the steady state characterized by no variations of currents and speed $\left(\frac{di_d}{dt} = \frac{di_q}{dt} = \frac{d\Omega_m}{dt} = 0\right)$ we deduce the following equations of P_g and the constraint of its maximization:

$$P_g = C_p\left(\frac{R\Omega_m}{V_w}\right)P_w - R_s\left(i_d^2 + i_q^2\right) - f\Omega_m^2, \quad (8)$$

$$C_p\left(\frac{R\Omega_m}{V_w}\right)P_w - f\Omega_m^2 = -T_e\Omega_m = \\ = \sqrt{\frac{3}{2}}p\Phi_{sf}\Omega_m i_q. \quad (9)$$

A popular method allowing to realise this is the Lagrange multipliers whose first step is to define the Lagrange function as:

$$\zeta = P_g + \lambda_\zeta \cdots \\ \cdot\left(C_p\left(\frac{R\Omega_m}{V_w}\right)P_w - f\Omega_m^2 - \sqrt{\frac{3}{2}}p\Phi_{sf}\Omega_m i_q\right). \quad (10)$$

The second step is to find a stationary point of this function by finding solution of the partial derivatives of this function with respect to its four variables, results of this are given as follows:

$$\frac{\partial \zeta}{\partial i_d} = 0 \Rightarrow i_d = 0. \tag{11}$$

$$\frac{\partial \zeta}{\partial \lambda_\zeta} = 0 \Rightarrow i_q = \frac{C_p \left(\frac{R\Omega_m}{V_w} \right) P_w - f\Omega_m^2}{\sqrt{\frac{3}{2}} p \Phi_{sf} \Omega_m}. \tag{12}$$

$$\frac{\partial \zeta}{\partial i_q} = 0 \Rightarrow \lambda_\zeta = \frac{4R_s}{3p^2\Phi_{sf}^2} \left(f - \frac{P_w}{\Omega_m^2} C_p \left(\frac{R\Omega_m}{V_w} \right) \right). \tag{13}$$

$$\frac{\partial \zeta}{\partial \Omega_m} = 0 \Rightarrow \frac{R\Omega_m}{V_w} \frac{\mathrm{d}C_p \left(\frac{R\Omega_m}{V_w} \right)}{\mathrm{d}\lambda} =$$
$$= \frac{\lambda_\zeta C_p \left(\frac{R\Omega_m}{V_w} \right) + (\lambda_\zeta + 2)f\Omega_m^2/P_w}{\lambda_\zeta + 1}. \tag{14}$$

From the first derivative Eq. (11) we deduce that P_g is maximised when $i_d = 0$, for that reason the FOC is selected as control of the implemented rectifier.

2)   FOC Structure

The structure of the FOC [10] can be found in the Fig. 1 consisting of two current regulation loops based on the two similar transfer functions whose form is:

$$G(s) = \frac{i}{v} = \frac{-1}{R + Ls}. \tag{15}$$

Then

$$i = \frac{-1}{R + Ls} \left(K_{pc} i - \frac{K_{ic}}{s} (i_{ref} - i) \right), \tag{16}$$

$$\frac{i}{i_{ref}} = \frac{\frac{K_{ic}}{L}}{s^2 + \frac{R + k_{pc}}{L} s + \frac{K_{ic}}{L}}. \tag{17}$$

Finally, we extract a second order equation with the following form:

$$\frac{i}{i_{ref}} = \frac{\omega_n^2}{s^2 + 2\xi\omega_n s + \omega_n^2}. \tag{18}$$

We consider the same parameters for both currents i_d and i_q, then coefficients $K_{ic} = L\omega_n^2$ and $K_{pc} = 2\xi\omega_n L - R$ are the same for the two control loops. As illustrated, Park and inverse Park transformations in this control require the angle θ estimated by the SRF-PLL. From v_d and v_q expressions, we deduce that a decoupling of the two currents control loops is necessary. Then the decoupling terms are:

$$DT_d = \omega_e L_q i_q, \tag{19}$$

$$DT_q = -\omega_e L_d i_d + \omega_e \sqrt{\frac{3}{2}} \Phi_{sf}. \tag{20}$$

3.   Sensorless MPPT

As outlined in the introduction, the aim of this work is to build a complete MPPT strategy for the proposed WECS detailed in Fig. 1 from which we can observe that the MPPT algorithm has two levels the first is a power regulation loop generating at its output the i_{qref} for the FOC and the second is an extremum seeking system giving the value of the K_{opt}. Generator rotor speed needed is estimated using the SRF-PLL.

3.1.   SRF-PLL as Angle and Speed Estimator

The SRF-PLL [11] is based on aligning the output frequency with the d axis in the dq frame by using a PI controller to force the q component voltage to zero. Referring to Fig. 1 which shows the basic structure of the SRF-PLL, the voltages in dq frame V_d and V_q are deduced from the three phase voltages V_a, V_b and V_c using Park's transform including the estimated phase angle θ:

$$\begin{bmatrix} V_d \\ V_q \\ V_0 \end{bmatrix} = \frac{2}{3} \begin{bmatrix} \sin(\theta) & \sin(\theta - \frac{2\pi}{3}) & \sin(\theta - \frac{4\pi}{3}) \\ \cos(\theta) & \cos(\theta - \frac{2\pi}{3}) & \cos(\theta - \frac{4\pi}{3}) \\ \frac{1}{2} & \frac{1}{2} & \frac{1}{2} \end{bmatrix} \begin{bmatrix} V_a \\ V_b \\ V_c \end{bmatrix}. \tag{21}$$

To align the SRF-PLL output with d axis, the PI controller forces the component to zero. When this output becomes in-phase with the supply voltage, the PI output will be equal to ω. Then the angle θ can be obtained by integrating the PI output as shown.

3.2.   MPPT Level 1: Power Regulation Loop

We consider the expression of the turbine generated power to extract the area where the points of maximum power are located. That is the curve described using the following expression:

$$P_{opt} = \frac{1}{2}\rho A \left(\frac{R\Omega_m}{\lambda_{opt}} \right)^3 C_{popt} = K_{opt}\Omega_m^3,$$
$$\text{with } K_{opt} = \frac{\rho AR^3 C_{popt}}{2\lambda_{opt}^3}. \tag{22}$$

As we have shown, the MPPT principle must be applied to the entire conversion chain. Then we constructed a power control loop based on the knowledge of the optimum power value, the reference power is obtained from the estimated rotational speed and the value of the K_{opt} parameter. The structure of this loop is illustrated in Fig. 1.

3.3.   MPPT Level 2: Seeking K_{opt}

For our case, we suppose that characteristics of the wind turbine are undefined, so the value of K_{opt} is undetermined. To determine this value continuously an automatic system is implemented, this system is a method of extremum seeking [12] presenting a second level of the MPPT. This block, as Fig. 1 describes, determines an average value K_{mean} and adds a very slow sinusoidal perturbation to this value to generate the value of K_{opt} used as an input for the power control loop. Finally, we obtain variations in the average value of instantaneous power P_{inst} according to variations of the K_{opt} value at the output of a high pass filter. The low pass filter rejects the frequency of the disturbance signal at the output of the multiplier. Gain a is the disturbance signal amplitude and ε adjusts the integrator gain which defines K_{mean} value.

4.   Experimental Results

The proposed control system was implemented using real time digital controller dSPACE 1104. This controller and the hardware test setup consisting of a 5 kW PMSG, wind turbine emulator, three phase rectifier and a DC load $R = 115.1\ \Omega$ are presented in Fig. 2.

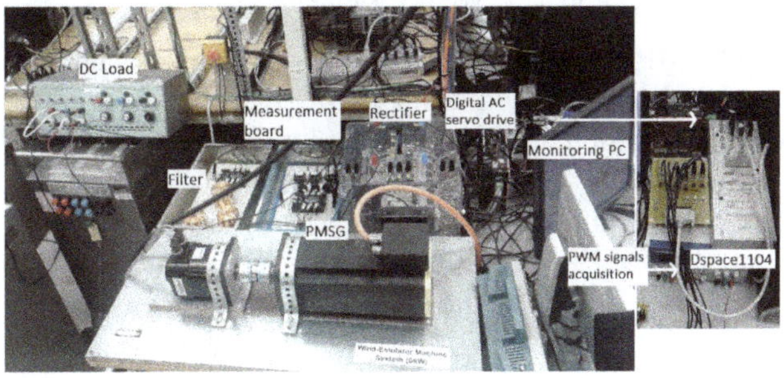

Fig. 2: Experimental test bed setup.

This paper considers a variable speed wind conversion system for investigation. To evaluate its performance and as illustrated in Fig. 3, a variable wind speed profile in the form of a repeating sequence of different values of the wind speed is considered.

According to Fig. 4, where the estimated position by the SRF-PLL and one of the three phase voltages precisely V_a are presented, the SRF-PLL is operating accurately, and this is clearer in the zoom.

The second estimated parameter by the SRF-PLL needed in the system control of our sensorless MPPT structure is the generator speed which is presented with the real generator speed in Fig. 5. It appears that values of the two speeds are close and also vary simultaneously.

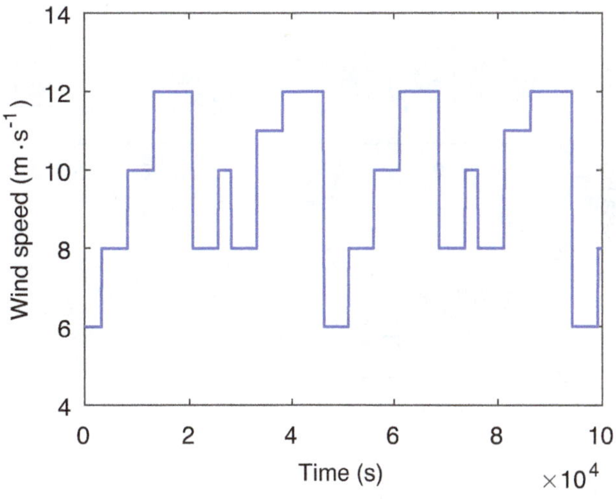

Fig. 3: Variable wind speed profile.

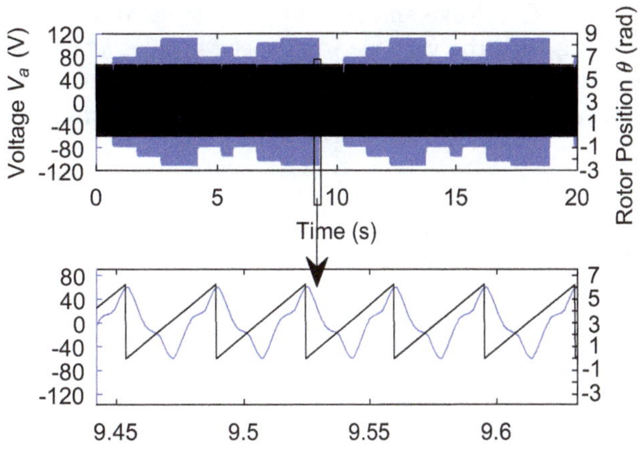

Fig. 4: Voltage V_a (V) and the enstimated Rotor position (rad).

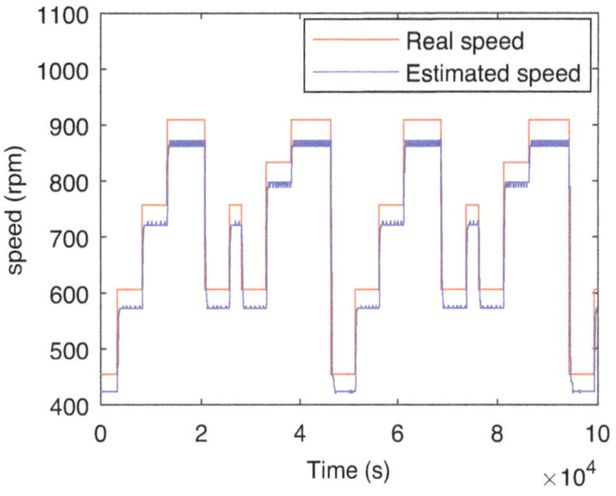

Fig. 5: Real generator speed (rpm) and estimated generator speed (rpm).

Figure 6 shows the generator output three phase voltages. To better visualise variations of this three phase voltages, a zoom in on a smaller section of the time line was applied.

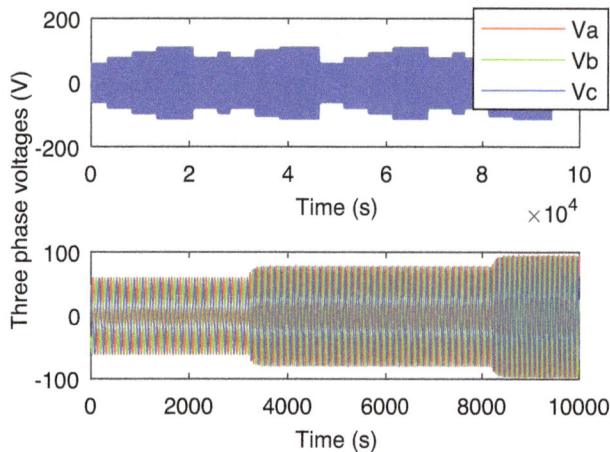

Fig. 6: Generator output three phase voltages.

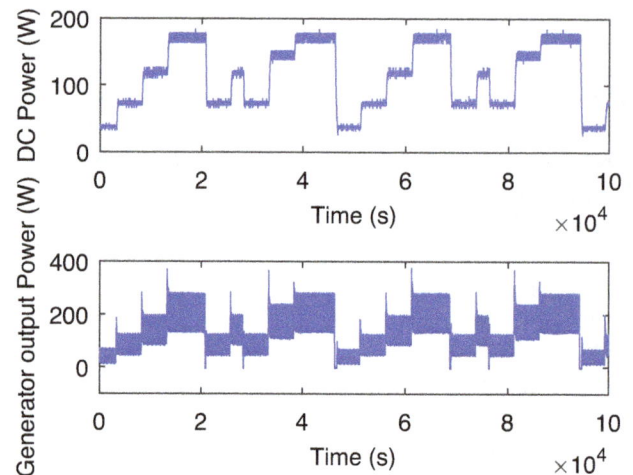

Fig. 8: Power at the DC load (W) and Power at the generator output (W).

The DC voltage and current are presented in Fig. 7 varying with the variable wind speed profile. Values of this two outputs are in accordance with the measurements carried out in real time during the experiments.

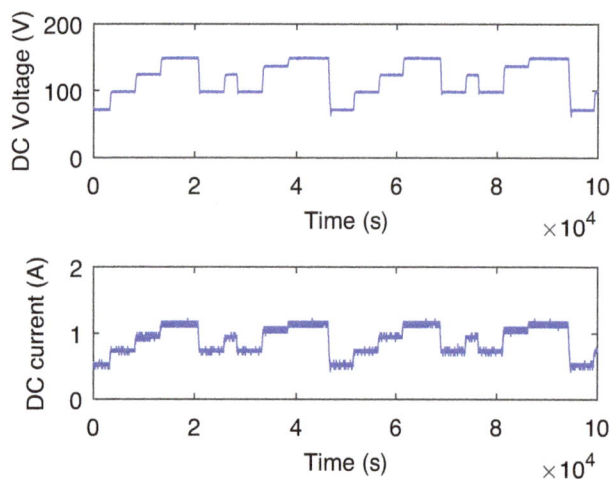

Fig. 7: DC voltage (V) and DC current (A).

Finally, Fig. 8 presents variations of both the DC power at the resistive load and the active generator output power.

5.   Conclusion

In this paper, a sensorless two-level MPPT strategy was applied to a variable speed wind energy conversion system composed of a wind turbine, a PMSG and an active rectifier connected to a DC load. The complete structure is presented in Fig. 1.

An analysis of the generator power maximization was discussed previously justifying the use of the FOC generating the PWM signals to the active rectifier.

The approach that has been used in this work aims to avoid the mechanical sensors for their aforementioned disadvantages, for this reason, an SRF-PLL was employed as a speed and angle estimator which are inputs of the FOC and the MPPT blocks.

The MPPT strategy adopted has two levels; the first level is a power regulation loop generating the value of i_{qref} needed in the FOC and taking as input the value of K_{opt} the coefficient that includes the turbine parameters and estimated by the second level which is a method of the extremum seeking.

An experimental setup with a 5 kW PMSG including a real time digital controller dSPACE 1104 was designed in order to validate the described approach.

From the last section where the experimental results are analysed and shown, it is apparent that the estimated speed by the SRF-PLL follows the real generator speed and then the main aim of employing the SRF-PLL is reached. Also from figures illustrating the power at the DC load and power at the generator output, we can deduce that the MPPT strategy adopted gives optimal extracted power from the studied WECS. As can be seen from the other obtained results, the whole introduced strategy is achievable in real time applications and can be replaced by other MPPT algorithms according to the needs and objectives of the system control.

Future work will investigate the comparison between the different estimators used in our previous works as outlined in the introduction, and the estimator employed in this work.

Acknowledgment

The authors gratefully acknowledge the support of the Energy Systems Research Laboratory, Department of Electrical and Computer Engineering, in Florida International University, Miami, FL 33174 USA, in which the experimental test setup was built and the experimental results were validated.

References

[1] MARUF HOSSAIN, M. and H. ALI. Future research directions for the wind turbine generator system. *Renewable and Sustainable energy reviews*. 2015, vol. 49, iss. 1, pp. 481–489. ISSN 1364-0321. DOI: 10.1016/j.rser.2015.04.126.

[2] LI, S., T. A. HASKEW, R. P. SWATLOSKI and W. GATHINGS. Design, Optimal and direct-current vector control of direct-driven PMSG wind turbines. *IEEE Transactions on Power Electronics*. 2012, vol. 27, iss. 5, pp. 2325–2337. ISSN 0885-8993. DOI: 10.1109/TPEL.2011.2174254.

[3] POLINDER, H., F. F. A. VAN DER PIJL, G. J. DE VILDER and P. J. TAVNER. Comparison of direct-drive and geared generator concepts for wind turbines. *IEEE Transactions on Energy Conversion*. 2006, vol. 21, iss. 3, pp. 725–733. ISSN 0885-8969. DOI: 10.1109/TEC.2006.875476.

[4] MERZOUG, M. S. and F. NACERI. Comparison of Field-Oriented Control and Direct Torque Control for Permanent Magnet Synchronous Motor (PMSM). *International Journal of Electrical, Computer, Energetic, Electronic and Communication Engineering*. 2008, vol. 2, no. 9, pp. 1797–1802. ISSN 5372-7291.

[5] KALA, H. and K. S. SANDHU. Effect of change in power coefficient on the performance of wind turbines with different dimensions. In: *International Conference on Microelectronics, Computing and Communications*. Durgapur: IEEE, 2016, pp. 1–4. ISBN 978-1-4673-6622-9. DOI: 10.1109/MicroCom.2016.7522487.

[6] ECHCHAACHOUAI, A., S. EL HANI, A. HAMMOUCH and S. GUEDIRA. A new Sensorless Maximum Power Point Tracking Technologies od Wind Conversion Chain based on a PMSG. In: *International Renewable and Sustainable Energy Conference*. Ouarzazate: IEEE, 2014, pp. 1–6. ISBN 978-1-4799-7337-8. DOI: 10.1109/IRSEC.2014.7059789.

[7] ECHCHAACHOUAI, A., S. EL HANI, A. HAMMOUCH and S. GUEDIRA. Extended Kalman Filter used to estimate speed rotation for sensorless MPPT of wind conversion chain based on a PMSG. In: *International Conference on Electrical and Information Technologies*. Marrakech: IEEE, 2015, pp. 172–177. ISBN 978-1-4799-7479-5. DOI: 10.1109/EITech.2015.7162989.

[8] TAN, Y., W. MOASE, C. MANZIE, D. NESIC and I. MAREELS. Extremum seeking from 1922 to 2010. In: *Proceedings of the 29th Chinese Control Conference*. Beijing: IEEE, 2010, pp. 14–26. ISBN 978-1-4244-6263-6.

[9] BHENDE, C. N., S. MISHRA and S. G. MALLA. Permanent Magnet Synchronous Generator-Based Standalone Wind Energy Supply System. *IEEE Tansactions on Sustainable Energy*. 2011, vol. 2, iss. 4, pp. 361–373. ISSN 1949-3029. DOI: 10.1109/TSTE.2011.2159253.

[10] KIRAN, Y. and P. S. PUTTASWAMY. Field Oriented Control of a Permanent Magnet Synchronous Motor using a DSP. *International Journal of Advanced Research in Electrical, Electronics and Instrumentation Engineering*. 2014, vol. 3, iss. 10, pp. 12364–12378. ISSN 2320-3765.

[11] YOUNG, K. and R. A. DOUGAL. SRF-PLL with dynamic center frequency for improved phase detection. In: *International Conference on Clean Electrical Power*. Capri: IEEE, 2009, pp. 212–216. ISBN 978-1-4244-2543-3. DOI: 10.1109/ICCEP.2009.5212055.

[12] AUBREE, R., F. AUGER and R. DAI. A new low-cost sensorless MPPT algorithm for small wind turbines. In: *First International Conference on Renewable Energies and Vehicular Technology*. Hammamet: IEEE, 2012, pp. 305–311. ISBN 978-1-4673-1168-7. DOI: 10.1109/REVET.2012.6195288.

About Authors

Amina ECHCHAACHOUAI was born in Morocco in 1990. She received the university degree in Electrical Engineering before being an engineer Networks and Telecommunications from the National School of Applied Sciences, Fes, Morocco. She is currently working towards the Ph.D. degree at The Department of Electrical Engineering in the Mohammed V Souissi University, ENSET-Rabat and her main research interest is in the control of a wind conversion chain based on a PMSG.

Soumia EL HANI Professor of Electrical Machine and drives at the Ecole Normale Superieure de

Enseignement Technique (ENSET), Mohammed V University Rabat, Morocco since 1992. She obtained her engineering degree From Higher school of Mines-Rabat, and her Ph.D. in Automatic from Mohammedia School of Engineering, Rabat Morocco in 2003. Her research interests are in the area of Robust Control, Monitoring and Diagnosis of Electromechanical Systems. She in charge of the research team 'Energy Optimisation, Diagnosis and Control' EODIC, research Laboratory in Energy, Electrotechnic, Robotic and Automatic. She is Author of several publications in the field of electrical engineering, including robust control systems, diagnosis and control systems of wind electric conversion. She has supervised Ph.D. and Masters Theses dealing with different research topics concerned with her research interests. Soumia El Hani is Prof. Dr. IEEE senior member, she was a member of the Organizing and the Scientific Committees of several international conferences dealing with topics related to Renewable Energy, Electrical Machines and Drives. Since the year 2015, Soumia El Hani is the co-founder and general co- chair of 'The International Conference on Electrical and Information Technologies'. Also, she is the co-editor of ICEIT 2015 and ICEIT 2016 Proceedings.

Ahmed HAMMOUCH received the master degree and the Ph.D. in Automatic, Electrical, Electronic by the Haute Alsace University of Mulhouse (France) in 1993 and the Ph.D. in Signal and Image Processing by the Mohammed V University of Rabat in 2004. From 1993 to 2013 he was a professor in the Mohammed V University in Morocco. Since 2009 he manages the Research Laboratory in Electronic Engineering. He is an author of several papers in international journals and conferences. His domains of interest include multimedia data processing and telecommunications. He is with National Center for Scientific and Technical Research in Rabat.

Imad ABOUDRAR was born in Agadir, Morocco. He received the M.Sc. degree in electrical engineering in 2016 from the Mohammed V University, Rabat Morocco- where he is currently working toward the Ph.D. degree in the department of electrical engineering. Since 2016, His research interests are related to renewable energy, His current activities include the improvement of energy quality of an integrated PV and Wind Hybrid System connected to the grid.

Appendix A
Experimentation PMSG Parameters

- Output power = 0–5 kW,
- Rated voltage = 208 Vrms,
- Rated speed = 1200 rpm,
- Resistance = 0.46 Ω,
- Inductance = 4.86 mH.

18

Ground Receiving Station Reference Pair Selection Technique for a Minimum Configuration 3D Emitter Position Estimation Multilateration System

Abdulmalik Shehu YARO[1,2], Ahmad Zuri SHA'AMERI[1], Nidal KAMEL[3]

[1]Department of Electronic and Computer Engineering, Faculty of Electrical Engineering,
Universiti Teknologi Malaysia, UTM Johor Bahru, 81310 Johor, Malaysia
[2]Department of Electrical and Computer Engineering, Faculty of Engineering,
Ahmadu Bello University, Sokoto Road, PMB 06 Zaria, Nigeria
[3]Department of Electrical and Electronic Engineering, Faculty of Engineering,
Universiti Teknologi Petronas, 32610 Seri Iskandar, Malaysia

yaroabdulmalik@yahoo.com, zuri@fke.utm.my, nidalkamel2@hotmail.com

Abstract. *Multilateration estimates aircraft position using the Time Difference Of Arrival (TDOA) with a lateration algorithm. The Position Estimation (PE) accuracy of the lateration algorithm depends on several factors which are the TDOA estimation error, the lateration algorithm approach, the number of deployed GRSs and the selection of the GRS reference used for the PE process. Using the minimum number of GRSs for 3D emitter PE, a technique based on the condition number calculation is proposed to select the suitable GRS reference pair for improving the accuracy of the PE using the lateration algorithm. Validation of the proposed technique was performed with the GRSs in the square and triangular GRS configuration. For the selected emitter positions, the result shows that the proposed technique can be used to select the suitable GRS reference pair for the PE process. A unity condition number is achieved for GRS pair most suitable for the PE process. Monte Carlo simulation result, in comparison with the fixed GRS reference pair lateration algorithm, shows a reduction in PE error of at least 70 % for both GRS in the square and triangular configuration.*

Keywords

Condition number, lateration algorithm, minimum configuration, multilateration, reference selection.

1. Introduction

Passive wireless positioning and navigation systems utilize aircraft transponder emission which is detected with the support of antenna of a Ground Receiving Station (GRS) for determining the position of the aircraft. Determining the position of an aircraft is a two-stage process [1]. The first stage involves the estimation of the position dependent signal parameter from the received aircraft transponder emission. Some examples of position dependent signal parameters are the Angle Of Arrival (AOA), the Time Of Arrival (TOA), the Time Difference Of Arrival (TDOA) and Receive Signal Strength (RSS). In the second stage, the position dependent signal parameter estimated at the first stage is input into a Position Estimation (PE) algorithm to determine the position of the aircraft. This is known as the PE process. Examples of the PE algorithm used at the PE process are angulation, fingerprinting and lateration. Multilateration system is an example of a wireless positioning system. The system estimates TDOA from the received aircraft transponder emission as its position dependent signal parameter and uses the lateration algorithm to determine the aircraft position [1], [2] and [3]. It consists of several specially placed GRSs all connected to a central processing unit. Depending on the number of GRS deployed, 2-Dimension (2D) or 3-Dimension (3D) position of the aircraft is resolved. For 3D PE, there is a minimum of four GRSs needed [3]. Many studies have described methods for TDOA estimation [4], [5], [6], [7] and [8]. For example, [4] compares TDOA estimation using cross-correlation and fast cross-correlation to determine which method is faster for practical and theoretical implementation. In [8], Signal-to-Noise Ratio (SNR) is used for the benchmark to compare the performance of five different TDOA estimation techniques.

The PE process is the scope of this work. Depending on the number of GRS (N), $N-1$ nonlinear hyperbolic equations are generated [9]. Several approaches developing the lateration algorithms have been proposed in articles [9], [10], [11], [12], [13], [14] and [15] which can be grouped into two as a linear and nonlinear approach [1] and [11]. The non-linear approach involves the use of linear approximation and iterative methods such as Taylor's series expansion to perform PE [9], [10] and [11]. The linear algorithm involves algebraically manipulating the hyperbolic equations to directly set an inverse problem that linearly relates the unknown aircraft position to the known TDOA measurements as described in [12], [13], [14] and [15]. Due to the convergence issue and the use of initial position estimated for the non-linear approach [10], this study focuses on using the linear approach to developing the lateration algorithm.

The linear lateration algorithm has been characterized with high PE error. Numerous researchers have proposed techniques such as weighting functions [16], total least squares [17] and Tikhonov regularization [9] for improving the PE accuracy of the lateration algorithm. These techniques efficiently to use but require at least five GRSs to be deployed. The minimum GRS deployed for 3D PE is four. Thus, these techniques cannot be used to improve PE accuracy of the lateration algorithm for 3D minimum configuration. The use of more than one GRS as a reference is suggested for improving the accuracy of the lateration algorithm [6] and [18]. The choice of the reference GRS has been reported to improve the PE accuracy of the lateration algorithm [19], [20] and [21]. In [20], a TDOA residual-based method was proposed to select the suitable GRS as a reference for PE with lateration algorithm in an active system. It was assumed that the emitter position is known but there was the need to continuously track the position of the system using another system. Using the known emitter position, each of the deployed GRS is used as a reference and the GRS that resulted in the least TDOA residual is chosen as a reference for subsequent estimation of the emitter position. An SNR based GRS reference selection method was proposed in [21]. With the assumption that all noise power at the GRS remains constant, the GRS with the highest received SNR is the closest to the emitter and it is the most suitable GRS to be used as a reference for the PE with the lateration algorithm. Using GRS pair as a reference for the lateration algorithm, this study suggests a technique to choose the suitable GRS reference pair for the lateration algorithm. The lateration algorithm

considered is for the 3D minimum configuration multilateration system. The suggested technique involves calculating the condition number of a derived matrix and choosing the GRS pair with the least condition number from the derived matrix. The proposed GRS reference pair selection technique is validated by comparison with the fixed GRS reference pair lateration algorithm, used in [12], with selected emitter position and the GRSs deployed in square and triangular configurations.

The rest of the paper is organized as follows. Section 2. presents the multilateration PE methodology and PE condition number analysis. The proposed GRS reference pair selection technique is described in Sec. 3. The result and discussion are presented in Sec. 4. followed by the conclusion in Sec. 5.

2. Multilateration PE Methodology and Condition Number Analysis

This section describes the variable GRS reference pair PE lateration algorithm followed by the condition number analysis of the multilateration PE mathematical model for different GRS reference pair.

2.1. Variable GRS Reference Pair Multilateration PE Methodology

Let $\mathbf{x} = (x, y, z)$ be the coordinate of a stationary emitter in 3D Euclidean space and $S_i = (x_i, z_i, z_i)$. the coordinate of the i-th GRS The distance travelled by the electromagnetic emission from the emitter position to the i-th GRS is calculated as:

$$d_i = c \cdot \tau_i = \sqrt{(x - x_i)^2 + (y - y_i)^2 + (z - z_i)^2}, \quad (1)$$

where $c = 3 \cdot 10^8$ m·s^{-1} is the speed of light and τ_i is the propagation time of the signal from the emitter to the i-th GRS.

The Path Difference (PD) between i-th and m-th GRS pair is obtained as:

$$d_{i,m} = \sqrt{(x - x_i)^2 + (y - y_i)^2 + (z - z_i)^2} - \\ + \sqrt{(x - x_m)^2 + (y - y_m)^2 + (z - z_m)^2}. \quad (2)$$

GRS pair is used as a reference for the lateration algorithm. Let the i-th and j-th GRSs to be chosen as reference pair with coordinates (x_i, j_i, k_i) and (x_j, j_j, k_j) respectively while the non-reference GRSs are labelled the m-th and n-th with coordinates

(x_m, j_m, k_m) and (x_n, j_n, k_n) respectively. Using the i-th GRS as a reference, two independent PD equations are obtained [12] and expressed in the following:

$$d_{i,n} = d_i - d_n, \qquad (3)$$

$$d_{i,m} = d_i - d_m, \qquad (4)$$

while for the j-th reference GRS, the two independent PD equations are obtained as:

$$d_{j,n} = d_j - d_n, \qquad (5)$$

$$d_{j,m} = d_j - d_m. \qquad (6)$$

Combining Eq. (3) and Eq. (4) after further simplification results into 3D plane equation which is presented in [12] as follows:

$$A_{i,n,m} = xB_{i,n,m} + yC_{i,n,m} + zD_{i,n,m}, \qquad (7)$$

where the coefficients of Eq. (7) are:

$$A_{i,n,m} = 0.5 \left(d_{i,m} - d_{i,n} + \frac{k_{i,m}}{d_{i,m}} - \frac{k_{i,m}}{d_{i,n}} \right), \qquad (8)$$

$$B_{i,n,m} = \frac{X_{n,i}}{d_{i,n}} - \frac{X_{m,i}}{d_{i,m}}, \qquad (9)$$

$$C_{i,n,m} = \frac{Y_{n,i}}{d_{i,n}} - \frac{Y_{m,i}}{d_{i,m}}, \qquad (10)$$

$$D_{i,n,m} = \frac{Z_{n,i}}{d_{i,n}} - \frac{Z_{m,i}}{d_{i,m}}, \qquad (11)$$

$$k_{i,w} = \left(x_i^2 + y_i^2 + z_i^2 \right) - \left(x_w^2 + y_w^2 + z_w^2 \right), \qquad (12)$$

$$X_{i,w} = x_i - x_w, Y_{i,w} = y_i - y_w, \qquad (13)$$

$$Z_{i,w} = z_i - z_w, w \in [m, n]. \qquad (14)$$

In addition, combining Eq. (5) and Eq. (6) after further simplification results into another 3D plane equation as follows:

$$A_{j,n,m} = xB_{j,n,m} + yC_{j,n,m} + zD_{j,n,m}, \qquad (15)$$

where the coefficients of Eq. (15) are:

$$A_{j,n,m} = 0.5 \left(d_{j,m} - d_{j,n} + \frac{k_{j,m}}{d_{j,m}} - \frac{k_{j,m}}{d_{j,n}} \right), \qquad (16)$$

$$B_{j,n,m} = \frac{X_{n,j}}{d_{j,n}} - \frac{X_{m,j}}{d_{j,m}}, \qquad (17)$$

$$C_{j,n,m} = \frac{Y_{n,j}}{d_{j,n}} - \frac{Y_{m,j}}{d_{j,m}}, \qquad (18)$$

$$D_{j,n,m} = \frac{Z_{n,j}}{d_{j,n}} - \frac{Z_{m,j}}{d_{j,m}}, \qquad (19)$$

$$k_{j,w} = \left(x_j^2 + y_j^2 + z_j^2 \right) - \left(x_w^2 + y_w^2 + z_w^2 \right), \qquad (20)$$

$$X_{j,w} = x_j - x_w, Y_{j,w} = y_j - y_w, \qquad (21)$$

$$Z_{j,w} = z_j - z_w, w \in [m, n]. \qquad (22)$$

Equation (7) and Eq. (15) when represented in matrix form is:

$$\begin{bmatrix} B_{i,n,m} & C_{i,n,m} & D_{i,n,m} \\ B_{j,n,m} & J_{j,n,m} & D_{j,n,m} \end{bmatrix} \cdot \begin{bmatrix} x \\ y \\ z \end{bmatrix} = \begin{bmatrix} A_{i,n,m} \\ A_{j,n,m} \end{bmatrix}, \qquad (23)$$

$$\mathbf{Q}_{ij} \cdot \mathbf{x} = \mathbf{a}_{ij}. \qquad (24)$$

Equation (23) is known as the multilateration 3D PE mathematical model for minimum GRS configuration. The subscript "i, n, m" and "j, n, m" for the entries of the matrices $\mathbf{Q}_{ij}$ and $\mathbf{a}_{ij}$ indicate that the entry is obtained using the i-th and j-th GRS as a reference respectively with the m-th and n-th GRS as non-reference. The location of the aircraft (x, y, z) is obtained by finding the inverse matrix solution of Eq. (23) with TDOA or PD measurements and GRSs coordinates as input.

2.2. Multilateration PE Mathematical Condition Number Analysis

In the practical application, the PD measurements are obtained with errors which affect the solution obtained using matrix Eq. (23). The effect of the PD measurement error on the solution of matrix Eq. (23) is determined by the sensitivity of matrix $\mathbf{Q}_{ij}$ defined by the condition number value. The condition number of a square matrix indicates on how the error in the input variables is amplified to the solution obtained using the system. Matrix $\mathbf{Q}_{ij}$ in Eq. (24) is a rectangular matrix whose condition number cannot be determined. Assuming that the GRSs have insignificant height difference, that is:

$$Z_{i,w} = z_i - z_w \approx 0, \qquad (25)$$

$$Z_{j,w} = z_j - z_w \approx 0, \qquad (26)$$

Tab. 1: Matrix $\mathbf{A}_{ij}$ condition number for different GRS pair as reference. Yellow shade indicates the GRS pair with the least $K(\mathbf{A}_{ij})$.

Range (km)	Bearing (°)	Altitude (km)	GRS reference pair condition number					
			$i=1$ & $j=2$	$i=1$ & $j=3$	$i=1$ & $j=4$	$i=2$ & $j=3$	$i=2$ & $j=4$	$i=3$ & $j=4$
5	60	1	17	21	4	19	10	30
50		1	87	59	17	64	50	99
5	120	7	37	21	31	8	30	49
50		7	89	50	66	17	60	101

for all $1 \leq i \leq 4$, $1 \leq j \leq 4$, $1 \leq w \leq 4$ and $i \neq j \neq w$, matrix $\mathbf{Q}_{ij}$ can be reduced to a square matrix written as:

$$\mathbf{A}_{ij} = \begin{bmatrix} B_{i,n,m} & C_{i,n,m} \\ B_{j,n,m} & C_{j,n,m} \end{bmatrix}, \quad (27)$$

where $D_{i,n,m} = D_{j,n,m} = 0$.

Using the matrix $\mathbf{A}_{ij}$, the condition number can be obtained and used in determining the effect of PD measurement error on the PE accuracy of the lateration algorithm. The condition number of matrix $\mathbf{A}_{ij}$ in Eq. (27) denoted as $K(\mathbf{A}_{ij})$ is obtained as:

$$K(\mathbf{A}_{ij}) = \|\mathbf{A}_{ij}\|_2 \cdot \|\mathbf{A}_{ij}^{-1}\|_2, \quad (28)$$

where $\|\mathbf{A}_{ij}\|_2$ and $\|\mathbf{A}_{ij}^{-1}\|_2$ are the 2-norm of matrix $\mathbf{A}_{ij}$ and its inverse respectively.

The 2-norm of the matrix $\mathbf{A}_{ij}$ and its inverse are defined with respect to entries in Eq. (27) which have been expressed in [22].

$$\|\mathbf{A}_{ij}\|_2 = \\ = \sqrt{|B_{i,m,n}|^2 + |C_{i,m,n}|^2 + |B_{j,m,n}|^2 + |C_{j,m,n}|^2}, \quad (29)$$

$$\|\mathbf{A}_{ij}^{-1}\|_2 = \\ = \frac{\sqrt{|B_{i,m,n}|^2 + |C_{i,m,n}|^2 + |B_{j,m,n}|^2 + |C_{j,m,n}|^2}}{\det(\mathbf{A}_{ij})}, \quad (30)$$

where $\det(\mathbf{A}_{ij})$ is a determinant of matrix $\mathbf{A}_{ij}$ expressed mathematically as:

$$\det \mathbf{A}_{ij} = (B_{i,m,n} \cdot C_{j,m,n}) - (B_{j,m,n} \cdot C_{i,m,n}). \quad (31)$$

Substituting Eq. (29) and Eq. (30) into Eq. (28), the condition number the matrix $\mathbf{A}_{ij}$ as function of its entries can be written as Eq. (32).

Equation (32) represents the condition number of matrix $\mathbf{A}_{ij}$ in Eq. (27) whose entries are obtained using the i-th and j-th GRS pair as a reference with the m-th and n-th as non-reference GRSs. For an emitter at a stationary position with a fixed GRS configuration, different GRS pair (i, j) will produce different entries of matrix $\mathbf{A}_{ij}$. This will result in different condition

number value in Eq. (32). Higher condition number values indicate greater error in the solution obtained using Eq. (23). Table 1 shows the condition number of the matrix $\mathbf{A}_{ij}$ using Eq. (32) for different GRS pair (i, j) at four emitter positions with GRS in the square configuration. Emitter positions are given in cylindrical coordinate system. The condition number differs for different emitter positions and GRS reference pairs. At fixed emitter position, different GRS pair produces different condition numbers. At emitter position (5 km, 60°, 1 km), GRS pair $i = 1$ and $j = 4$ has the least condition number value while GRS pair $i = 3$ and $j = 4$ has the highest condition number value. At emitter position (50 km, 120°, 7 km), GRS pair $i = 2$ and $j = 3$ has the least condition number value while $i = 3$ and $j = 4$ has the highest condition number value. For each emitter position, the pair with the least condition number value, used as a reference for the lateration algorithm, will result in the emitter position estimated with the least error. This means that for emitter positions (5 km, 60°, 1 km) and (50 km, 60°, 1 m), the suitable GRS pair as a reference are the $i = 1$ and $j = 4$. For emitter positions (5 km, 120°, 7 km) and (50 km, 120°, 7 km), the suitable GRS pairs as a reference are $i = 2$ and $j = 3$.

GRS reference selection for PE is carried out prior to the actual PE process. The available parameters related to the emitter position which can be used for selecting the suitable GRS pair as a reference are the PD measurements only. Thus, the condition number obtained from matrix $\mathbf{A}_{ij}$ in Eq. (27) cannot be used since it is a function of both PD measurements and GRS coordinate. In next section, the approach is developed to determine the suitable GRS reference pair to be selected for the PE process.

3. Proposed GRS Reference Pair Selection Technique

In this section, the technique for selection of the suitable GRS reference pair for the PE process is presented. In Subsec. 2.2. , it was concluded that using matrix $\mathbf{A}_{ij}$ to determine the suitable GRS pair as a reference for PE is not possible. Matrix $\mathbf{A}_{ij}$ can

$$K(\mathbf{A}_{ij}) = \frac{\left(|B_{i,m,m}|^2 + |C_{i,m,m}|^2 + |B_{j,m,m}|^2 + |C_{j,m,m}|^2\right)}{(B_{i,m,m} \cdot C_{j,m,m}) - (B_{j,m,m} \cdot C_{i,m,m})}. \qquad (32)$$

$$\mathbf{A}_{ij} = \begin{bmatrix} \left(\dfrac{X_{m,i}}{d_{i,m}} - \dfrac{X_{n,i}}{d_{i,n}}\right) & \left(\dfrac{Y_{m,i}}{d_{i,m}} - \dfrac{Y_{n,i}}{d_{i,n}}\right) \\ \left(\dfrac{X_{m,j}}{d_{j,m}} - \dfrac{X_{n,j}}{d_{j,n}}\right) & \left(\dfrac{Y_{m,j}}{d_{j,m}} - \dfrac{Y_{n,j}}{d_{j,n}}\right) \end{bmatrix} = \begin{bmatrix} \dfrac{(X_{m,i}d_{i,n} - X_{n,i}d_{i,m})}{d_{i,m}d_{i,n}} & \dfrac{(Y_{m,i}d_{i,n} - Y_{n,i}d_{i,m})}{d_{i,m}d_{i,n}} \\ \dfrac{(X_{m,j}d_{j,n} - X_{n,j}d_{j,m})}{d_{j,m}d_{j,n}} & \dfrac{(Y_{m,j}d_{j,n} - Y_{n,j}d_{j,m})}{d_{j,m}d_{j,n}} \end{bmatrix} =$$

$$= \begin{bmatrix} (d_{i,m} \cdot d_{i,n})^{-1} & 0 \\ 0 & (d_{j,m} \cdot d_{j,n})^{-1} \end{bmatrix} \cdot \begin{bmatrix} (X_{m,i}d_{i,n} - X_{n,i}d_{i,m}) & (Y_{m,i}d_{i,n} - Y_{n,i}d_{i,m}) \\ (X_{m,j}d_{j,n} - X_{n,j}d_{j,m}) & (Y_{m,j}d_{j,n} - Y_{n,j}d_{j,m}) \end{bmatrix}, \qquad (33)$$

be split into two matrices while one of the matrices is having only the PD measurements as its entries. From Eq. (27), the matrix $\mathbf{A}_{ij}$ is written as Eq. (33). Let

$$\mathbf{M}_{ij} = \begin{bmatrix} (d_{i,m} \cdot d_{i,n})^{-1} & 0 \\ 0 & (d_{j,m} \cdot d_{j,n})^{-1} \end{bmatrix}. \qquad (34)$$

$$\mathbf{N}_{ij} = \begin{bmatrix} (X_{m,i}d_{i,n} - X_{n,i}d_{i,m}) & (Y_{m,i}d_{i,n} - Y_{n,i}d_{i,m}) \\ (X_{m,j}d_{j,n} - X_{n,j}d_{j,m}) & (Y_{m,j}d_{j,n} - Y_{n,j}d_{j,m}) \end{bmatrix}. \qquad (35)$$

Then

$$\mathbf{A}_{ij} = \mathbf{M}_{ij} \cdot \mathbf{N}_{ij}. \qquad (36)$$

Matrix $\mathbf{M}_{ij}$ and $\mathbf{N}_{ij}$ are both square matrices. The matrix $\mathbf{M}_{ij}$ is having only PD measurements obtained using any possible GRS pair (i, j) as its entries. This matrix can be used instead of matrix $\mathbf{A}_{ij}$ for condition number calculation to determine the suitable GRS pair as a reference for the PE process. The condition number of matrix $\mathbf{M}_{ij}$ as a function of the PD measurement is obtained as Eq. (37).

Further simplification of Eq. (37) will result in:

$$K(\mathbf{M}_{ij}) = \left(\frac{d_{j,m} \cdot d_{j,n}}{d_{i,m} \cdot d_{i,n}}\right) + \left(\frac{d_{i,m} \cdot d_{i,n}}{d_{j,m} \cdot d_{j,n}}\right). \qquad (38)$$

Using Eq. (38), the condition numbers for all the possible GRS pairs are obtained. The pair with the least condition number is chosen as a reference for the PE process with the lateration algorithm. The summary of the procedure for selection of GRS reference pair for four numbers of GRSs is described as follows:

- Obtain the PD measurement set using Eq. (39) for each of the possible GRS pair (i, j) as references as shown below.

$$\mathbf{d}_{i,j,m,n} = [d_{i,m}, d_{i,n}, d_{j,m}, d_{j,n}]. \qquad (39)$$

- Using the PD measurement set from (i) for each GRS pair, substitute into Eq. (38) and solve for $K(\mathbf{M}_{ij})$.

- Choose the GRS pair with the least $K(\mathbf{M}_{ij})$ value from step (ii) as the reference pair for the PE process with the lateration algorithm.

4.   Results and Discussion

In this section, the technique for the selecting the suitable GRS reference pair based on the condition number which is calculated using Eq. (38) for the PE using the lateration algorithm is validated. This is done with a comparison of the condition number which has been obtained using Eq. (38) with the support of Eq. (32). Validation is carried out for some selected emitter positions with GRSs in the square and triangular configuration as shown in Fig. 1. It has been established

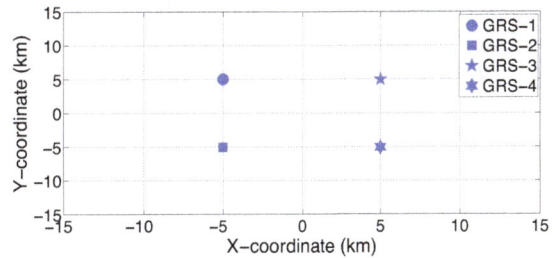

(a) Square GRS configuration.

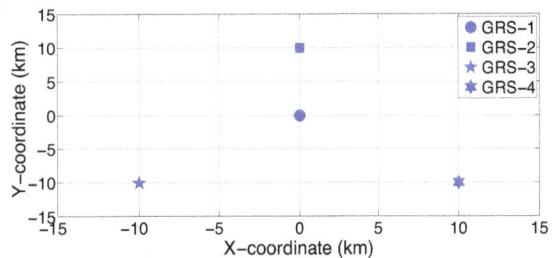

(b) Triangular GRS configuration.

Fig. 1: Square and triangular GRS configuration with GRS separation of 10 km.

$$KM_{ij} = \frac{1}{\det(\mathbf{M}_{ij})} \cdot \|\mathbf{M}_{ij}\|_2^2 = \frac{\left\| \begin{bmatrix} (d_{i,m} \cdot d_{i,n})^{-1} & 0 \\ 0 & (d_{j,m} \cdot d_{j,n})^{-1} \end{bmatrix} \right\|_2^2}{\det\left(\begin{bmatrix} (d_{i,m} \cdot d_{i,n})^{-1} & 0 \\ 0 & (d_{j,m} \cdot d_{j,n})^{-1} \end{bmatrix} \right)} = \tag{37}$$

$$= \frac{\left(\sqrt{(d_{i,m} \cdot d_{i,n})^{-2} + (d_{j,m} \cdot d_{j,n})^{-2}} \right)^2}{(d_{i,m} \cdot d_{i,n} \cdot d_{j,m} \cdot d_{j,n})^{-1}} = (d_{i,m} \cdot d_{i,n} \cdot d_{j,m} \cdot d_{j,n}) \cdot \left(\frac{1}{(d_{i,m} \cdot d_{i,n})^2} + \frac{1}{(d_{j,m} \cdot d_{j,n})^2} \right).$$

that the GRS pair with the least condition number is the suitable GRS pair as a reference for the lateration algorithm.

For each of the GRS configurations, the validation of the proposed GRS reference pair selection technique is carried out for the emitter positions which are defined in Tab. 2.

Tab. 2: Emitter positions for validation.

No.	Emitter position	Range (km)	Altitude (km)	Bearing (°)
1	A			30
2	B	5	7	120
3	C			220
4	D			320

Table 3 shows the condition number comparison of the matrix $\mathbf{A}_{ij}$ in Eq. (27) and $\mathbf{M}_{ij}$ in Eq. (34) using Eq. (32) and Eq. (37) respectively for the square GRS configuration. For the selected emitter positions considered, it is seen that the GRS pair with the least KA_{ij} also has the least KM_{ij}. At emitter position A, the GRS pair with the least $KA_{ij} = 8$ and least $KM_{ij} = 1$ is the pair $i = 1$ and $j = 4$. At emitter location B, the GRS pair with the least $\mathbf{A}_{ij} = 8$ and least $KM_{ij} = 1$ is the pair $i = 2$ and $j = 3$. It is also seen that the GRS pair with the least $\mathbf{A}_{ij}$ will have $KM_{ij} = 1$. This means that the GRS pair suitable as a reference for PE process with the lateration algorithm at any given emitter position will have $KM_{ij} = 1$.

Table 4 shows the condition number comparison of matrix $\mathbf{A}_{ij}$ in Eq. (27) and matrix $\mathbf{M}_{ij}$ in Eq. (34) using Eq. (32) and Eq. (37) respectively for the triangular GRS configuration. The same conclusion for the square GRS configuration is deduced for the triangular configuration. Even thought, at the emitter positions A and D the least $K(\mathbf{A}_{ij})$ is obtained by more than one GRS reference pair. One of the GRS pairs was chosen as the most suitable for the PE process of the emitter at the selected position which has $K(\mathbf{M}_{ij}) = 1$.

4.1.   PE Accuracy Improvement

In this section, the PE accuracy of the lateration algorithm with the proposed reference selection technique in Sec. 3. is compared with the fixed GRS reference pair approach (GRS 1 and GRS 2) used in [12]. PE Root Mean Square Error (RMSE) is used as the performance measure for the comparison. Mathematically, the PE RMSE is obtained as:

$$PE_{rmse} = \\ = \sqrt{\frac{\sum\limits_{i=1}^{N} \left[(\hat{x} - x)^2 + (\hat{y} - y)^2 + (\hat{z} - z)^2 \right]}{N}}, \tag{40}$$

where (x, y, z) are the known emitter coordinates and $(\hat{x}_i, \hat{y}_i, \hat{z}_i)$ are the estimated emitter coordinates at the i-th Monte Carlo simulation realization. The Monte Carlo simulation results were obtained after 500 realizations. The PD Estimation (PDE) error was modelled as $N(0, \sigma^2)$ and it was assumed to be the same at all the spatially placed GRSs.

By varying the PDE error standard deviation from 0 m to 2 m, the PE RMSE of the lateration algorithm with the proposed reference selection technique and that of the fixed GRS reference pair were obtained and compared. Figure 2 and Fig. 3 show the PE RMSE comparison between two approaches for emitter at position B using the square and triangular configuration respectively. The PE RMSE increases with increase in the PDE error standard deviation from 0 m to 2 m. Comparison between the PE RMSE of the lateration algorithm with the proposed reference selection technique for both square and triangular configuration shows that there is an improvement in the PE accuracy by the reduction in the PE RMSE. From Fig. 2, at PDE error standard deviation of 1 m, the PE RMSE of the lateration algorithm with the proposed technique is 6.25 m and that of using the fixed GRS reference pair is 21.78 m. This means a reduction in the PE RMSE of about 15.53 m (~ 71 %) was achieved with the proposed technique at emitter position A with the GRS in the square configuration. Extending the analysis to the triangular configuration, at PDE error standard devi-

Tab. 3: Square GRS configuration condition number comparison. Yellow shade indicates the GRS pair with the least $K\mathbf{A}_{ij}$ value while green shade indicates the GRS pair with the least $K\mathbf{M}_{ij}$ value.

Emitter position		GRS reference pair					
		$i=1$ & $j=2$	$i=1$ & $j=3$	$i=1$ & $j=4$	$i=2$ & $j=3$	$i=2$ & $j=4$	$i=3$ & $j=4$
A	$K(\mathbf{A}_{ij})$	21	49	8	31	37	30
	$K(\mathbf{M}_{ij})$	3	5	1	2	3	6
B	$K(\mathbf{A}_{ij})$	37	21	31	8	30	49
	$K(\mathbf{M}_{ij})$	3	3	2	1	6	5
C	$K(\mathbf{A}_{ij})$	85	73	7	80	98	60
	$K(\mathbf{M}_{ij})$	17	9	1	2	16	9
D	$K(\mathbf{A}_{ij})$	85	98	80	7	73	60
	$K(\mathbf{M}_{ij})$	17	16	2	1	9	9

Tab. 4: Triangular GRS configuration condition number comparison. Yellow shade indicates the GRS pair with the least $K(\mathbf{A}_i j)$ value while green shade indicates the GRS pair with the least $K(\mathbf{M}_i j)$ value.

Emitter position		GRS reference pair					
		$i=1$ & $j=2$	$i=1$ & $j=3$	$i=1$ & $j=4$	$i=2$ & $j=3$	$i=2$ & $j=4$	$i=3$ & $j=4$
A	$K(\mathbf{A}_i j)$	14	2	2	2	2	15
	$K(\mathbf{M}_{ij})$	2	4	1	5	3	2
B	$K(\mathbf{A}_i j)$	3	3	8	5	1	2
	$K(\mathbf{M}_{ij})$	3	2	3	3	1	2
C	$K(\mathbf{A}_i j)$	13	44	13	8	4	7
	$K(\mathbf{M}_{ij})$	15	3	13	5	1	5
D	$K(\mathbf{A}_i j)$	10	1	2	1	4	10
	$K(\mathbf{M}_{ij})$	2	1	3	1	2	2

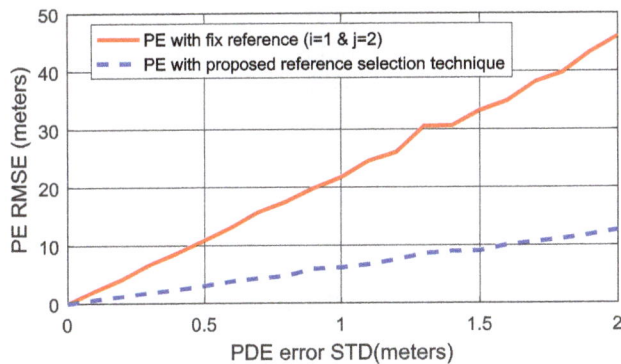

Fig. 2: PE RMSE comparison with the square GRS configuration.

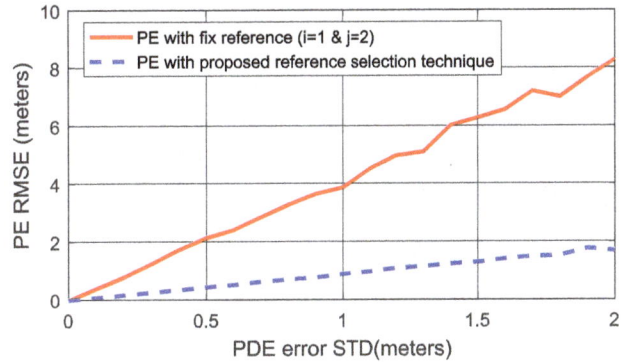

Fig. 3: PE RMSE comparison with the triangular GRS configuration.

ation of 1 m, the reduction in PE RMSE of about 3 m ($\sim 77\,\%$) was obtained.

Furthermore, comparing the PE RMSE for the square and triangular GRS configuration, the triangular GRS configuration resulted in the least PE RMSE. This is due to the low condition number values obtained with the triangular configuration as shown in Tab. 4 compared to the square configuration as shown in Tab. 3. Thus, the triangular GRS configuration will result in higher PE accuracy compared to the square GRS configuration.

5. Conclusion

This research has accomplished a method to select the suitable GRS reference pair which is to be used for improving PE accuracy of the lateration algorithm for a minimum configuration 3D multilateration system. The technique was validated by condition number calculation and PE RMSE estimation comparison with a fixed GRS reference pair lateration algorithm. Condition number calculation results indicate that the most suitable GRS pair, used as a reference for the lateration algorithm, has the least condition number value. PE RMSE Monte Carlo simulation results comparison shows that the proposed reference selection technique improved the PE accuracy of the lateration algorithm by a reduction in the PE RMSE of at least 70 % for both square and triangular GRS configuration. Further work will focus on the extension of the technique for more than 4 GRSs.

Acknowledgment

The authors thank Universiti Teknologi Malaysia (UTM) for providing the resources and support for this research.

References

[1] FALLETTI, E., M. LUISE and D. DARDARI. *Satellite and terrestrial radio positioning techniques: A signal processing perspective.* 1st ed. Amsterdam: Academic Press, 2011. ISBN 978-0-123-82085-3.

[2] PETROCHILOS, N., G. GALATI and E. PIRACCI. Separation of SSR Signals by Array Processing in Multilateration Systems Sign In or Purchase. *IEEE Transactions on Aerospace and Electronic Systems.* 2009, vol. 45, iss. 3, pp. 965–982. ISSN 0018-9251. DOI: 10.1109/TAES.2009.5259177.

[3] NEVEN, W. H. L., T. J. QUILTER, R. WEEDON and R. A. HOGENDOORN. Wide area multilateration. In: *Eurocontrol* [online]. 2005. Available at: https://www.eurocontrol.int/sites/default/files/publication/files/surveilllance-report-wide-area- multilateration-200508.pdf.

[4] YAN, H. and W. LIU. Design of time difference of arrival estimation system based on fast cross correlation. In: *2nd International Conference on Future Computer and Communication (ICFCC).* Wuha: IEEE, 2010, pp. V2-464–V2-466. ISBN 978-1-4244-5824-0. DOI: 10.1109/ICFCC.2010.5497484.

[5] DOU, H., Q. LEI, W. LI and Q. XING. A new TDOA estimation method in Three-satellite interference localisation. *International Journal of Electronics.* 2015, vol. 102, iss. 5, pp. 839–854. ISSN 1362-3060. DOI: 10.1080/00207217.2014.942886.

[6] MARMAROLI, P., X. FALOURD and H. LISSEK. A Comparative Study of Time Delay Estimation Techniques for Road Vehicle Tracking. In: *11th French Congress of Acoustics and 2012 Annual IOA Meeting.* Nantes: IEEE, 2012, pp. 4135–4140. Available at: https://hal.archives-ouvertes.fr/hal-00810981/document.

[7] KNAPP, C. and G. CARTER. The generalized correlation method for estimation of time delay. *IEEE Transactions on Acoustics, Speech, and Signal Processing.* 1976, vol. 24, iss. 4, pp. 320–327. ISSN 0096-3518. DOI: 10.1109/TASSP.1976.1162830.

[8] ZHANG, Y. and W. H. ABDULLA. A comparative study of time-delay estimation techniques using microphone arrays: School of Engineering Report no. 619. In: *Semanticscholar* [online]. 2005. Available at: https://pdfs.semanticscholar.org/f32c/b5fb2d0205d7c9b35e8d48edeef6c3d354db.pdf

[9] MANTILLA-GAVIRIA, I. A., M. LEONARDI, G. GALATI and J. V. BALBASTRE-TEJEDOR. Time-difference-of-arrival regularised location estimator for multilateration systems. *IET Radar, Sonar & Navigation.* 2014, vol. 8, iss. 5, pp. 479–489. ISSN 1751-8792. DOI: 10.1049/iet-rsn.2013.0151.

[10] CHAITANYA, D. E., M. N. V. S. S. KUMAR, G. S. RAO and R. GOSWAMI. Convergence issues of taylor series method in determining unknown target location using hyperbolic multilateration. In: *International Conference on Science Engineering and Management Research (ICSEMR).* Chennai: IEEE, 2014, pp. 1–4. ISBN 978-1-4799-7613-3. DOI: 10.1109/ICSEMR.2014.7043670.

[11] RUI, L. and K. HO. Bias analysis of maximum likelihood target location estimator. *IEEE Transactions on Aerospace and Electronic Systems.* 2014, vol. 50, iss. 4, pp. 2679–2693. ISSN 0018-9251. DOI: 10.1109/TAES.2014.130318.

[12] SHA'AMERI, A. Z., Y. A. SHEHU and W. ASUTI. Performance analysis of a minimum configuration multilateration system for airborne emitter position estimation. *Defence S and T Technical Bulletin.* 2015, vol. 8, iss. 1, pp. 27–41. ISSN 1985-6571.

[13] GILLETTE, M. D. and H. F. SILVERMAN. A Linear Closed-Form Algorithm for Source Localization From Time-Differences of Arrival. *IEEE Signal Processing Letters.* 2008, vol. 15, iss. 1, pp. 1–4. ISSN 1558-2361. DOI: 10.1109/LSP.2007.910324.

[14] BUCHER, R. and D. MISRA. A Synthesizable VHDL Model of the Exact Solution for Three-dimensional Hyperbolic Positioning System. *VLSI Design.* 2002,

vol. 15, iss. 2, pp. 507–520. ISSN 1563-5171. DOI: 10.1080/1065514021000012129.

[15] DUDA, R. O., P. E. HART and D. G. STORK. *Pattern Classification.* 2nd ed. New York: Wiley, 2002. ISBN 978-0-471-05669-0.

[16] LIN, L., H. C. SO, F. K. W. CHAN, Y. T. CHAN and K. C. HO. A new constrained weighted least squares algorithm for TDOA-based localization. *Signal Processing.* 2013, vol. 93, iss. 11, pp. 2872–2878. ISSN 0165-1684. DOI: 10.1016/j.sigpro.2013.04.004.

[17] HO, K. C. Bias Reduction for an Explicit Solution of Source Localization Using TDOA. *IEEE Transactions on Signal Processing.* 2012, vol. 60, iss. 5, pp. 2101–2114. ISSN 1941-0476. DOI: 10.1109/TSP.2012.2187283.

[18] FARD, H. T., H. F. H. KASHANI, Y. NOROUZI and M. ATASHBAR. Multi Reference CTLS Method for Passive Localization of Radar Targets. *Journal of Advanced Defence Science and Technology.* 2013, vol. 3, iss. 3, pp. 179–185. ISSN 2228-5865. Available at: http://adst.ir/article-1-140-en.html.

[19] DELOSME, J., M. MORF and B. FRIEDLANDER. Source location from time differences of arrival: Identifiability and estimation. In: *IEEE International Conference on Acoustics, Speech, and Signal Processing (ICASSP 80).* Denver: IEEE, 1980, pp. 818–824. ISBN 978-1-4799-5835-1. DOI: 10.1109/ICASSP.1980.1170965.

[20] XU, Q., Y. LEI, J. CAO and H. WEI. An improved algorithm based on reference selection for time difference of arrival location. In: *7th International Congress on Image and Signal Processing (CISP).* Dalian: IEEE, 2015, pp. 953–957. ISBN 978-1-4799-5835-1. DOI: 10.1109/CISP.2014.7003916.

[21] RENE, J. E., D. ORTIZ, P. VENEGAS and J. VIDAL. Selection of the reference anchor node by using SNR in TDOA-based positioning. In: *IEEE Ecuador Technical Chapters Meeting (ETCM).* Guayaquil: IEEE, 2016, pp. 1–4. ISBN 978-1-5090-1629-7. DOI: 10.1109/ETCM.2016.7750813.

[22] GOLUB, G. H. and C. F. VAN LOAN. *Matrix Computations.* 4th ed. Baltimore: Johns Hopkins University Press, 2013. ISBN 0-8018-5414-8.

About Authors

Abdulmalik Shehu YARO was born in Kaduna, Nigeria. He received his B.Eng. degree in Electrical Engineering from Ahmadu Bello University, Zaria in 2012 and his M.Eng. degree in Electrical-Electronics and Telecommunications from University Teknologi Malaysia in 2014. He is currently pursuing his Ph.D. degree on Surveillance system at the Department of Electrical and Computer Engineering at Universiti Teknologi Malaysia.

Ahmad Zuri SHA'AMERI obtained his B.Sc. in Electrical Engineering from the University of Missouri-Columbia, USA in 1984, and M.Eng. Electrical Engineering and Ph.D. both from Universiti Teknologi Malaysia in 1991 and 2000 respectively. At present, he is an associate professor, Coordinator for the Digital Signal and Image Processing (DSIP) Research Group and Academic Coordinator for the DSP Lab, Electronic and Computer Engineering Department, Faculty of Electrical Engineering, Universiti Teknologi Malaysia. His research interest includes signal theory, signal processing for radar and communication, signal analysis and classification, and information security. The subjects taught at both undergraduate and postgraduate levels include digital signal processing, advance digital signal processing, advance digital communications and information security. He has also conducted short courses for both government and private sectors. At present, he has published 160 papers in his areas of interest at both national and international levels in conferences and journals.

Nidal KAMIL received the Ph.D. degree (Hons.) from the Technical University of Gdansk, Poland, in 1993. His Ph.D. work was focused on the subspace-based array signal processing for direction-of-arrival estimation. Since 1993, he has been involved in the research projects related to estimation theory, noise reduction, optimal filtering, and pattern recognition. He developed SNR estimator for antenna diversity combining and introduced the data glove for online signature verification. Currently, he is an Associate Professor at the Petronas University of Technology, Tronoh, Perak, Malaysia. His research interests include brain signal processing for epilepsy assessment and seizure prediction.

19

Multiple Time-Instances Features of Degraded Speech for Single Ended Quality Measurement

Rajesh Kumar DUBEY[1], *Arun KUMAR*[2]

[1]Department of Electronics and Communication Engineering, Jaypee Institute of Information Technology, Sector-62, 201309 Noida, Uttar Pradesh, India
[2]Centre for Applied Research in Electronics, Indian Institute of Technology Delhi, Huaz Khas, 110016 New Delhi, India

rajesh.dubey@jiit.ac.in, arunkm@care.iitd.ac.in

Abstract. *The use of single time-instance features, where entire speech utterance is used for feature computation, is not accurate and adequate in capturing the time localized information of short-time transient distortions and their distinction from plosive sounds of speech, particularly degraded by impulsive noise. Hence, the importance of estimating features at multiple time-instances is sought. In this, only active speech segments of degraded speech are used for features computation at multiple time-instances on per frame basis. Here, active speech means both voiced and unvoiced frames except silence. The features of different combinations of multiple contiguous active speech segments are computed and called multiple time-instances features. The joint GMM training has been done using these features along with the subjective MOS of the corresponding speech utterance to obtain the parameters of GMM. These parameters of GMM and multiple time-instances features of test speech are used to compute the objective MOS values of different combinations of multiple contiguous active speech segments. The overall objective MOS of the test speech utterance is obtained by assigning equal weight to the objective MOS values of the different combinations of multiple contiguous active speech segments. This algorithm outperforms the Recommendation ITU-T P.563 and recently published algorithms.*

Keywords

Auditory feature, degraded speech, speech quality.

1. Introduction

The speech processing algorithms and codecs are used in modern telecommunication systems and thus the monitoring and maintaining the quality of speech is important from customer satisfaction point of view to maintain and improve the quality of service. One aspect of this requirement for the automated system is to evaluate the speech quality objectively and continuously. If the quality of speech is not up to the mark, the proper bandwidth allocation or other speech enhancement techniques can be utilized to improve the quality of speech and thus the quality of service. There are two methods for signal based speech quality measurement: Double ended (Intrusive technique) and single ended (Non-intrusive technique). Double ended (Intrusive technique) requires original clean speech signal along with the received degraded speech signal to compute the quality rating called objective MOS, while single ended (Non-intrusive technique) uses only received degraded speech signal to compute the quality rating [1]. The non-intrusive method of speech quality measurement is suitable for system automation and real-time applications where the original clean speech signal is practically impossible to obtain such as mobile communications, telephonic communication, Direct-to-Home (DTH) signal of television (TV), Voice over Internet Protocol (VoIP) signal, etc. The Recommendation ITU-T P.563 (May 2004) is the standard for single ended (non-intrusive) speech quality measurement [2]. The subjective measurement is the ideal way to obtain the speech quality rating of degraded speech signal where the speech signal is played and average value of opinions of about 16–20 listeners is treated as quality rating for a particular speech utterance and called the subjective MOS as per the Recommendation ITU-T P.800-Aug.1996 [3]. The measurement of

Multiple Time-Instances Features of Degraded Speech for Single Ended Quality Measurement

159

speech quality has been done using different types of features obtained from speech encoder and GMM mapping in [4], without considering any degradation model. The human auditory system modelled explicitly or implicitly as Lyon's cochlear model is used in this work. Reference [5], which takes into account for the critical band and different auditory phenomenon such as masking the effect of human auditory system. The functional role of the human auditory system and the articulator system characteristics in the form of temporal envelope representation of speech have been utilized in the Auditory Non-Intrusive Quality Estimation (ANIQUE) model [6]. The Lyon's auditory features computed for entire speech let us call as "single time-instance features" and their mapping to the speech quality score by GMM has been given in [7]. The combination of different single time-instance speech features including auditory features and features related to vocal-tract resonances are used for GMM mapping and speech quality evaluation in [8]. The method given in [9] is assessing dimensions of perceptual quality space using linear regression and the dimension used is the loudness of speech which describes a non-optimal sound level. Estimating the quality and intelligibility of speech degraded by additive noise and distortions associated with telecommunication networks, based on a data driven framework of feature extraction and tree based regression, is given in [10].

The limitations of current research in the literature are that the features used for speech quality measurement are single time-instance, where the entire speech utterance is used for the computation of features, and these features are mapped to the objective quality rating score. In this work, the features are computed at multiple time-instances which capture the presence of noise at different locations of the speech utterance instead of averaging the effect over the entire speech utterance. Thus, the use of single time-instance features is not accurate and adequate in capturing the time localized information of short-time transient distortions and their distinction from plosive sounds of speech, particularly degraded by impulsive noise. The Voice Activity Detection (VAD) algorithm is employed to get the active speech segments of different speech utterances [11]. Here, active speech means both voiced and unvoiced frames except silence. Now, the combinations of multiple contiguous active speech segments of speech utterance are made in increasing order till all the active speech segments are accounted for. These combinations of active segments are divided into frames and features are computed on per frame basis using Lyon's auditory model. These per frame features are combined over the frames to give features of the different combinations of multiple contiguous active speech segments. In similar manner, Mel-Frequency Cepstral Coefficients (MFCC) [12] and [13] and Line Spectral Frequencies (LSF) features [14] are computed at multiple

time-instances and concatenated to obtain the feature vector. The subjective MOS of the speech utterance is taken as the subjective MOS for each of the different multiple time scale estimates (the combination of multiple contiguous active speech segments) during GMM training. The objective MOS values for each of the multiple time scale estimates are computed using the GMM parameters and different multiple time-scale features of test speech utterance. The overall objective MOS of the test speech utterance is computed by assigning equal weights to the objective MOS values of different multiple time scale estimates. The results are compared with Recommendation ITU-T P.563, the standard for non-intrusive technique of speech quality measurement, and different state-of-art recently published works [13], [15], [16], [17] and [18], which are using single time-instance features approach in terms of Pearson's correlation coefficient and RMSE between the subjective MOS and the overall objective MOS of speech utterances. The proposed algorithm using the combination of Lyon's auditory features, MFCC and LSF features, all computed at multiple time-instances, outperforms the state-of-art recent works.

2. Multiple Time-Instances Auditory Features

The more detailed statistical information of local features, particularly for contiguous speech segments, can be captured in multiple time-instances estimates, if non-stationary noise is present in the speech utterance. Thus, it is expected that the correlation between the subjective and the objective MOS in speech quality measurement problem will improve in multiple time-instances features approach. The degraded speech is input to the multiple time-instance auditory feature computation modules. At the very first stage, it will have to pass through VAD algorithm to remove silence region and find out the different active speech regions present in the speech utterances. For a speech utterance having three active speech segments, the output of VAD algorithm is schematically shown in Fig. 1.

The active speech segments at the output of the VAD algorithm are used in increasing order to make the different combinations of multiple time duration active speech segments till all the active speech segments are accounted for. The method of making concatenation to obtain different multiple time-instances estimates as the combinations of active speech segments for a speech utterance having three active speech segments is shown in Fig. 2. It will be continued till all the active segments are accounted for.

The first active segment is, say SEG1. Next, the combinations of the first and second active speech seg-

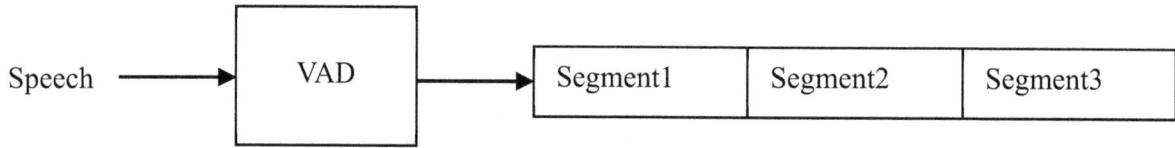

Fig. 1: Active speech segments and their concatenation.

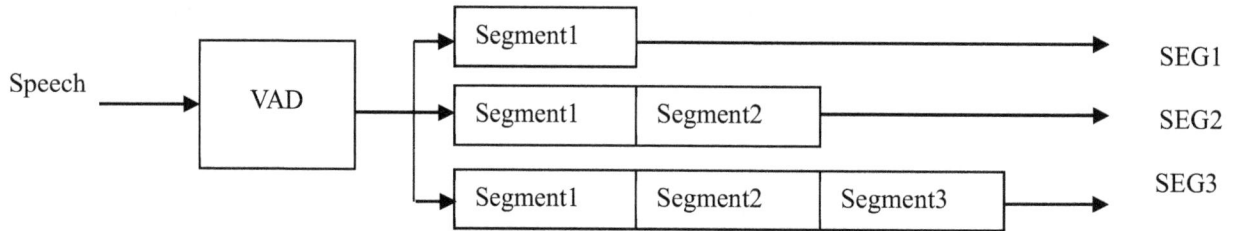

Fig. 2: Combinations of three active speech segments for different time-instances estimates for illustration.

ments is, say, SEG2. The combinations of the first, second and third active speech segments is, say, SEG3 and so on. In a similar manner, for K number of active speech segments in a speech utterance, there will be K different combinations of multiple contiguous active speech segments, on the lines of SEG1, SEG2, ... up to SEGK. These combinations of multiple contiguous active speech segments such as SEG2, SEG3, ... up to SEGK are divided into frames of fixed duration of 16 ms and passed through 64-channel Lyon's auditory model to compute 64 auditory features on frame-by-frame basis after windowing with a Hamming window of 16 ms duration with 50 % overlap. The mean, variance, skewness and kurtosis over the frames of 64 auditory features are computed and concatenated to obtain 256-dimensional Lyon's feature vector. The dimensionality of the feature vector is reduced from 256 to 30 by using Principal Component Analysis (PCA) to preserve more than 98 % of the energy. In the multiple time-instances features approach, the duration of active speech segments is varying over time.

In a similar manner, 13-dimensional multiple time-instances MFCC and 10-dimensional multiple time-instances LSF feature vectors are also computed on per frame basis. All these feature vectors are now concatenated to obtain a 53-dimensional feature vector. In a similar manner, 53-dimensional feature vectors are computed for all multiple time-instances estimates such as SEG2, SEG3 and so on up to SEGK. For the training of joint GMM according to Expectation Maximization (EM) algorithm [19], the 53-dimensional feature vectors are appended with the subjective MOS values of the corresponding speech utterance. The subjective MOS for each of the multiple time-instances estimates is taken as the subjective MOS of the speech utterance, as shown in Fig. 3, because no separate subjective MOS will be available for the multiple time-instances estimates in any database. The objective MOS of each of the multiple time-instances estimates is computed us-

ing GMM parameters namely mean, mixture weight, and covariance matrix and 53-dimensional feature vectors of the corresponding multiple time-instances estimates. The objective MOS value of ith multiple-time scale estimate $\hat{\theta}_i$ as a function of 53-dimensional multiple time-instances feature vector ψ is obtained using the Minimum Mean Square Error (MMSE) criterion [4]:

$$\hat{\theta}_i = \hat{\theta}_i(\psi) = arg\min_{\hat{\theta}_i(\psi)} E\left\{(\theta - \hat{\theta}_i(\psi))^2\right\} = \\ = E\left\{\theta_i/\psi\right\}, \qquad (1)$$

where θ is the subjective MOS of corresponding speech utterance. The three databases are randomized to use leave-one-out10-fold cross validation process for training and testing. That is, 90% data are used for training and 10 % data are used for testing. The process is repeated 10-times to obtain the objective MOS values for all the multiple time-instances estimates. In this work, GMM with 12 mixture components are used and all the GMM training parameters are computed offline and stored in a library. In real-time monitoring, only test speech will be used but there will be an algorithmic buffering delay corresponding to one sentence speech utterance before the multiple time-instances speech quality evaluation algorithms are applied.

The averaging of the objective MOS values of the multiple time-instances estimates is done i.e. equal weights are assigned to the objective MOS values of the different multiple time-instances estimates to compute the overall objective MOS of the corresponding speech utterance. If $\hat{\theta}$ is the objective MOS of speech utterance, then it is computed by taking the average of the objective MOS values of K SEGs, $\hat{\theta}_i$ is given by:

$$\hat{\theta} = \frac{1}{K}\sum_{i=1}^{K}\hat{\theta}_i, \qquad (2)$$

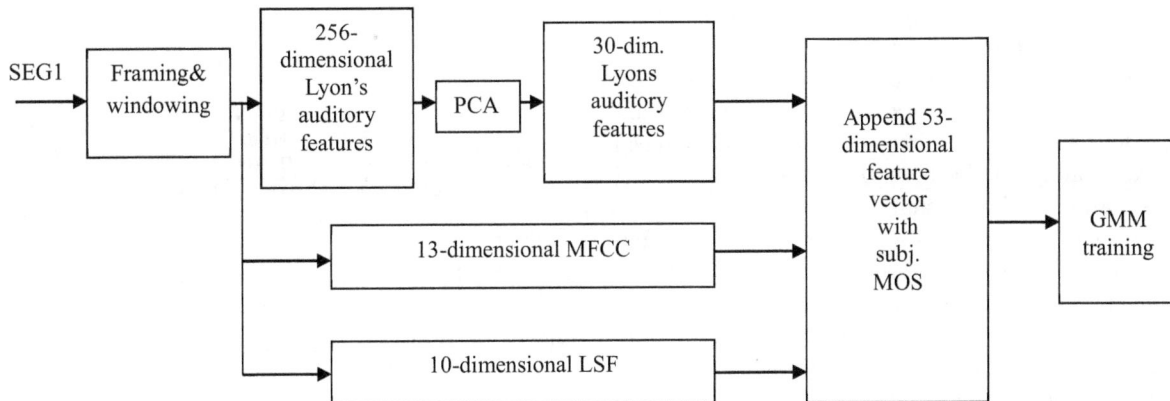

Fig. 3: Computation of 53-dimensional feature vector, and appending with the subjective MOS for GMM training.

where K is the number of active speech segments which will be equal to the number of combinations of multiple contiguous active speech segments.

3. Description of Databases

In this work, three databases are used namely ITU-T Supplement-23 database [20], NOIZEUS-2240 and NOIZEUS-960 [21]. The first database of 1328 speech utterances constitute Expt.-1 (A, D, O) having a total of 528 speech utterances with 8 kbps ITU & ETSI standard CODECS interworking and Expt.-3 (A, C, D, O) is having a total of 800 speech utterances with channel errors and background noises. All these 1328 speech utterances degraded at 332 different degradation conditions and are of 8 second duration each, sampled at 8 kHz and subjective MOS labelled according to Absolute Category Rating (ACR). The second database is having a total of 2240 degraded speech utterances of 3 second duration each, sampled at 8kHz and degraded by 4 different types of noise namely babble, car, street and train noise at two different SNR levels, 5 dB and 10 dB. A total of 20 clean speech utterances are degraded at 112 different conditions of degradation. The third one, NOIZEUS-960 database, which is taken from NOIZEUS database of noisy speech corpus of 960 speech sentences of 30 clean speech signals are sampled at 8 kHz and of 3 seconds duration each. The clean signals are degraded by 8 different types of noise namely airport, babble, car, exhibition, restaurant, station, street and suburban train at 4 different SNR levels (0 dB, 5 dB, 10 dB and 15 dB). The NOIZEUS-2240 and NOIZEUS-960 speech utterances are not having subjective MOS associated with them and thus subjective listening test was conducted to obtain the subjective MOS in our laboratory. The statistical analysis of the subjective MOS rating is presented in [22] to ensure the high degree inter-and-intra-rater reliability.

4. Results and Analysis

The Pearson's correlation coefficient and RMSE between the subjective MOS score θ and estimated overall objective MOS score $\hat{\theta}$, both computed as condition averaged value, are used as figure of merit in most of the literatures of single ended speech quality measurement algorithms. In this work, unconditioned values of the subjective and objective MOS are also used for the computation of Pearson's correlation coefficient and RMSE [8], where MOS values of speech sentence-by-sentence are used, because it will be more realistic. Results are given and compared in Tab. 1 for condition averaged MOS values and Tab. 3 for unconditioned MOS values using three databases. The comparison of results between single time-instance [8] and multiple time-instances approaches is presented along with Recommendation ITU-T P.563. The overall weighted average of the correlation using multiple time-instances estimates is 0.980 as against single time-instance features approach which is 0.960 [8], whereas the correlation is 0.934 using the ITU-T Rec. P.563 algorithm over the three databases for condition averaged MOS case as given in Tab. 1.

In [8], on same databases 37-dimensional feature vectors formed by combining 14-dimensional reduced size

Tab. 1: Correlation coefficients and RMSE between the subjective and the estimated overall objective MOS for the condition averaged MOS case.

Data-base	No. of speech utteran-ces	ITU-T Rec. P.563		Proposed model	
		Corre-lation	RM SE	Corre-lation	RM SE
ITU-T Supp. 23	1328	0.815	0.450	0.966	0.168
NOIZEUS-960	960	0.951	0.250	0.995	0.039
NOIZEUS-2240	2240	0.954	0.422	0.986	0.068

Lyon's auditory model features, 13-dimensional MFCC features, and 10-dimensional LSF features, all computed at single time-instance for entire speech utterances are used. In this work, 53-dimensional feature vectors formed by combining 30-dimensional reduced size Lyon's auditory features, 13-dimensional MFCC features, and 10-dimensional LSF features, all computed at multiple time-instances are used. The basis for dimensionality reduction of Lyon's auditory features using PCA from 256 to 14 in the case of single time-instance is preservation of 98 % energy. According to this criterion, the dimensionality of multiple time-instances Lyon's auditory features is reduced from 256 to 30 using PCA. Moreover, the MFCC features are 13-dimensional and LSF features are 10-dimensional. Thus, single time-instance feature vectors are 37-dimensional and multiple time-instances feature vectors are 53-dimensional.

Tab. 2: Comparison of single time-instance and multiple time-instances features approach using equal weights with ITU-T Rec. P.563 taking condition averaged subjective MOS and estimated objective MOS.

Data of Different Expts.	No. of Speech Utterances	ITU-T Rec. P.563	Single time-instance features Lyon's, MFCC & LSF	Multiple time-instances features Lyon's MFCC, & LSF with equal weight
Exp.1(A) -French	176	0.885	0.912	0.967
Exp.1(D) -Japanese	176	0.842	0.933	0.975
Exp.1(O) -A.English	176	0.902	0.946	0.988
Exp.3(A) -French	200	0.867	0.887	0.949
Exp.3(C) -Italian	200	0.854	0.851	0.954
Exp.3(D) -Japanese	200	0.929	0.908	0.948
Exp.3(O) -A.English	200	0.918	0.891	0.961
NOIZEUS -960	960	0.951	0.993	0.995
NOIZEUS -2240	2240	0.955	0.980	0.985
Weighted Average	**934**	**0.960**	**0.980**
Std. Dev.		0.041	0.046	0.018
Confidence Interval (95 %)		0.027	0.030	0.012

Tab. 3: Correlation coefficients and RMSE between the unconditioned subjective MOS and the unconditioned estimated overall objective MOS.

Database	No. of speech utterances	ITU-T Rec. P.563		Proposed model	
		Correlation	RMSE	Correlation	RMSE
ITU-T Supp. 23	1328	0.7168	0.580	0.9233	0.335
NOIZEUS -960	960	0.7169	0.856	0.9180	0.277
NOIZEUS -2240	2240	0.3057	0.998	0.7007	0.379

Tab. 4: Comparison of single time-instances and multiple time-instances features approach using equal weights with ITU-T Rec. P.563 taking unconditioned subjective and estimated objective MOS.

Data of Different Expts.	No. of Speech Utterances	ITU-T Rec. P.563	Single time-instance features Lyon's, MFCC & LSF	Multiple time-instances features Lyon's MFCC, & LSF with equal weight
Exp.1(A) -French	176	0.759	0.837	0.921
Exp.1(D) -Japanese	176	0.701	0.828	0.934
Exp.1(O) -A.English	176	0.790	0.828	0.956
Exp.3(A) -French	200	0.768	0.773	0.889
Exp.3(C) -Italian	200	0.762	0.753	0.903
Exp.3(D) -Japanese	200	0.801	0.806	0.901
Exp.3(O) -A.English	200	0.788	0.745	0.91
NOIZEUS -960	960	0.717	0.859	0.918
NOIZEUS -2240	2240	0.306	0.690	0.695
Weighted Average		**0.529**	**0.756**	**0.807**
Std. Dev.		0.155	0.054	0.076
Confidence Interval (95 %)		0.101	0.035	0.050

The results in terms of Pearson's correlation coefficient and RMSE for condition averaged MOS are also compared in Tab. 5 with the published results of recent works in [13], [15] and [16] which were using a database of 1792 speech utterances that was a subset of NOIZEUS-2240 database of 2240 speech utterances used in this work. The comparison is also shown by bar graph in Fig. 4. Here, we have conducted subjective listening tests to obtain the subjective MOS for 2240 speech utterances, while in [13], [15] and [16] they have used their own respective subjective scores. The value of correlation reported in [13] for the condition averaged case is 0.9002 and the RMSE to be 0.33, whereas in this proposed work the correlation obtained is 0.986 and the RMSE to be 0.068 respectively for the NOIZEUS-2240 database. In [15], the maximum value of Pearson's correlation coefficients obtained is 0.910 in test-1 which uses 8-fold cross validation process, whereas 10-fold cross-validation process has been used

Tab. 5: Comparison of results in terms of Pearson's correlation coefficient and RMSE with recently published works [11], [13] and [14] on NOIZEUS-2240 database.

Methods		Correlation	RMSE
Ref [11]		0.9002	0.33
Ref [13]	Test-1	0.910 (Estimated from given bar chart)	0.190 (Estimated from given bar chart)
	Test-2	0.886	194
	Test-3	0.842	248
Ref [14]	Mean	0.77	0.29
	Variance	0.83	0.25
	Mean + Variance	0.90	0.20
Proposed Work		0.986	0.068

in this proposed work. In [16], the mean and variance statistics of Gabor PCA features gives the best performance for speech quality assessment on NOIZEUS-2240 database. It uses 80 % of data for training and 20 % for testing. The maximum Pearson's correlation coefficients obtained to be 0.90 and RMSE 0.20 in [16].

Tab. 6: Comparison of results in terms of Pearson's correlation coefficient with recently published works [16] on ITU-T P.Supplement-23 database.

Database	ITU-T Rec. P.563	Bag-of-Words Representation Algorithm	Multiple time-instances features Lyon's, MFCC & LSF
Exp.1(A) -French	0.885	0.933	0.967
Exp.1(D) -Japanese	0.842	0.902	0.975
Exp.1(O) -A.English	0.902	0.949	0.988
Exp.3(A) -French	0.867	0.925	0.949
Exp.3(C) -Italian	0.854	0.849	0.954
Exp.3(D) -Japanese	0.929	0.888	0.948
Exp.3(O) -A.English	0.918	0.902	0.961
Average	**0.885**	**0.893**	**0.963**

The comparison of results in terms of Pearson's correlation coefficient for NOIZEUS-960 database has also been done with [17] for condition averaged MOS, which is the same speech database used in this work. Here, we have conducted subjective listening tests to obtain the subjective MOS for 960 speech utterances, while in [17] they have used their own respective subjective scores. The Pearson's correlation coefficients obtained in [17] was 0.933 as against 0.995 in this proposed work. In [17], 70 % of data has been used for training while 30 % for testing. The comparison of results in terms of Pearson's correlation coefficient for ITU-T P. Supplement-23 database has also been done with recent work [18] in Tab. 6 for seven sub-databases for condition averaged MOS values. In these comparisons, it is observed that

the proposed work performs better than these recently published works.

5. Inferences Drawn from Results

From the overall results expressed in tabular form and different comparisons, the following inferences are made:

- The multiple time-instances estimates to compute the objective MOS score of the overall speech utterance gives higher correlation as compared to the single time-instances features approach.

- For both, the condition averaged MOS case or unconditioned MOS case, correlation coefficients and RMSE are significantly better for multiple time-instances estimates as compare to single time-instances estimates over the different databases.

- In this algorithm, the combination of reduced size Lyon's auditory features with MFCC and LSF features are used as feature vectors in the study. In this, even there will be some duplicity of information in the features, but the combination of features gives better result in terms of correlation and RMSE between the subjective MOS and the estimated overall objective MOS for speech on sentence-by-sentence basis. By combining these feature vectors, the correlation coefficient, in both the cases of unconditioned and condition averaged MOS increases significantly.

6. Conclusion

Lyon's auditory features, MFCC and LSF features are computed for multiple time-instances for the different combinations of multiple contiguous active speech segments. These multiple time-instances features are combined for a speech utterance for single ended speech quality measurement. The overall objective MOS of the speech utterance is computed by assigning equal weights (averaging) of the MOS values of the multiple time-instances estimates. The results in terms of correlation coefficients between the subjective and the estimated overall objective MOS for different types of noisy speech databases are obtained and compared with the different single time-instances approaches recently published and Rec. ITU-T P.563 and found that multiple time-instances approach outperforms.

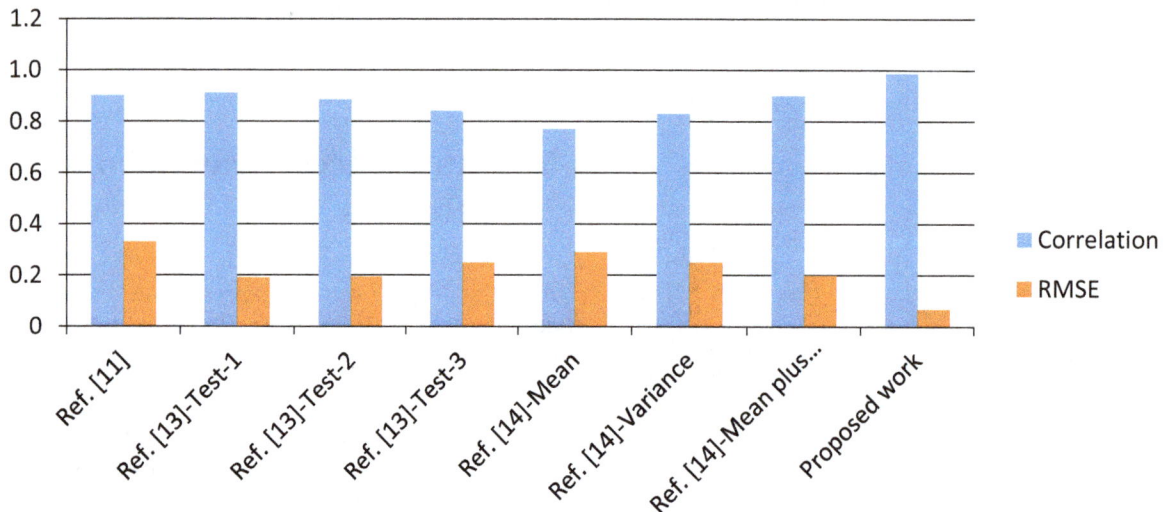

Fig. 4: Bar graph comparison with recent published work [11], [13] and [14].

Acknowledgment

The authors would like to express their gratitude to Mr. Yi Hu and Dr. Philipos C. Loizou of Department of Electrical Engineering, The University of Texas at Dallas, Richardson, TX75083-0688, USA for providing the NOIZEUS-2240 database of 2240.

References

[1] MALFAIT, L., J. BERGER and M. KAST-NER. P.563-The ITU-T standard for single-ended speech quality assessment. *IEEE Transactions on Audio, Speech and Language Processing.* 2006, vol. 14, iss 6, pp. 1924–1934. ISSN 1558-7924. DOI: 10.1109/TASL.2006.883177.

[2] ITU-T RECOMMENDATION P. 563. *Single ended method for objective speech quality assessment in narrow-band telephony applications.* Geneva: ITU-T, 2005.

[3] ITU-T RECOMMENDATION P. 800. *Methods for subjective determination of transmission quality.* Geneva: ITU-T, 1996.

[4] GRANCHAROV, V., D. Y. ZHAO, J. LIND-BLOM and W. B. KLEIJN. Low-complexity, non-intrusive speech quality assessment. *IEEE Transactions on Audio, Speech and Language Processing.* 2006, vol. 14, iss. 6, pp. 1948–1956. ISSN 1558-7924. DOI: 10.1109/TASL.2006.883250.

[5] LYON, R. F. A computational model of filtering, detection, and compression in the cochlea. In: *IEEE International Conference on Acoustics, Speech and Signal Processing.* Paris: IEEE,

1982, pp. 1282–1285. ISBN 978-82-80125-127. DOI: 10.1109/ICASSP.1982.1171644.

[6] KIM, D. S. ANIQUE: An auditory model for single ended speech quality estimation. *IEEE Transactions on Audio, Speech and Language Processing.* 2005, vol. 13, iss. 5, pp. 821–831. ISSN 1063-6676 DOI: 10.1109/TSA.2005.851924.

[7] AUDHKHASI, K. and A. KUMAR. Two scale auditory features based non-intrusive speech quality evaluation. *IETE Journal of Research.* 2010, vol. 56, iss. 2, pp. 111–118. ISSN 0377-2063.

[8] DUBEY, R. K. and A. KUMAR. Non-intrusive speech quality assessment using several combinations of auditory features. *International Journal of Speech Technology.* 2013, vol. 16, iss. 1, pp. 89–101. ISSN 1381-2416. DOI: 10.1007/s10772-012-9162-4.

[9] KOSTER, F., V. CERCOS-LLOMBART, G. Gabriel MITTAG and S. MOLLER. Non-intrusive estimation model for the speech-quality dimension loudness. In: *Proceedings of Speech Communication, 12. ITG Symposium.* Berlin: IEEE, 2016, pp. 175–179. ISBN 978-3-8007-4275-2.

[10] SHARMA, D., Y. WANG, P. A. NAYLOR and M. BROOKES. A data-driven non-intrusive measure of speech quality and intelligibility. *Speech Communication.* 2016, vol. 80, iss. C, pp. 84–94. ISSN 0167-6393. DOI: 10.1016/j.specom.2016.03.005.

[11] RABINER, L. R. and M. R. SAMBUR. Voiced-unvoiced-silence detection using the Itakura LPC distance measure. In: *IEEE International Conference on Acoustics, Speech and Signal Processing.* New Jersey: IEEE, 1977, pp. 323–326. DOI: 10.1109/ICASSP.1977.1170330.

[12] HASAN, M. R., M. JAMIL, M. G. RABBANI and M. S. RAHMAN. Speaker identification using mel-frequency cepstral coefficient. In: *3rd International Conference on Electrical & Computer engineering*. Bangladesh: ICECE, 2004, pp. 565–568. ISBN 984-32-1804-4.

[13] NARWARIA, M., W. LIN, I. V. MCLOUGH-LIN, S. EMMANUEL and L. T. CHIA. Non-intrusive quality assessment of noise suppressed speech with mel-filtered energies and support vector regression. *IEEE Transactions on Audio, Speech and Language Processing*. 2012, vol. 20, iss. 4, pp. 1217–1232. ISSN 1558-7924. DOI: 10.1109/TASL.2011.2174223.

[14] BOZKURT, E., E. ERZIN, C. E. ERDEM and A. T. ERDEM. Use of line spectral frequencies for emotion recognition from speech. In: *IEEE International Conference on Pattern Recognition*. Istanbul: IEEE, 2010, pp. 3708–3711. ISBN 978-1-4244-7541-4. DOI: 10.1109/ICPR.2010.903.

[15] SONI, M. H. and H. A. PATIL. Novel deep autoencoder features for non-intrusive speech quality assessment. In: *Proceedings of the 24th European Signal Processing Conference (EUSIPCO)*. Budapest: IEEE, 2016, pp. 2315–2319. ISBN 978-1-5090-1891-8. DOI: 10.1109/EUSIPCO.2016.7760662.

[16] LI, Q., Y. FANG, W. LIN and D. THALMANN. Non-intrusive quality assessment for enhanced speech signals based on spectro-temporal features. In: *International Conference on Multimedia and Expo Workshops (ICMEW)*. Chengdu: IEEE, 2014, pp. 1–6. ISBN 978-1-4799-4716-4. DOI: 10.1109/ICMEW.2014.6890561.

[17] ISLAM, M. R., M. A. RAHMAN, M. N. HASAN, A. S. HOSSAIN, A. N. UDDIN and M. A. HAQUE. Non-intrusive objective evaluation of speech quality in noisy condition. In: *9th International Conference on Electrical and Computer Engineering (ICECE)*. Dhaka: IEEE, 2016, pp. 586–589. ISBN 978-1-5090-2964-8. DOI: 10.1109/ICECE.2016.7853988.

[18] LI, Q., Y. FANG, W. LIN and D. THALMANN. Bag-of-words representation for non-intrusive speech quality assessment. In: *China Summit and International Conference on Signal and Information Processing (ChinaSIP)*. Chengdu: IEEE, 2015, pp. 616–619. ISBN 978-1-4799-1947-5. DOI: 10.1109/ChinaSIP.2015.7230477.

[19] DEMPSTER, A. P., N. LAIRD and D. B. RUBIN. Maximum likelihood from incomplete data via the EM algorithm. *Journal of the Royal Statistical Society. Series B (Methodological)*. 1977, vol. 39, iss. 1, pp. 1–38. ISSN 1369-7412.

[20] ITU-T RECOMMENDATION SUPPLEMENT 23. *Methods for subjective determination of transmission quality*. Geneva: ITU-T, 1998.

[21] NOIZEUS: A noisy speech corpus for evaluation of speech enhancement algorithms. In: *University of Texas at Dallas* [online]. 2009. Available at: http://ecs.utdallas.edu/loizou/speech/noizeus/.

[22] DUBEY, R. K. and A. KUMAR. Comparison of subjective and objective speech quality ass assessment for different degradations/noise conditions. In: *International Conference on Signal Processing and Communication (ICSC)*. Noida: IEEE, 2015, pp. 261–266. ISBN 978-1-4799-6762-9. DOI: 10.1109/ICSPCom.2015.7150659.

About Authors

Rajesh Kumar DUBEY received the B.Tech. degree in Electrical Engineering from the National Institute of Technology Hamirpur (HP), India in 1999, the M.Tech. degree in Electrical Engineering in 2002, and the Ph.D. degree from the Center for Applied Research in Electronics, in 2014, both from the Indian Institute of Technology Delhi, India. Since 2006, he is working as an Assistant Professor (Sr. Grade) in the department of Electronics and Communication Engineering, Jaypee Institute of Information Technology (JIIT), Noida, India. His research interests span the areas of Digital Signal and Speech Processing.

Arun KUMAR received the B.Tech, M.Tech, and Ph.D. degrees in Electrical Engineering from the Indian Institute of Technology Kanpur, India, in 1988, 1990, and 1995, respectively. He was a visiting researcher at the University of California, Santa Barbara, from 1994 to 1996. Since 1997, he is working as professor at the Center for Applied Research in Electronics, Indian Institute of Technology Delhi, India. His research interests span the areas of Digital Signal Processing, Speech Processing, Underwater Acoustics, Air Acoustics and Communications. In these areas, he has introduced new courses at the Masters level, supervised more than 40 funded research projects, and several Masters and Ph.D theses at IIT Delhi. Dr. Kumar is a recipient of the Young Scientist Award of the International Union of Radio Science (URSI).

20

The Method of the Sensitivity Comparison of the Tin Dioxide Gas Sensor in Periodic Steady State

Libor GAJDOSIK

Department of Telecommunications, Faculty of Electrotechnical Engineering and Computer Science,
VSB–Technical university of Ostrava, 17. listopadu 15, 708 33 Ostrava, Czech Republic

libor.gajdosik@vsb.cz

Abstract. *The formulas for the time dependency of the electrical conductivity of the sensor in thermal periodic steady state in the clean air atmosphere were derived herein. The created model of the sensor was experimentally verified and enables to compare the sensitivity to the tested substance at the frequencies at which the tests were carried out. The experiments were carried out with the sensors MQR 1003, SP 11, and TGS813. The sensors were tested in the clean air atmosphere and subsequently in the presence of ethanol, acetone and toluene vapour in the air at three different frequencies.*

Keywords

Semiconductor gas sensor, temperature modulation, thermal cycling, thermal inertia, tin dioxide.

1. Introduction

Sensors are often operated in periodic time variable heating regimes, because of better detection properties compared to the constant heating regime [1], [2], [3], [4], [5], [6], [7], [8], [9], [10], [11] and [12]. The overview of used types of time dependent heating of the sensor can be found in [13]. The theoretical models of the sensor response at the thermal modulation were presented in [14], [15], [16], [17], [18] and [19]. But in these works thermal inertia properties of the sensor were not considered. This is carried out herein. A heating system of the sensor has thermal inertia properties and that is why the detection properties of the sensor are dependent on the frequency of the heating voltage as well. The simple model of the behaviour of the sensor described herein can help to decide if the sensor is more

or less sensitive to the given compound at given frequency.

2. Experiments

The commercial sensors of type MQR 1003, SP 11, and TGS 813 were operated in the clean air of 31 % relative humidity and then in the vapour of the single compound of concentration 100 ppm in the clean air. Ethanol, acetone, and toluene were used in the experiments. The sensor response, which is the electrical conductivity of the detection layer, was sensed in periodic steady state. The heating voltage was realized by a function in the following way:

$$U(t) = \sqrt{U_0 + U_A \sin(\omega t)}, \qquad (1)$$

$$U_0 = \frac{U_M^2 + U_m^2}{2}, \qquad U_A = \frac{U_M^2 - U_m^2}{2}, \qquad (2)$$

where t is the time, ω is the angular frequency, U_M is the maximum value of heating voltage and U_m is the minimum value of the heating voltage. For the experiments $U_M = 5$ V, $U_m = 2$ V. The response of the sensor was sampled in 32 points in every period of the heating voltage. The sampling rate t_s was chosen by experimental experience $t_s = 1000$ ms, $t_s = 500$ ms and $t_s = 200$ ms. Theses values are related to the frequencies of the sine component of the heating voltage $f = 0.0312$ Hz, $f = 0.0625$ Hz, $f = 0.156$ Hz, respectively.

3. Theory

A sensor consists of a heating system made of a heating resistance conductor covered by a ceramic insulator layer. The ceramic layer is covered by a metal oxide layer on which a pair of measuring electrodes is placed to measure the electrical conductivity of the metal oxide layer. The conductor is heated by an electrical voltage and heat penetrates through the ceramic layer into the detection layer. A gas detection does not occur until a specific temperature of the detection layer is reached, which is manifested by a change of the electrical conductivity of the detection layer. Each of these layers has its specific thermal properties. All these layers can be considered as one fictive layer of the thermal diffusivity a to simplify the mathematical solution. The heating voltage U generates the electrical power:

$$P = \frac{U^2}{R}, \tag{3}$$

where P is the electrical active power, R is the electrical resistance of the heating system of the sensor and U is the electrical voltage. This power is converted according to Joule - Lenz law into heat, which heats up the detection layer of the sensor to the temperature ϑ by the following formula:

$$Q = C\,\vartheta, \tag{4}$$

where C is the heat capacity of the system and Q is the Joule heat. It follows that the temperature ϑ is directly proportional to the Joule heat and consequently to the square power of the heating voltage U by the following formula:

$$\vartheta \approx \frac{1}{CR}U^2 = kU^2, \tag{5}$$

where k is the constant of proportionality. Substituting U from Eq. (1), we find:

$$\vartheta = kU^2 = k[U_0 + U_A \sin(\omega t)]. \tag{6}$$

Equation (6) can be rewritten into the following form:

$$\vartheta = \vartheta_0 + \vartheta_A \sin(\omega t)\,, \tag{7}$$

where ϑ_0 is the constant component and ϑ_A is the amplitude of the sine component of the driving temperature of the detection system of the sensor. Joule heat is conducted through the sensor and affects the temperature of the detection layer in the point x at the time t. This temperature can be designated as $\vartheta(x,t)$. But we do not measure the temperature $\vartheta(x,t)$ directly, we measure its manifestation, which is the electrical conductivity of the detection layer. The electrical conductivity of the detection layer is related to the temperature of the sensor according to [20] by the following formula:

$$G = G_0 e^{-\frac{r}{\vartheta(x,t)}}, \tag{8}$$

where G_0 and r represent coefficients which are dependent on the material of the sensor and on the tested gas.

Heat conduction inside the layer can be solved by the Fourier - Kirchhoff equation. Heat conduction is considered in the halfplane $x > 0$ in the direction of the axis x:

$$a\frac{\partial^2 \vartheta}{\partial x^2} = \frac{\partial \vartheta}{\partial t}, \tag{9}$$

where a is the thermal diffusivity of the material $(\mathrm{m^2 \cdot s^{-1}})$, ϑ is the temperature and t is the time. The heating system is supposed to be placed in the origin $x = 0$. We assume the initial condition in the following form to calculate the solution of Eq. (9) for sine component of Eq. (7):

$$\vartheta(0,t) = \vartheta_A \sin \omega t. \tag{10}$$

The periodical steady state solution is desired. The solution is supposed in the form as the product of two functions:

$$\vartheta(x,t) = g(x) \cdot f(t), \qquad \text{where} \quad f(t) = e^{j\omega t}. \tag{11}$$

Substituting Eq. (11) into Eq. (9) we obtain:

$$a\frac{\partial^2 g}{\partial x^2}e^{j\omega t} = \frac{\partial}{\partial t}\left(ge^{j\omega t}\right) = j\omega g e^{j\omega t}. \tag{12}$$

Since the term $\exp(j\omega t)$ cancels on both sides of Eq. (12), we obtain:

$$a\frac{\mathrm{d}^2 g}{\mathrm{d}x^2} = j\omega g. \tag{13}$$

We rewrite Eq. (13) into the form:

$$\frac{\mathrm{d}^2 g}{\mathrm{d}x^2} - \frac{1}{a}j\omega g = 0. \tag{14}$$

We find the solution with the use of the relevant characteristic equation:

$$\lambda^2 - \frac{1}{a}j\omega = 0, \qquad \rightarrow \qquad \lambda^2 = j\frac{\omega}{a}. \tag{15}$$

The number λ equals:

$$\lambda_{1/2} = \pm\sqrt{j\frac{\omega}{a}} = \pm(1+j)\sqrt{\frac{\omega}{2a}} = \pm(1+j)h. \tag{16}$$

The solution of Eq. (13) equals:

$$g(x) = K_1 e^{\lambda_1 x} + K_2 e^{\lambda_2 x}. \tag{17}$$

Since $\vartheta(x,t) < \infty$ is valid, we consider the solution only in the form:

$$g(x) = K_1 e^{\lambda_1 x}, \qquad \text{where} \qquad \lambda_1 = -(1+j)h\,. \tag{18}$$

Considering Eq. (11) the solution is written in the form:

$$\vartheta(x,t) = g(x) \cdot f(t) = K_1 e^{-(1+j)hx}e^{j\omega t}, \tag{19}$$

$$\vartheta(x,t) = K_1 e^{-hx}e^{j(-hx+\omega t)}. \tag{20}$$

Since the driving temperature $\vartheta(0,t)$ is expressed by a sine function, we consider only the sine component in the solution i.e. the imaginary part of Eq. (15). We can write:

$$\vartheta(x,t) = K_1 e^{-hx} \sin(-hx + \omega t). \qquad (21)$$

Regarding Eq. (10) in case that

$$K_1 = \vartheta_A, \qquad (22)$$

it is necessary to modify the solution described by Eq. (21) for used experimental conditions. If $\omega = \infty$ in Eq. (21), it leads to $\exp(-hx) = 0$ and then $\vartheta(x,t) = 0$. The sensor heated by the voltage described by Eq. (1) reaches non-zero temperature at the frequency $\omega = \infty$. This is why Eq. (21) is to be modified by introducing the constant ϑ_0. We obtain:

$$\vartheta(x,t) = \vartheta_0 + \vartheta_A e^{-hx} \sin(-hx + \omega t). \qquad (23)$$

Equation (23) is also the solution of Eq. (9), because Eq. (23) contains only additional constant ϑ_0 compared to Eq. (21). If the heating voltage is described by Eq. (1), the temperature of the detection layer in the place x at the time t is described by Eq. (23). With the use of Eq. (7) we obtain the conductivity of the detection layer in the following form:

$$G(t) = G_0 e^{-\frac{r}{\vartheta(x,t)}}, \qquad \text{where}$$
$$\vartheta(x,t) = \vartheta_0 + \vartheta_A e^{-hx} \sin(-hx + \omega t). \qquad (24)$$

The term hx can be modified with the use of Eq. (16) and Eq. (23) into the following form:

$$hx = x\sqrt{\frac{\omega}{2a}} = x\sqrt{\frac{\pi f}{a}} = h_1\sqrt{f}, \qquad h_1 = x\sqrt{\frac{\pi}{a}}. \quad (25)$$

The coefficient h_1 is dependent on the construction of the sensor, x is the thickness of the heated fictive layer. Then it is possible to write the electrical conductivity of the sensor in the following form:

$$G(t) = e^{-\frac{1}{\vartheta(x,t)}}, \qquad \text{where}$$
$$\vartheta(t) = A + Be^{-h_1\sqrt{f}} \sin(-h_1\sqrt{f} + 2\pi\sqrt{f}t), \qquad (26)$$

where the constants A, B and h_1 can be determined from experimental data. This function can be used for approximation of experimental values of the electrical conductivity of the detection layer. The choice of the constants $G_0 = r = 1$ is admissible from the mathematical point of view to reach the equality of $G(t)$ between Eq. (26) and Eq. (24).

The heating voltage $U(t)$ and the electrical conductivity $G(t)$ are periodical functions, whose periodicity is determined by the sine function. Maxima and minima of $G(t)$ and $U(t)$ appear when the sine equals ± 1. When concidering following in Eq. (1):

$$\omega t_i = (2n+1)\frac{\pi}{2}, \qquad n = 0, 1, 2, \ldots, \qquad (27)$$

then the maxima of the heating voltage $U(t)$ occur in the time t_i. When concidering following in Eq. (24):

$$-h_1\sqrt{f} + \omega t_j = (2n+1)\frac{\pi}{2}, \qquad n = 0, 1, 2, \ldots, \qquad (28)$$

then the maxima of the electrical conductivity $G(t)$ occur at the time t_j. It follows from Eq. (27) and Eq. (28), that the following equation must be satisfied for two matching maxima of $U(t_i)$ and $G(i_j)$:

$$-h_1\sqrt{f} + \omega t_j = \omega t_i. \qquad (29)$$

We express from Eq. (29):

$$h_1 = \frac{\omega t_j - \omega t_i}{\sqrt{f}} = 2\pi\sqrt{f}(t_j - t_i) = 2\pi\Delta t\sqrt{f}, \quad (30)$$

where Δt is the time delay between the heating voltage and the electrical conductivity. Δt can be calculated from the experimental values of $G(t)$ and $U(t)$. Then h_1 can be calculated from Eq. (30).

We designate in Eq. (26), that:

$$u = e^{-h_1\sqrt{f}}. \qquad (31)$$

With the use of Eq. (26) and Eq. (31), the following expressions can be defined as:

$$G_M = e^{-\frac{1}{A+Bu}}, \qquad G_m = e^{-\frac{1}{A-Bu}}, \qquad (32)$$

where G_M represents the maximum and G_m represents the minimum value of the electrical conductivity. The constants A and B can be expressed from Eq. (32). Substituting u from Eq. (31) we obtain:

$$A = \frac{1}{2}\left(\frac{1}{-\ln G_M} + \frac{1}{-\ln G_m}\right), \qquad (33)$$

$$B = \frac{1}{2}\left(\frac{1}{-\ln G_M} - \frac{1}{-\ln G_m}\right)e^{h_1\sqrt{f}}. \qquad (34)$$

Equation (33) and Eq. (34) enable to calculate A and B from the experimental values of G_M and G_m. The coefficient h_1 is calculated by Eq. (30).

The percent deviation δ was defined by the following formula:

$$\delta = \frac{S_e - S_a}{S_a} \cdot 100, \qquad (35)$$

where S_e is the difference $S_e = G_M - G_m$ taken from the experimental values and S_a is the difference $S_a = G'_M - G'_m$, where G'_M and G'_m designate the theoretical values given by Eq. (26). The quantity S_e we can also call the swing of the experimental sine component whereas S_a the swing of the theoretical sine component. The quantity δ enables to compare magnitudes of the alternating sine components of the sensor response at given frequency.

4.   Results

The values of the time difference Δt were found in the experimental values of the measurement carried out in the clean air. It was found, that the maximum of G is delayed behind the maximum of $U(t)$ as expected according to heat conduction theory. The value h_1 was calculated from Eq. (30). The overview of the calculated values is in Tab. 1.

Tab. 1: The experimental values of the sensor in the clean air.

f (Hz)	Δt (s)	h_1 ($m^2 \cdot s^{-1} \cdot K^{-1}$)	Average of h_1
		MQR 1003	
0.0312	4.0	4.442	
0.0625	3.0	4.712	4.209
0.156	1.4	3.474	
		SP 11	
0.0312	1.0	1.110	
0.0625	1.5	2.356	1.982
0.156	1.0	2.481	
		TGS 813	
0.0312	3.0	3.332	
0.0625	2.5	3.926	3.577
0.156	1.4	3.474	

The constants A and B were calculated from Eq. (33) and Eq. (34) at the frequency $f = 0.0312$ Hz with the use of the average value of h_1 from Tab. 1. When two near maxima of G_M or two near minima of G_m occurred, their average value was used instead them. The calculated values are in Tab. 2.

Tab. 2: The coefficients of the sensor in the clean air at the frequency $f = 0.0312$ Hz.

G_M (μS)	G_m (μS)	$A \cdot 10^2$ (1)	$B \cdot 10^2$ (1)
	MQR 1003		
10.12	0.29	7.668	2.159
	SP 11		
9.82	0.30	7.665	1.429
	TGS 813		
10.07	0.49	7.810	1.745

The data in Tab. 2 were considered as input data used in Eq. (26) and only the frequency f was changed. Then Eq. (26) was used as the model of behaviour of the sensor.

An example of the results is in Fig. 1, Fig. 2 and Fig. 3. At the beginning the theoretical curves were calculated at the frequency $f = 0.0312$ Hz. The curves fit the experimental data well. The example is in Fig. 1. The accordance is a bit worse in Fig. 2, where the frequency $f = 0.0625$ Hz was used. The curve is even shifted compared to the measured data in Fig. 3, where the frequency $f = 0.156$ Hz was used. Heat is conducted from the heating system faster into the sensor compared to convection of heat away of the sensor into its surrounding at higher frequency. This phenomenon can elevate the temperature of the sensor and shift the theoretical curve compared to measured data. The

heat exchange between the sensor and its surrounding was not considered in the derivation of Eq. (26). Despite this the neglecting of the phenomenon is acceptable for the purpose of which Eq. (26) is used herein.

Table 3 includes the deviation δ calculated for the clean air. From it follows that the deviation δ does not exceed 23 %.

The sensors were further tested in the concentration $x = 100$ ppm of the single tested vapour in the air at the same heating voltage as in the clean air case. The conductivities G_M and G_m were found out in the experimental data of the tested vapour and average of h_1 from Tab. 1 was used to calculate the constants A and B with the use of Eq. (33) and Eq. (34) at the frequency

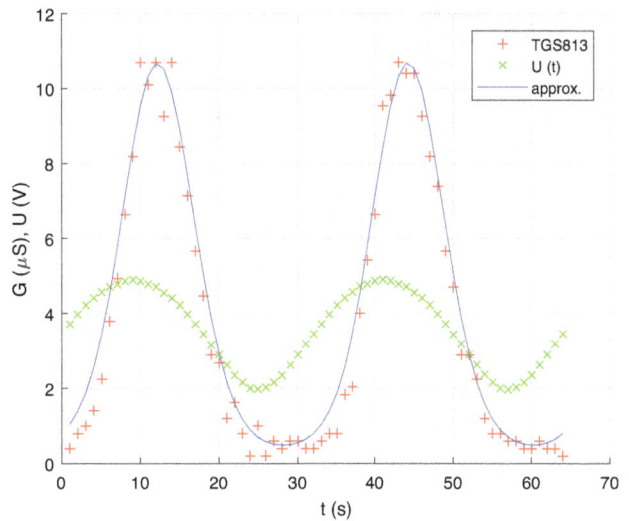

Fig. 1: The electrical conductivity $G(t)$ of the sensor TGS 813 at frequency $f = 0.0312$ Hz in the clean air. $U(t)$ is the heating voltage.

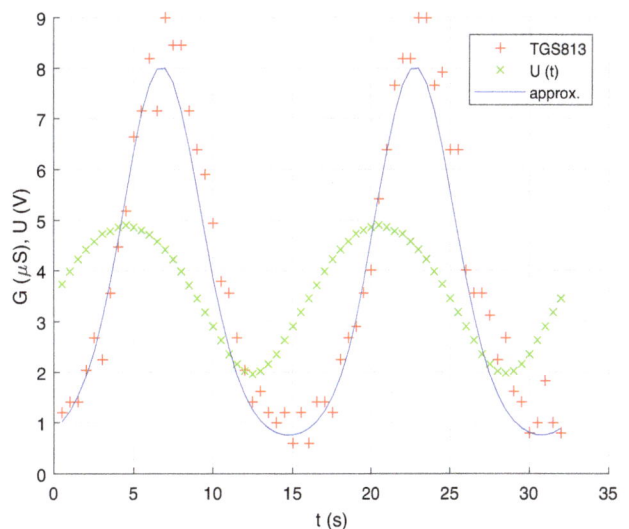

Fig. 2: The electrical conductivity $G(t)$ of the sensor TGS 813 at frequency $f = 0.0625$ Hz in the clean air. $U(t)$ is the heating voltage.

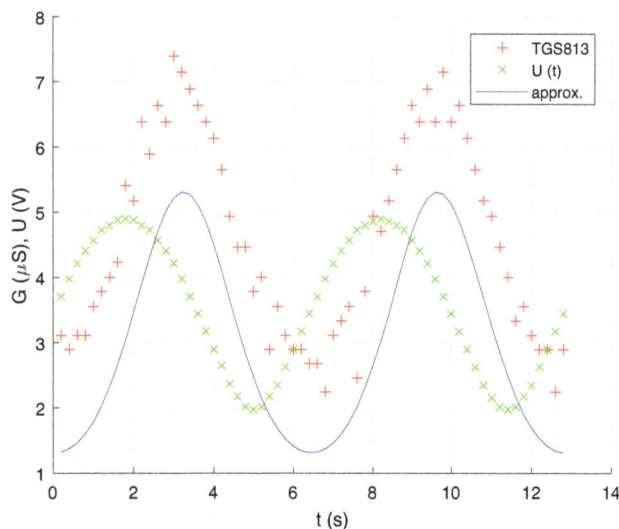

Fig. 3: The electrical conductivity $G(t)$ of the sensor TGS 813 at frequency $f = 0.156$ Hz in the clean air. $U(t)$ is the heating voltage.

Tab. 3: The deviations of δ in % for the sensors in clean air at the used frequencies.

f=0.0312 Hz	f=0.0625 Hz	f=0.156 Hz
MQR 1003		
3.04	−1.8	2.5
SP 11		
2.0	2.0	14.5
TGS 813		
1.5	17.2	23.0

$f = 0.0312$ Hz. These constants were used to calculate the reference curve by Eq. (26) and subsequently the deviation δ by Eq. (35) at all the frequencies. Table 4 includes the results.

Tab. 4: The values of δ in % for the sensor in the tested vapours.

	f=0.0312 Hz	f=0.0625 Hz	f=0.156 Hz
acetone			
MQR 1003	0.5	12.9	59.9
SP 11	2.7	−51.6	−62.4
TGS 813	4.0	−19.5	−27.4
ethanol			
MQR 1003	0.8	46.3	53.8
SP 11	5.4	-14.6	−42.2
TGS 813	0.9	6.7	−25.3
toluene			
MQR 1003	0.5	10.9	−45.7
SP 11	7.6	−54.7	−79.7
TGS 813	7.5	−37.8	−42.1

The deviations δ in the column at the frequency $f = 0.0312$ Hz are small, because of the constants A and B which were calculated from the experimental data at this frequency. This represents the reference column. The results in Tab. 4 show that the swing of the conductance of the sensor MQR 1003 is greater by 59.9 % at the frequency $f = 0.156$ Hz for the acetone case than expected. Similarly it is for the case when ethanol was used at the frequencies $f = 0.625$ Hz and

$f = 0.156$ Hz. On the other hand the sensor SP 11 indicates for the acetone case and for the toluene case the swing smaller by −51.6 % or even less at the frequencies $f = 0.0625$ Hz and $f = 0.156$ Hz then expected. It follows that the parameter δ can serve as a decision criterion if the sensor is more or less sensitive to the compound at given frequency compared to the expected value. The curve calculated by Eq. (26) can be considered as a reference curve for which the parameter δ is calculated by Eq. (35).

An example of measured data and the dependency of Eq. (26) is in Fig. 4, Fig. 5, and Fig. 6. For the measured data, the maximum of G is not always delayed behind the maximum of $U(t)$ as expected when a compound is detected. That is why the method of the calculation of h_1 by Eq. (30) fails here. The coefficient h_1 is related to the thermal properties of the sensor and it would be independent on a detected compound. The correct value of h_1 can be calculated when the experimental data of the clean air are taken.

It is general knowledge [21], [22] and [23], that during the detection of ethanol the maximum of the electrical conductance occurs at specific heating voltage between the values $U = 2$ V and 5 V. Hence the maximum of $G(t)$ for the ethanol case occurs earlier than the maximum of $G(t)$ for the clean air case. The corollary of this is less or even a negative delay Δt between $G(t)$ and $U(t)$. The phenomenon can be seen in Fig. 4, Fig. 5 and Fig. 6. Ethanol is not only one compound with maximum electrical conductance at specific heating voltage and the negative delay Δt can occur also during the test of other substances.

It follows from Eq. (26) that for $f = \infty$, the magnitude of the sine component tends to zero. The value

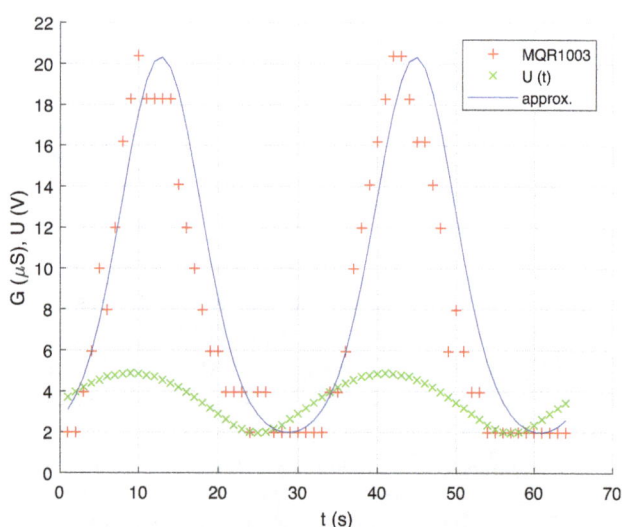

Fig. 4: The electrical conductivity $G(t)$ of the sensor MQR 1003 for ethanol of the concentration $x = 100$ ppm at frequency $f = 0.0312$ Hz of the heating voltage. $U(t)$ is the heating voltage.

5.   Conclusion

The electrical conductivity of the tin dioxide gas sensor was derived herein. The model of the sensor is based on thermal inertia and it is valid for specific periodical heating voltage. The model was verified with the experimental data. It was found that the model can be used in the determination of sensitivity of the sensor to the tested substance at given frequency.

Acknowledgment

This study was enabled due to Project TA04021263 entitled The Intelligent Modules for Communication and Lightening provided by TACR (The Technology Czech Science Foundation) and due to Project FV10369 SIDAS entitled The System of Inelligent Detection and Signaling of Clashing States for Improvement the Truck Safety provided by The Ministry of Industry and Trade. Further this article was prepared within the frame of sustainability of the project No. CZ.1.07/2.3.00/20.0217 "The Development of Excellence of the Telecommunication Research Team in Relation to International Cooperation" within the frame of the operation programme "Education for competitiveness" that was financed by the Structural Funds and from the state budget of the Czech Republic.

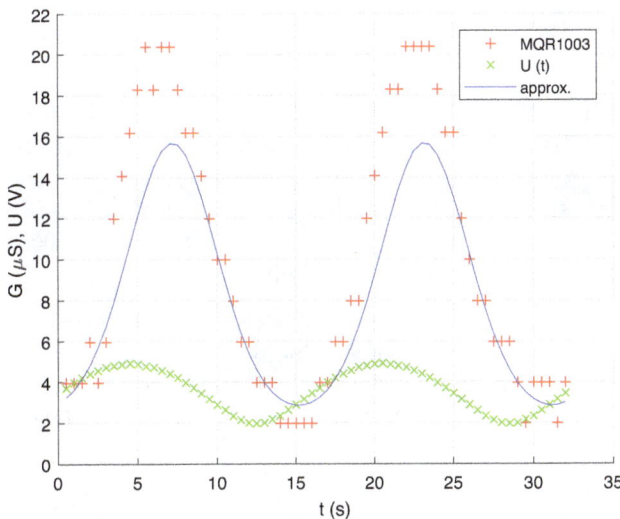

Fig. 5: The electrical conductivity $G(t)$ of the sensor MQR 1003 for ethanol of the concentration $x = 100$ ppm at frequency $f = 0.0625$ Hz of the heating voltage. $U(t)$ is the heating voltage.

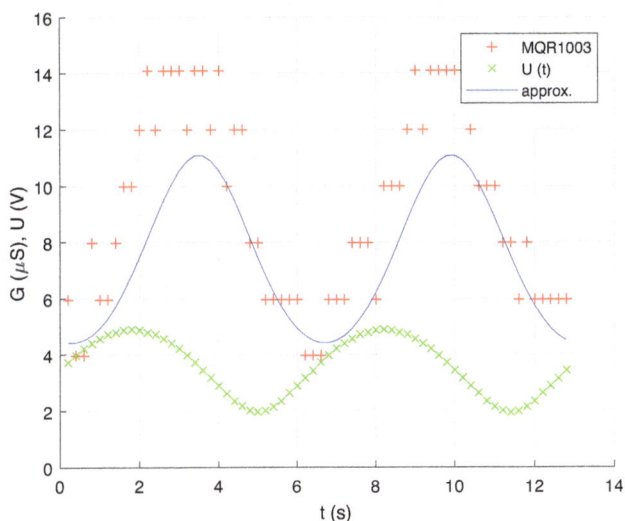

Fig. 6: The electrical conductivity $G(t)$ of the sensor MQR 1003 for ethanol of the concentration $x = 100$ ppm at frequency $f = 0.156$ Hz of the heating voltage. $U(t)$ is the heating voltage.

of the electrical conductivity $G(t)$ becomes time independent despite the time dependent heating. The electrical conductivity tends to the value as if the sensor were heated by the specific constant voltage. It is possible, using Eq. (31), Eq. (32), Eq. (33) and Eq. (34), to calculate the frequency at which the amplitude of the sine component of the electrical conductivity reaches the prescribed value. The values of δ in Tab. 4 indicate great positive or negative deviations compared to the expected value. It means that the decay of the swing of the sine component is not monotonous in the presence of the detected substance as theoretically expected and the prediction of such a frequency can be carried out only experimentally for the tested compound.

References

[1] XINGJIU, H., J. S. DONGLIANG, P. ZONXIN and Y. ZENGLIANG. Rectangular mode of operation for detecting pesticide residue by using a single SnO$_2$-based gas sensor. *Sensors and Actuators B: Chemical*. 2003, vol. 96, iss. 3, pp. 630–635. ISSN 0925-4005. DOI: 10.1016/j.snb.2003.07.006.

[2] XINGJIU, H., M. FANLI, P. ZONGXIN, X. WEIHONG and L. JINHUAI. Gas sensing behavior of a single tin dioxide sensor under dynamic temperature modulation. *Sensors and Actuators B: Chemical*. 2004, vol. 99, iss. 2–3, pp. 444–450. ISSN 0925-4005. DOI: 10.1016/j.snb.2003.12.013.

[3] HERRERO-CARON, F., D. J. YANEZ and F. DE BORJA RODRIGUEZ. An active, inverse temperature modulation strategy for single sensor odorant classification. *Sensors and Actuators B: Chemical*. 2015, vol. 206, iss. 1, pp. 555–563. ISSN 0925-4005. DOI: 10.1016/j.snb.2014.09.085.

[4] NAKATA, S., H. NAKAMURA and K. YOSHIKAWA. New strategy for the development of a gas sensor based on the dynamic

characteristics: principle and preliminary experiment. *Sensors and Actuators B: Chemical.* 1992, vol. 8, iss. 2, pp. 187–189. ISSN 0925-4005. DOI: 10.1016/0925-4005(92)80179-2.

[5] NAKATA, S., K. TAKEMURA and K. NEYA. Non-linear dynamic responses of a semiconductor gas sensor: Evaluation of kinetic parameters and competition effect on the sensor response. *Sensors and Actuators B: Chemical.* 2001, vol. 76, iss. 1–3, pp. 436–441. ISSN 0925-4005. DOI: 10.1016/S0925-4005(01)00652-9.

[6] NAKATA, S., H. OKUNISHI and Y. NAKASHIMA. Distinction of gases with a semiconductor sensor under a cyclic temperature modulation with second-harmonic heating. *Sensors and Actuators B: Chemical.* 2006, vol. 119, iss. 2, pp. 556–561. ISSN 0925-4005. DOI: 10.1016/j.snb.2006.01.009.

[7] JAEGLE, M., J. WOLLENSTEIN, T. MEISINGER, H. BOTTNER, G. MULLER, T. BECKER and C. BOSCH-BRAUNMUHL. Micromachined thin film SnO_2 gas sensors in temperature-pulsed operation mode. *Sensors and Actuators B: Chemical.* 1999, vol. 57, iss. 1–3, pp. 130–134. ISSN 0925-4005. DOI: 10.1016/S0925-4005(99)00074-X.

[8] HEILIG, A., N. BARSAN, U. WEIMAR, M. SCHWEIZER-BERBERICH, J. W. GARDNER and W. GOPEL. Gas identification by modulating temperatures of SnO_2-based thick film sensors. *Sensors and Actuators B: Chemical.* 1997, vol. 43, iss. 1–3, pp. 45–51. ISSN 0925-4005. DOI: 10.1016/S0925-4005(97)00096-8.

[9] VERGARA, A., E. LLOBET, J. BREZMES, P. IVANOV, C. CANE, I. GRACIA, X. VILANOVA and X. CORREIG. Quantitative gas mixture analysis using temperature-modulated micro-hotplate gas sensors: Selection and validation of the optimal modulating frequencies. *Sensors and Actuators B: Chemical.* 2007, vol. 123, iss. 2, pp. 1002–1016. ISSN 0925-4005. DOI: 10.1016/j.snb.200.

[10] ILLYASKUTTY, N., J. KNOBLAUCH, M. SCHWOTZER and H. KOHLER. Thermally modulated multi sensor arrays of SnO_2/additive/electrode combinations for enhanced gas identification. *Sensors and Actuators B: Chemical.* 2015, vol. 217, iss. 1, pp. 2–12. ISSN 0925-4005. DOI: 10.1016/j.snb.2015.03.018.

[11] HOSSEIN-BABAEI, F. and A. AMINI. A breakthrough in gas diagnosis with a temperature-modulated generic metal oxide gas sensor. *Sensors and Actuators B: Chemical.* 2012, vol. 166–167, iss. 1, pp. 419–425. ISSN 0925-4005. DOI: 10.1016/j.snb.2012.02.082.

[12] IONESCU, R. and E. LLOBET. Wavelet transform-based fast feature extraction from temperature modulated semiconductor gas sensors. *Sensors and Actuators B: Chemical.* 2002, vol. 81, iss. 2–3, pp. 289–295. ISSN 0925-4005. DOI: 10.1016/S0925-4005(01)00968-6.

[13] LEE, A. P. and B. J. REEDY. Temperature modulation in semiconductor gas sensing. *Sensors and Actuators B: Chemical.* 1999, vol. 60, iss. 1, pp. 35–42. ISSN 0925-4005. DOI: 10.1016/S0925-4005(99)00241-5.

[14] IONESCU, R., E. LLOBET, S. AL-KHALIFA, J. W. GARDNER, X. VILANOVA, J. BREZMES and X. CORREIG. Response model for thermally modulated tin-oxide absed microhotplate gas sensors. *Sensors and Actuators B: Chemical.* 2003, vol. 95, iss. 1–3, pp. 203–211. ISSN 0925-4005. DOI: 10.1016/S0925-4005(03)00420-9.

[15] FORT, A., S. ROCCHI, M. B. SERRANO-SANTOS, M. MUGNAINI, V. VIGNOLI, A. ATREI and R. SPINICCI. CO sensing with SnO_2 based thick film sensors: Surface state model for conductance responses during thermal modulation. *Sensors and Actuators B: Chemical.* 2006, vol. 116, iss. 1–2, pp. 43–48. ISSN 0925-4005. DOI: 10.1016/j.snb.2005.11.070.

[16] DING, J., T. J. MCAVOY, R. E. CAVICCHI and S. SEMANCIK. Surface state trapping models for SnO_2-based microhotplate sensors. *Sensors and Actuators B: Chemical.* 2001, vol. 77, iss. 3, pp. 597–613. ISSN 0925-4005. DOI: 10.1016/S0925-4005(01)00765-1.

[17] KUNT, T., T. MCAVOY, R. E. CAVICCHI and S. SEMANCIK. Optimization of temperature programmed sensing for gas identification using micro-hotplate sensors. *Sensors and Actuators B: Chemical.* 1998, vol. 53, iss. 1–2, pp. 24–43. ISSN 0925-4005. DOI: 10.1016/S0925-4005(98)00244-5.

[18] GOSANGI, R. and R. GUTIERREZ-OSUNA. Active temperature modulation of metal-oxide sensors for quantitative analysis of gas mixtures. *Sensors and Actuators B: Chemical.* 2013, vol. 185, iss. 1, pp. 201–210. ISSN 0925-4005. DOI: 10.1016/j.snb.2013.04.056.

[19] MARTINELLI, E., D. POLESE, A. CATINI, A. D'AMICO and C. DI NATALE. Self-adapted temperature modulation in metal-oxide semiconductor gas sensors. *Sensors and Actuators B: Chemical.* 2012, vol. 161, iss. 1–3, pp. 534–541. ISSN 0925-4005. DOI: 10.1016/j.snb.2011.10.072.

[20] MORRISON, S. R. Semiconductor gas sensors. *Sensors and Actuators B: Chemical.* 1981–1982, vol. 2, iss. 1, pp. 329–341. ISSN 0925-4005. DOI: 10.1016/0250-6874(81)80054-6.

[21] SINGH, R. C., N. KOHLI, N. P. SINGH and O. SINGH. Ethanol and LPG sensing characteristics of SnO_2 activated Cr_2O_3 thick film sensor. *Bulletin of Materials Science.* 2010, vol. 33, iss. 5, pp. 575–579. ISSN 0250-4707.

[22] SCHMID, W., N. BARSAN and U. WEIMAR. Sensing of hydrocarbons with tin oxide sensors: possible reaction path as revealed by consumption measurements. *Sensors and Actuators B: Chemical.* 2003, vol. 89, iss. 3, pp. 232–236. ISSN 0925-4005. DOI: 10.1016/S0925-4005(02)00470-7.

[23] HAE-WON, C. and L. MAN-JONG. Sensing characteristics and surface reaction mechanism of alcohol sensors based on doped SnO_2. *Journal of Ceramic Processing Research.* 2006, vol. 7, iss. 3, pp. 183–191. ISSN 1229-9162.

About Authors

Libor GAJDOSIK graduated in telecommunication engineering from the Czech Technical University (Prague, Czech Republic) in 1983. He received his Ph.D. in 1998 from the VSB–Technical University of Ostrava (Ostrava, Czech Republic). His thesis was focused on the detection properties of tin dioxide gas sensors at the dynamic operating modes. He has been an assistant professor at the Technical University of Ostrava since 1989. His main interests are electronics, circuit theory and chemical sensors.

21

A Novel Digital Background Calibration Technique for 16 Bit SHA-less Multibit Pipelined ADC

Swina NARULA[1], *Munish VASHISHATH*[2], *Sujata PANDEY*[1]

[1]Amity School of Engineering and Technology, Amity University Uttar Pradesh,
Sector 125, Noida, Uttar Pradesh 201303, India
[2]YMCA University of Science and Technology Faridabad, Sector 6, Mathura Road,
Opp. Sanjay Memorial Industrial Estate, Faridabad, Haryana 121006, India

swinanarula@gmail.com, munishvashishath@ymcaust.ac.in, spandey@amity.edu

Abstract. *In this paper, a high resolution of 16 bit and high speed of 125 MS·s^{-1}, multibit Pipelined ADC with digital background calibration is presented. In order to achieve low power, SHA-less front end is used with multibit stages. The first and second stages are used here as a 3.5 bit and the stages from third to seventh are of 2.5 bit and last stage is of 3-bit flash ADC. After bit alignment and truncation of total 19 bits, 16 bits are used as final digital output. To remove linear gain error of the residue amplifier and capacitor mismatching error, a digital background calibration technique is used, which is a combination of Signal Dependent Dithering (SDD) and butterfly shuffler. To improve settling time of residue amplifier, a special circuit of voltage separation is used. With the proposed digital background calibration technique, the Spurious-Free Dynamic Range (SFDR) has been improved to 97.74 dB @ 30 MHz and 88.9 dB @ 150 MHz, and the signal-to-noise and distortion ratio (SNDR) has been improved to 79.77 dB @ 30 MHz, and 73.5 dB @ 150 MHz. The implementation of the Pipelined ADC has been completed with technology parameters of 0.18 μm CMOS process with 1.8 V supply. Total power consumption is 300 mW by the proposed ADC.*

Keywords

Butterfly, CMOS, digital background calibration, op-amp, pipelined ADC, SHA-less front-end, Signal Dependent Dithering.

1. Introduction

Various digital applications, such as communication base stations, cable head end, professional HDTV cameras and video digitizers require Analog to Digital Converters (ADCs) of high-resolution, high-speed and low cost [1], [2] and [3]. Moreover, the modern IF-sampling super heterodyne communication systems require an DC of reduced receiver complexity, reduced overall system cost and should be able to sample input signals above 150 MHz. It's a big challenge for designers to implement a low power Pipelined ADC of high sampling rates with more than 16-bit resolution, which could be good for high speed data communication. Different types of errors like op-amp nonlinearity, capacitor mismatch and finite op-amp gain may come in MDAC, which degrades the performance and limit the resolution of Pipelined ADC. Linear errors like finite op-amp gain and capacitor mismatch could be removed by using various techniques [4], but to minimize nonlinear errors, very few techniques are available in the industry [5]. Many digital background calibration techniques have been developed due to limitations of analog circuits. Moreover, CMOS is the technology which is preferably used to implement switched capacitor circuits due to its low cost. Similar types of pipelined ADCs are the most popular in mobile communication systems.

A low power 16-bit, 125 MS·s^{-1} pipelined ADC is presented here for high resolution applications. To save power, the dedicated front-end Sample and Hold Amplifier (SHA) has been removed, which also permits ADC to use a smaller input sampling capacitor to facilitate its drivability. But the SHA-less architecture causes some problems like IF sampling front-end, which could be removed by using specific circuit designs [6]. By using a multi-bit front-end stage, additional power could be saved, but the circuit complexity will increase. The proposed Pipelined ADC adopts a multi-bit structure, where the first and second stages are of 3.5 bit, whereas other stages from 3 to 7 are of 2.5 bit each, and the 3-bit Flash ADC works as the last stage. To minimize the effects of nonlinear gain errors, a two-stage amplifier with a gain-boosted structure is implemented in the first stage to realize a high swing and high gain op-amp. The capacitor mismatches and linear gain errors are corrected by a combination of techniques, named as Signal Dependent Dithering (SDD) [7] and Butterfly Shuffler (BS) [8]. These techniques are useful for removing small signal linearity errors. Due to capacitor mismatching and other non-idealities of Op-amp, the ADC transfer function may be affected by discontinuities produced by DNL errors. By using dithering, the problem gets resolved as described in this paper. By using CMOS process technology, performance requirement gets reduced and simplifies the design of the analog circuit by inserting a digital algorithm. The calibration algorithm requires an extra circuit between sub-ADC and sub-DAC, which introduces the problem of an extra delay as well as reduces the settling time for residue amplifier. To remove these problem, a properly adjusted clock is used, which could control the switch to short differential outputs of Op-amp in hold phase. To remove the interference between different stages, separate voltage references are used. So, the first two stages are using different voltage reference than the other backend stages of Pipelined ADC. It has large transient current to charge and discharge capacitors in hold phase to backend stages, which is another reason to use different voltage references here.

Different sections of this paper is as follows. In Section 2. , front-end considerations are discussed. In Section 3. , the implementation details of the major circuit blocks of the proposed pipelined ADC are discussed. Both the dithering and butterfly calibration techniques used for the proposed pipelined ADC are discussed in Section 4. Section 5. presents the measurement results for the proposed ADC. The summary & conclusion of this paper is the part of Section 6.

2.   Front-End Considerations

It was a dream to accomplish high SNDR and low power dissipation simultaneously in early Pipelined ADC designs. To achieve high SNDR, it requires a large sampling capacitors, which increases power dissipation along with circuit complexity. By using the advantages of bipolar devices, BiCMOS technology could solve the problem of drivability of Pipelined ADC [9]. And, by using CMOS technology, the problem of large input sampling capacitor could be resolved by

the switched-capacitor CMOS pipelined ADCs where on-chip input buffer has been removed. In pipelined ADC, a dedicated front-end SHA is most commonly used, which highly contributes to noise, distortion and consume significant power for the complete Pipelined ADC unit. So, by removing SHA, power dissipation can be reduced along with noise reduction. Also, reducing the number of those stages which contribute to noise is the best solution to increase SNDR and SFDR without using large sampling capacitors.

In this way, other stages of Pipelined ADC can work well even with smaller sampling capacitors to minimize the overall circuit's noise and power. By using the smaller values of the input sampling capacitors, the sizes of switches can be reduced further to attain the same bandwidth, and hence the nonlinear parasitic capacitance in the input sampling network could be reduced [10]. Further the linearity of sampling circuit has improved, but the complexity of circuit gets increased with the multi bit ADC architecture [11] and [12].

A single stage amplifier has been considered here with an effective trans-conductance (gm) and feedback factor (β), whose results are also applicable to a two-stage amplifier. So, the Bandwidth (BW) for that is given by the Eq. (1):

$$BW = \beta \frac{gm}{CL}. \qquad (1)$$

By adding each bit in stage-1, feedback factor β becomes halved, so the load capacitor (CL), goes half to keep bandwidth constant, whereas power dissipation remains constant. Here, the load capacitor (CL) is actually the addition of the effective loading caused by the feedback capacitor (Cf), sampling capacitors (Cs) of stage-1, the overall sampling capacitance of stage-2 $(Cs2)$ and parasitic capacitances. If CL has been dominated by $Cs2$ then CL can be halved by making half of $Cs2$ to make BW constant. There are two major sources of noise, which contributes to the accumulative noise sampled by stage-2 $(\sigma2)$ and the first source of the noise in the stage-2 sampling circuit is due to the switches $(\sigma2\_ST)$. When the stage-1 amplifier is in hold phase, the noise occurred that time acts as another major source of noise $(\sigma2\_AP1)$. By referring to the input of the ADC, $\sigma2\_ST$ can be written as Eq. (2):

$$\sigma2\_ST \alpha \sqrt{\frac{KT}{CL}}/G. \qquad (2)$$

So, there is a reduction factor of $\sqrt{2}$ by which $\sigma2\_ST$ has been reduced. Equation (3) and Eq. (4) represent $\sigma1$ and $\sigma2\_AP1$ respectively, which stay moderately unaffected to first order when referred to input and stage-1 samples the track mode noise. Cp indicates

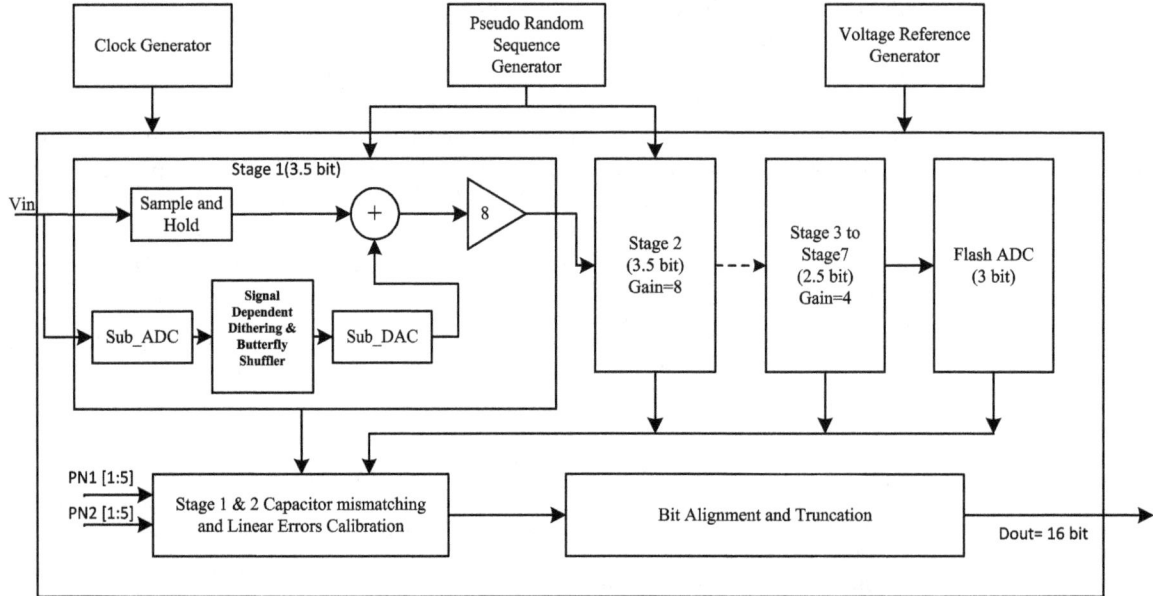

Fig. 1: Proposed 16 bit SHA-less multi-bit Pipelined ADC.

the summing node parasitic capacitance of stage 1.

$$\sigma 1\alpha\sqrt{\frac{KT(Cs+Cf+Cp)}{Cs^2}}=\sqrt{\frac{KT}{\beta GCL}}, \quad (3)$$

$$\sigma 2\_AP1\alpha\sqrt{\frac{KT}{\beta CL}}/G. \quad (4)$$

It's indicated from the equations above that by adding each quantized bit in the first stage, the gain will increase by two, at the same time some other factor will decrease by two, which is the feedback factor. So, there is the requirement to make a balance between power and number of bits. Here, SHA-less front-end has been chosen to achieve low power, high SNDR, high SFDR and minimum value of sampling capacitors but at the price of circuit complexity particularly for sampling frequencies near 200 MHz.

3. Proposed Pipelined ADC

In Fig. 1, the block diagram of 16-bit pipelined ADC is shown. The first two stages of the 16-bit Pipelined ADC are of 3.5 bits, the succeeding seven stages are of 2.5 bits each, and the 3-bit flash ADC has chosen for the last stage. To get a solution for high resolution of 16 bits, for low power consumption and for removing the capacitor mismatching problem, the first stage used here is of 3.5 bits. Multi-bit Pipelined ADC is much better than the Pipelined ADC, having all stages of the same kind. Sometimes, these multi-bit Pipelined ADCs are used even without calibration because these are able to save more power [22]. A problem that

may arise here is the complicated MDAC circuit designing, when resolution becomes higher than 4 bit. So, whereas the initial stages are of higher resolution, the succeeding stages have comparatively low resolution. As the number of stages are going to be scaled down along the pipelined ADC chain, the problem of capacitor mismatching and higher bias currents will be relaxed. Less area utilization is the additional benefit of multibit structured ADC because most of the stages are of low bit size.

The proposed architecture consists of a reference generator, pseudo random sequence generator, digital calibration block, bit alignment and truncation block, different stages of ADC of different bit size and clock generator. A voltage reference generator is used with low temperature coefficient and high Power Supply Rejection Ratio (PSRR). The clock generator is able to generate low jitter clocks for high input sampling frequencies. The capacitor mismatching and the residue amplifier's linear gain errors are corrected by digital background calibration which combines two techniques as SDD and butterfly shuffler. Two pseudo random noise sequences PN1 [1:5] and PN2 [1:5] are used to control Signal Dependent Dithering (SDD) and butterfly shuffler. The final 16-bit digital output can be achieved through digital bit alignment and truncation method.

3.1. SHA-Less Front-End

As SHA-less front-end has been used in this Pipelined ADC, both the multiplying DAC (MDAC) and the flash concurrently are able to sample the input signal.

The flash produces quantized output, which initiates MDAC to generate the residue signal for the next stage. The value of gain has been set to 8 for the MDAC of stage-1, and gain value set to 4 for stage 3 to stage 7. The stage-1 is able to quantize 4 bits and one bit out of those is a redundant bit, which is used to tolerate comparator offsets [15].

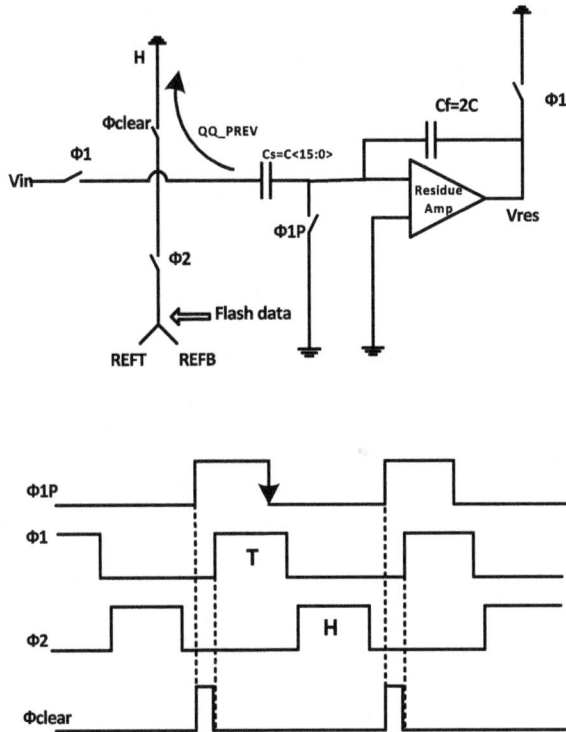

Fig. 2: Removing the nonlinear charge kickback to Vin, 4-bit stage-1 MDAC operation with Φ clear.

Figure 2 shows the stage-1 as a 4 bit MDAC with Φclear to eliminate the nonlinear charge kickback to Vin. The non-linear kickback could be settled, but the fact of how fast depends the value of Cs and sampling rate [14]. Another factor of the circuit which can be taken care of is effective impedance, as it drives the input. So, a large value of Cs is required for low noise coupled to stage-1 with its input buffer. This will avoid the kickback. The sampling flash architecture is used for the SHA-less front end to reduce the mismatch effects between the flash and the signal sampled by the MDAC [16] and [17].

Figure 3 shows the SHA-less stage-1 Switched-capacitor implementation of 4-bit without any dithering [22]. During the track phase Φ1, the flash sampling capacitors (C1s_fl) and the sampling capacitor (Cs) are concurrently sampling Vin. Also, sampling the reference ladder is completed by a different set of capacitors in the flash (C1ref_fl). To keep track of the input signal and to minimize sampling distortion, linear bootstrapped input switches are used to perform

this task [18] and [19]. C1s_fl and C1ref_fl are short together at the beginning of Φ2, to produce input for the pre-amplifier, and latch to produce valid flash data together with a small delay. The flash data is used to produce the residue voltage (Vres) which is driven by the MDAC reference switches used at the output of the residue amplifier. Here, during Φ2, the most challenging job is to achieve the total delay over the pre-amplifier and the latch, for the available settling time for MDAC. So, both the residue amplifier and flash comparators have to be faster in this SHA-less front-end, which consumes more power in SHA-less front-end as compared to a conventional one.

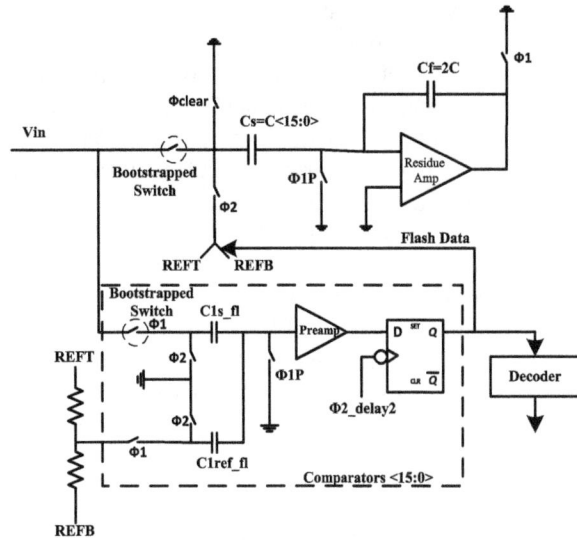

Fig. 3: SHA-less Stage-1, 4 bit Switched-capacitor implementation.

The choice between SHA-less front-end and conventional SHA front-end depends upon the application because by eliminating the SHA, power could be saved but for medium resolution and medium sampling rates only. For high resolution and very high sampling rate, power saving is difficult as flash comparators could use the maximum time offered for MDAC settling. So, the designing of fast and low-power comparators is practically a challenging job. Even for high sampling rates ADCs, SHAs are designed specifically for the application, which cannot be generalized. Here a precise design of SHA-less architecture with high sampling rate of 125 MS·s^{-1} is presented, which is capable to save power significantly. Since Cs and C1s_fl are simultaneously sample Vin, a problem that may arise in SHA-less front end is the difference between the values of flash and values sampled by the MDAC. It may use the available correction range, and sometimes it causes different sampled values than actual, due to timing and bandwidth mismatch between MDAC sampling networks and the flash ADC. Also for higher input frequencies, the difference becomes high and if frequencies increase further, stage-1 may cross the range of correction, which may result in missing codes. Hence, the important factor is to reduce the sampling mismatch and to increase the range of tolerance so that it always comes within the correction range. To get a complete range of codes, without any missing one, it should fulfil the following equation:

$$|\text{Sampling mismatch} + \text{Flash offsets}| < \text{VREF}/2^4. \quad (5)$$

To maximize the tolerance of sampling mismatch, their offsets should be minimized and offset cancellation could be used [20]. So, the fast comparators, used here for stage-1, are subject to offsets with high values to satisfy offset cancellation requirement. Due to the sampling flash architecture for the sampling capacitor (C1s_fl) and the reference capacitor (C1ref_fl), the capacitive reduction of 2 can be achieved in advance for pre-amplification, which doubles the flash offset and refers to input as associated to normal flash. In the stage-1, with increase in number of bits, the tolerance of sampling mismatch would double the available correction range and the design would be easy. The comparators are relaxed now from offset cancellation problem as matching requirement between MDAC sampling network and flash has been resolved but at the cost of high power consumption. To drive the sampling switches to minimize the timing skew concerning the MDAC sampling and flash clocks, the same buffered clock edge has been used here.

Since the two networks are different, matching their bandwidths is a tricky job as Cs is much larger than C1s_fl. The second reason is that the flash network is made up of two equal capacitors (C1s_fl and C1ref_fl), whereas the MDAC has two unequal capacitors, a large Cs and a small Cf. The third reason is that the parasitic capacitance of pre-amplifier is much lower than the residue amplifier. In-spite of all the differences, the bandwidths could be matched for two networks using post layout extracted simulations. The circuits could be cautiously by taking care of bandwidth, comparator offset cancellation and timing matching. SHA-less ADC can be sample inputs beyond 300 MHz and within the correction range itself.

3.2.   Flash Comparator

The flash comparator for stage-1 is similar to [17] but without cross-coupled PMOS pairs with the advantage of shorter regeneration time as shown in Fig. 4. In this comparator, PMOS are used as loads to make up a pre-amplifier. Here a diode is coupled to a latch and PMOS loads. When Φ2 begins, the pre-amplifier starts to amplify the input signal up to the peaks of different nodes.

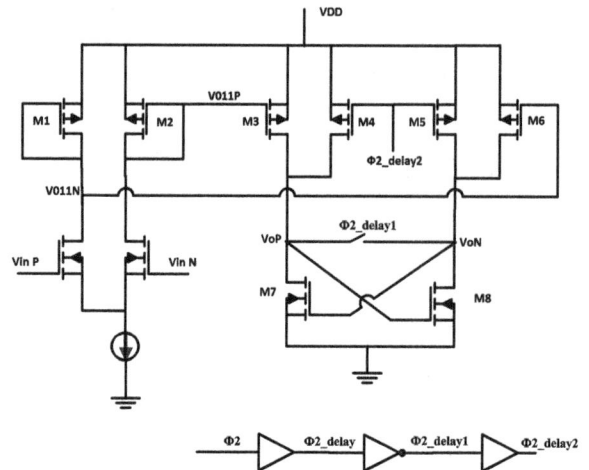

Fig. 4: Flash comparator circuit for Stage-1.

3.3.   Residue Amplifier (RA)

The residue amplifier used here is basically a Two-stage Miller-compensated design as shown in Fig. 5, and it is able to give high gain, large swing and excellent linearity [21] and [22]. The telescopic structure with PMOS inputs are used here for the first stage and NMOS differential pair with PMOS loads are used at the second stage. There is a large gm per unit current, so NMOS inputs should be in the first stage of the Residue Amplifier. But for CMOS process technology, there is an indication of greater value of 1/f noise, if NMOS devices are used comparatively to PMOS devices. A problem of low frequency nature of 1/f noise could be solved

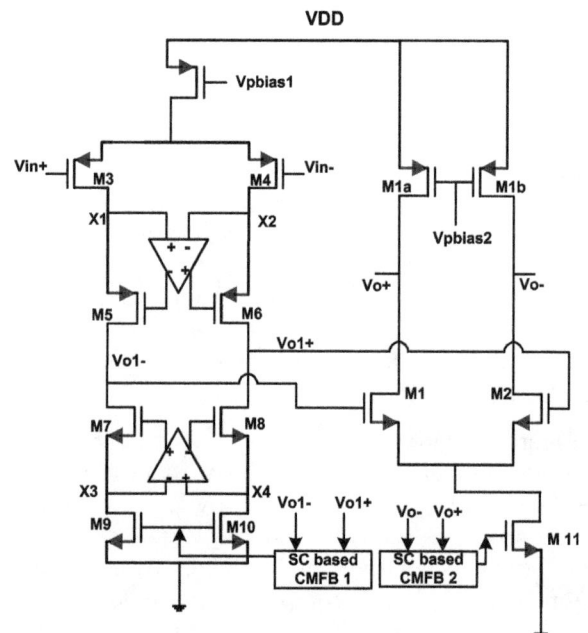

Fig. 5: Residue Amplifier circuit for Stage-1.

by growing the Miller capacitor, which can directly filter the noise. Therefore, a large Miller capacitor is used here. On the other side, an auto-zeroing amplifier could be used, which would help to reduce 1/f noise but at the price of increased circuit complexity and power consumption. To get the lowest power solution, PMOS inputs are preferably used for stage-1 due to their lower 1/f noise nature.

3.4. Voltage References

To supply stable voltages to all stages of Pipelined ADC, specific Voltage reference structure is used as mentioned by Keetal, [27]. Using this voltage referencing arrangement, there will be no effect on final digital output because digital calibration performed for the first two stages will be passed to backend stages as well. When hold phase comes, depending upon outputs of sub_ADC, the sampling capacitor may charge or discharge with sub_DAC's reference voltages. If just one reference voltage was used, a large current would be required by the capacitors of the first stage, and the transient output of other stages may lead to slow settling, hence two different voltages are used here. When the hold phase starts, the sampling capacitors also start, which supplies small current. By using separate voltages, the final outputs will be unaffected. In this way, the current utilization would be optimized and hence power could be saved.

4. Proposed Calibration Technique

High gain Op-amps are used here in this design at each stage to remove nonlinear errors and capacitor mismatching. The linear gain errors are treated together and could be corrected by the background digital calibration technique. For this design, Signal Depended Dithering (SDD) [7] is used with butterfly Shuffler [8]. For a 3.5 bit MDAC stage, the transfer function is changed for SDD with fifteen additional comparators a1 a15 as shown in Fig. 6.

Depending on the current PN code and the location of output residue voltage (Vres), dithering is injected here. For the calibration algorithm of 3.5 bit MDAC stage, the circuit modifications has to be done as per the residue plot's requirements, indicated clearly in Fig. 6.

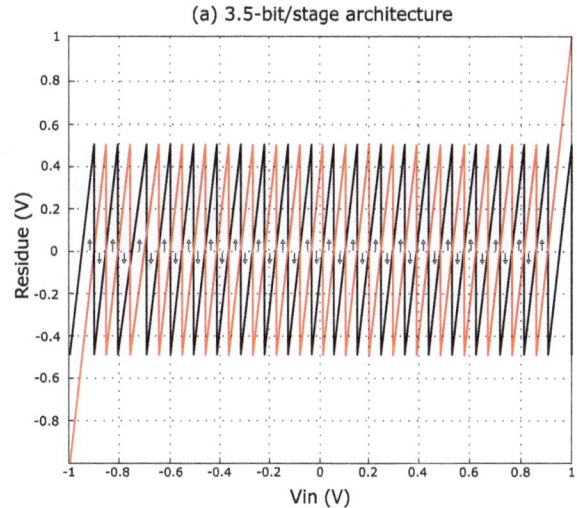

Fig. 6: Determination of Signal Dependent Dithering in Residue plot for a 3.5 bit MDAC stage.

For PN [2:5] =4'b0000, the transfer function can be expressed as per following equation:

$$Vres = G \cdot \left[Vin - \sum_{i=1}^{14} \left(Di \cdot Vref \frac{Ci}{Cs} \right), \right.$$

$$\left. -PNdither \cdot Vref \cdot \frac{C15}{Cs} \right],$$

$$\text{where } G = \frac{Cs}{Cf + (Cs + Cf + Cp)/A},$$

$$\text{and } Cs = \sum_{i=1}^{16} Ci.$$

(6)

The equivalent sampling capacitor is Cs, the feedback capacitor is Cf, the equivalent input parasitic capacitor of op-amp is Cp and the actual gain of the residue amplifier is G. When PN [2:5] is equal to 4'b0000, the dithering signal is injected to C15 and D1 D14 are the fourteen normal outputs of sub_ADC. PNdither is equal to (PN [1] +loc_res) 2, where PN sequences PN [1] = -1, 1 and loc_res decides the location of output residue Vres. If Vres is on the upper half plane of residue graph Fig. 7, loc_res=1 makes PNdither = 0, 1 which means the transfer curve can move down or remains unchanged. The transfer curve can move up or remains unchanged, when loc_res= -1, which makes PNdither = 0, -1. To extract the errors, Eq. (6) multiplies by -2 so,

$$Vres(-2) \cdot PN[1] = -2G\left[Vin - \sum_{i=1}^{14}\left(Di \cdot Vref\frac{Ci}{Cs}\right)\right.$$

$$\left(\frac{PN[1] + loc_res}{2}\right) \cdot Vref \cdot \left.\frac{C15}{Cs}\right] PN[1]$$

$$= PN\_modulate + G \cdot Vref \cdot \frac{C15}{Cs}$$

$$as\ PN[1]^2 = 1$$

$$where\ PN\_modulate = -2G\left[Vin - \sum_{i=1}^{14}\right.$$

$$\left(Di \cdot Vref\frac{Ci}{Cs}\right) - \left(\frac{loc\_res}{2}\right)Vref \cdot \left.\frac{C15}{Cs}\right] \cdot PN[1].$$
$$(7)$$

As loc_res is not related to PN [1] so, PN_modulate can be assumed as a noise that could be avoided with the help of a low-pass filter, which is ideally an accumulator and is able to average a large quantity of samples. So, the Eq. (8) is another way to write the Eq. (7):

$$WG, 15 = \frac{1}{N}\sum_{i=1}^{N} -2Vres(n) \cdot PN[1]|N \to \infty$$
$$= G \cdot Vref \cdot \frac{C15}{Cs}.$$
$$(8)$$

By considering infinite samples, whose average has been taken, WG,15 would include both capacitor mismatches and linear gain error. By using Butterfly Sampling technique for injecting dithering to each capacitor Ci (i= 1 to 16), which is controlled by four bit PN sequence PN [2:5] shown in Fig. 7. Total of 16 different WG (i=1 to 16) are obtained from Eq. (8) and saved to their respective registers, which are marked as WG,i (i=1 to 16) is shown in Tab. 1.

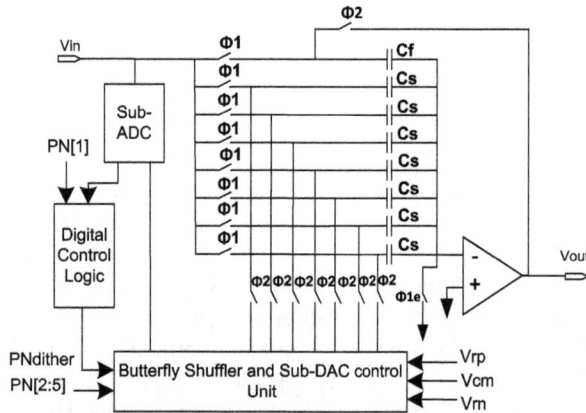

Fig. 7: 3.5 bit MDAC stage's circuit modifications for calibration.

The working of butterfly shuffler is illustrated in Fig. 8. The final digital output after calibration and by assuming backend stages are ideal, will be equal to G.Vin/4 which is linear.

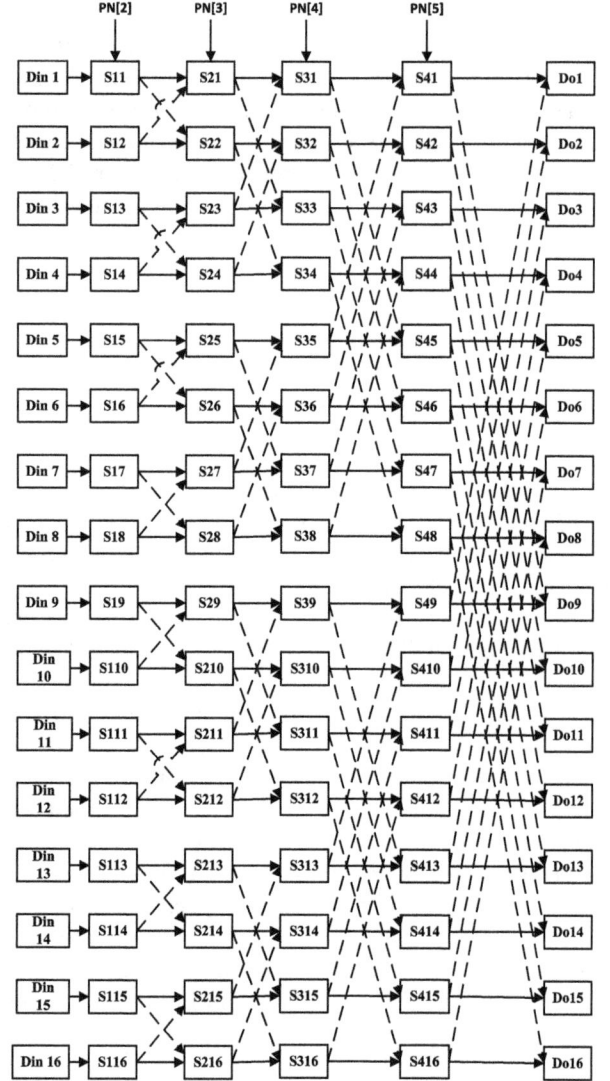

Fig. 8: Butterfly shuffler for 3.5-bit stage.

5.   Results and Measurements

A two-stage Miller compensated Op-amp topology has been used for the proposed 16-bit Pipelined ADC to attain a large output swing and a high gain. As in this design, the first stage requires a larger gm and a higher gain, so the first stage of Op-amp uses NMOS inputs with a telescopic and gain-boosting structure. With design margins and process variations, the first two stage's loop gains are slightly different to each other.

Figure 9 shows the SNDR and SFDR response of the ADC for w.r.t to input frequency range of 30 MHz to 150 MHz at sampling frequency at 125 MS·s^{-1}. With the increase in frequency SFDR decreases due to increase in parasitic capacitance and decrease in gain. Figure 10 shows the values obtained by applying the FFT at 125 MS·s^{-1} without calibration, and the values of SFDR and SNDR are 80 dB and 70.28 dB respectively. Five different frequencies ranging from 30 MHz

Tab. 1: Total of 16 different WG (i=1 to 16) has saved to the respective registers.

PN [2:5]	Do 1	Do 2	Do 3	Do 4	Do 5	Do 6	Do 7	Do 8	Do 9	Do 10	Do 11	Do 12	Do 13	Do 14	Do 15	Do 16
0000	D1	D2	D3	D4	D5	D6	D7	D8	D9	D10	D11	D12	D13	D14	*	0
0001	D9	D10	D11	D12	D13	D14	*	0	D1	D2	D3	D4	D5	D6	D7	D8
0010	D5	D6	D7	D8	D1	D2	D3	D4	D13	D14	*	0	D9	D10	D11	D12
0011	D13	D14	*	0	D9	D10	D11	D12	D5	D6	D7	D8	D1	D2	D3	D4
0100	D3	D4	D1	D2	D7	D8	D5	D6	D11	D12	D9	D10	*	0	D13	D14
0101	D11	D12	D9	D10	*	0	D13	D14	D3	D4	D1	D2	D7	D8	D5	D6
0110	D7	D8	D5	D6	D3	D4	D1	D2	*	0	D13	D14	D11	D12	D9	D10
0111	*	0	D13	D14	D11	D12	D9	D10	D7	D8	D5	D6	D3	D4	D1	D2
1000	D2	D1	D4	D3	D6	D5	D8	D7	D10	D9	D12	D11	D14	D13	0	*
1001	D10	D9	D12	D11	D14	D13	0	*	D2	D1	D4	D3	D6	D5	D8	D7
1010	D6	D5	D8	D7	D2	D1	D4	D3	D13	D13	0	*	D10	D9	D12	D11
1011	D14	D13	0	*	D10	D9	D12	D11	D6	D5	D8	D7	D2	D1	D4	D3
1100	D4	D3	D2	D1	D8	D7	D6	D5	D12	D11	D10	D9	0	*	D14	D13
1101	D12	D11	D10	D9	0	*	D17	D13	D4	D3	D2	D1	D8	D7	D6	D5
1110	D8	D7	D6	D5	D4	D3	D2	D1	0	*	D14	D13	D12	D11	D10	D9
1111	0	*	D14	D13	D12	D11	D10	D9	D8	D7	D6	D5	D4	D3	D2	D1

*Pndither

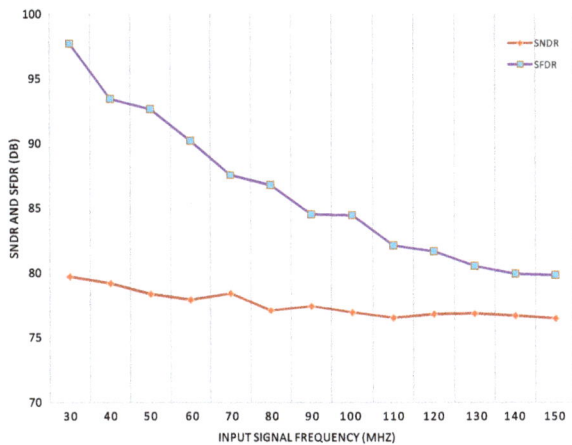

Fig. 9: Measured SFDR and SNDR at versus input signal frequency.

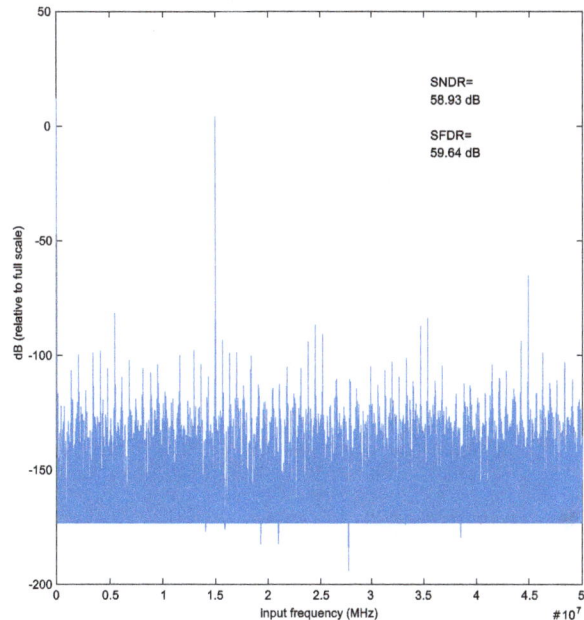

Fig. 10: Measured FFT before calibration.

to 150 MHz are given to the ADC to test its dynamic performance, which indicates that the SNDR and SFDR give better results for input signal below 150 MHz. Figure 11 shows the values of the SFDR and SNDR obtained by applying the FFT of sampling a 30 MHz full-scale input at 125 MS·s^{-1}, and is able to achieve SFDR and SNDR of 97.74 dB and 79.77 dB, respectively. Figure 12 shows the values of SNDR and SFDR after applying an FFT of sampling a 150 MHz input, and achieves an SFDR and SNDR of 88.9 dB and 73.5 dB respectively. Based on different results, ADC has been tested after subtracting out the jitter of the clock generator used. As per SNR measurements, the estimation of internal jitter of the ADC is 70 fs. The indication of low-jitter in the clock circuit is the slow drop of SNDR at high frequencies.

The DNL is improved from +1/-1 to 0.57/-0.48 and INL has improved from 8.7/-8.8 to 0.5/-0.5 after calibration as shown in Fig. 13 and Fig. 14. The total

power consumption of the proposed ADC is 300 mW, which includes the CMOS output drivers, operating at a 1.8 V supply. The calibration time for stage 1 and 2 is about 40s with sampling rate as 125 MS·s^{-1}. The performance comparison of proposed ADC, with other Pipelined ADCs [21], [22], [23], [24], [25] and [26] are presented in Tab. 2.

The ADC of [21] is a 14-bit Pipelined ADC, implemented in Nano technology of 90 nm, but due to high values of INL and DNL, it has limited scope. Because all digital post processing has been done off-chip with help of a PC, which hasn't integrated on a chip considered as another problem of this ADC. The ADC in [22] is a 16-bit 125 MS·s^{-1} SHA-less 4-bit front-end, which is able to achieve high values of SFDR and

Tab. 2: Comparison Table.

Parameter	[21]	[22]	[23]	[24]	[25]	[26]	This work
Technology (µm)	0.09	0.18	0.18	0.25BiCmos	0.18	0.18	0.18
Resolution	14	16	14	16	14	10	16
Sampling Rate	100 MS·s⁻¹	125 MS·s⁻¹	60 MS·s⁻¹	160 MS·s⁻¹	100 MS·s⁻¹	165 MS·s⁻¹	125 MS·s⁻¹
Supply (V)	1.2	2	1.6	3.3/5	1.8	1.8	1.8
INL (LSB)	1.3	1.5	0.6	4	3.86	0.6	0.57
DNL (LSB)	0.9	0.5	0.5	1.6	0.65	0.41	0.5
SNDR (dB)	73	78.6	73.3	76.16	69.1	56.1	79.77
SFDR (dB)	90	87	84	102	82.7	64	97.74
Power (mW)	250	385	67.8	1600	121	22	300

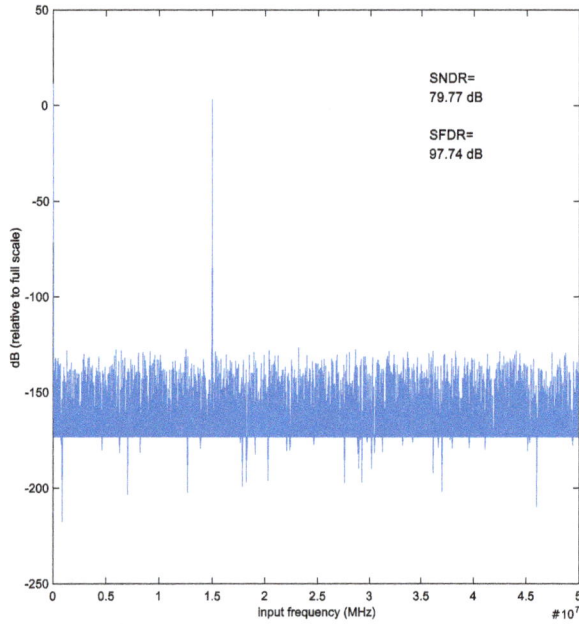

Fig. 11: Measured FFT when ADC is sampling a 30 MHz input at 125 MS·s⁻¹.

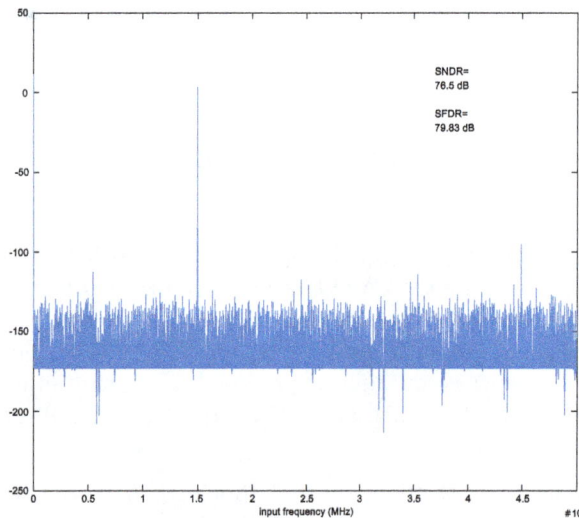

Fig. 12: Measured FFT when ADC is sampling a 150 MHz input at 125 MS·s⁻¹.

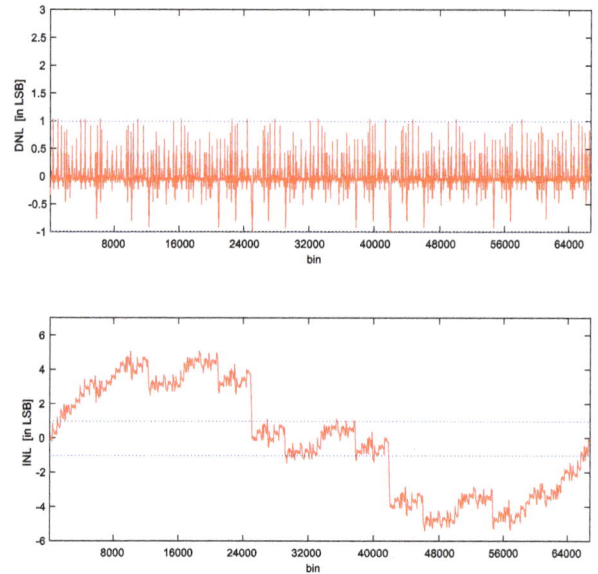

Fig. 13: Measured DNL and INL of ADC at 125 MS·s⁻¹ before Calibration.

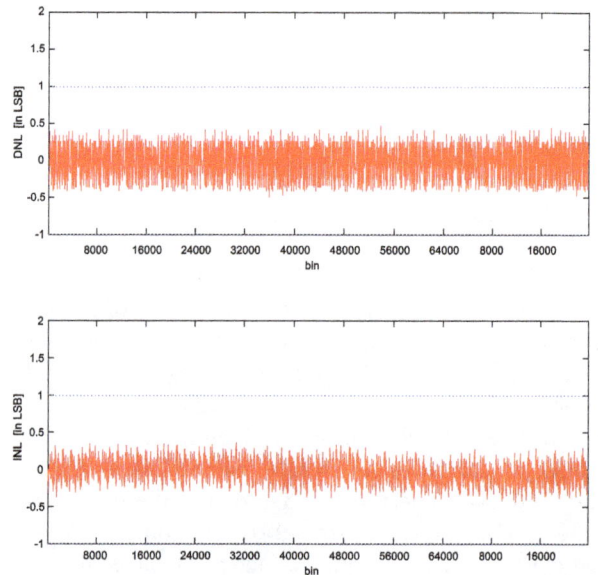

Fig. 14: Measured DNL and INL of ADC at 125 MS·s⁻¹ after Calibration.

SNDR, but this ADC has based on foreground factory digital calibration to correct for capacitor mismatches, that's a onetime process only. The ADC presented in [23] is based on a novel calibration technique with an auxiliary error amplifier and sign-sign LMS adaption. The problem comes with this ADC, when it has to operate on high sampling frequencies. The ADC used in [24] proved the viability of a low distortion 16-bit Pipelined ADC with sampling rate of 100 MS·s^{-1} to 160 MS·s^{-1} with SiGe BiCMOS process technology. The quite large power consumption and complicated implementation are major problems with this published ADC. The ADC presented in [25] is based on the study of the stage scaling theory. In this ADC, to implement a low power technique, residual amplifiers have been shared by two cascading MDACs. High value of INL and low value of SNDR are drawbacks of this Pipelined ADC. In [26], to reduce power consumption, double sampling MDAC and amplifier sharing techniques have been used. This low resolution 10-bit ADC with low values of SNDR and SFDR is only suitable for Video applications. From the comparison table, it's clear that the static and dynamic performance of the proposed Pipelined ADC are better than those published Pipelined ADCs, which operates from a supply voltage of 1.8 V, resolution of 16 bit and using CMOS process technology.

6. Conclusion

A 16-bit 125 MS·s^{-1} Pipelined ADC of Multi bit stage, without SHA in front-end, is described in this paper. The first and second stage of the 16-bit Pipelined ADC are of 3.5 bits, the succeeding stages from third to seventh are of 2.5 bits per stage, and a 3-bit flash ADC is used as the last stage of ADC. After bit alignment and truncation of 19 bits from all stages, the final digital output has achieved the 16-bit size. To improve ADC performance, the method of voltage reference separation has been used, which is really effective to prevent the kickback and settling within the available settling time. High gain Op-amp has been used in this design at each stage to remove nonlinear errors and capacitor mismatching. A combination of the two techniques of signal-dependent dithering and butterfly shuffler are used in the first two stages as a digital background calibration technique to remove linear errors. For the proposed ADC, the total power consumption is 300 mW. The DNL has improved from +1/-1 to 0.57/-0.48 and INL has improved from 8.7/-8.8 to 0.5/-0.5 after calibration. Due to the proposed calibration technique, SFDR has achieved a value of 97.74 dB and SNDR of 79.77 dB, when sampling with a 30 MHz full scale input at 125 MS·s^{-1}. The values of SFDR and SNDR achieved by sampling with a 150 MHz input are 88.9 dB and 73.5 dB respectively.

Acknowledgment

The authors of this paper would especially like to thank the reviewers for their precious time to give their expert comments. Per their valuable suggestions changes have been incorporated into the final version of this manuscript.

References

[1] CHOI, M. and A. A. ABIDI. A 6-b 1.3-Gsample s A D converter in 0.35-μm CMOS. *IEEE Journal of Solid-State Circuits.* 2001, vol. 36, iss. 12, pp. 1847–1858. ISSN 0018-9200. DOI: 10.1109/4.972135.

[2] WALDEN, R. H. Analog-to-digital converter survey and analysis. *IEEE Journal of Selected Areas in Communications.* 1999, vol. 17, iss. 4, pp. 539–550. ISSN 0733-8716. DOI: 10.1109/49.761034.

[3] MURMANN, B. and E. B. BERNHARD. *Digitally Assisted Pipeline ADCs Theory and Implementation.* 1st ed. Xi'an: Xi'an Jiaotong University Press, 2006. ISBN 1402078404.

[4] SIRAGURSA, E. and I. GALTON. A digitally enhanced 1.8-V 15-bit 40MSample/s CMOS pipelined ADC. *IEEE Journal of Solid-State Circuits.* 2004, vol. 39, iss. 12, pp. 2126–2138. ISSN 0018-9200. DOI: 10.1109/JSSC.2004.836230.

[5] GRACE, C. R., P. J. HURST and S. H. LEWIS. A 12b 80 MS/s pipelined ADC with bootstrapped digital calibration. *IEEE international Solid-State Circuits Conference.* 2004, vol. 1, iss. 1, pp. 460–539. ISSN 0193-6530. DOI: 10.1109/ISSCC.2004.1332793.

[6] CHANG, D. Y. Design techniques for pipelined ADC without using a front-end sample-and-Hold amplifier. *IEEE Transactions on Circuits and Systems I: Regular Papers.* 2004, vol. 51, iss. 11, pp. 2123–2132. ISSN 1549-8328. DOI: 10.1109/TCSI.2004.836842.

[7] SHU, Y. S., and B. S. SONG. A 15-bit Linear 20-MS/s Pipelined ADC Digitally Calibrated with Signal Dependent Dithering. *IEEE Journal of Solid-State Circuits.* 2008, vol. 43, iss. 2, pp. 342–350. ISSN 0018-9200. DOI: 10.1109/JSSC.2007.914260.

[8] WELZ, J. and I. GALTON. Necessary and sufficient conditions for mismatch shaping in a general class of multibit DACs. *IEEE Transactions*

on Circuits and Systems II: Analog and Digital Signal Processing. 2002, vol. 49, iss. 12, pp. 748–759. ISSN 1057-7130. DOI: 10.1109/TCSII.2002.807269.

[9] ALI, A. M. A., C. DILLON, R. SNEED, A. S. MORGAN, S. BARDSLEY, J. KORNBLUM, and L. WU. A 14-bit 125 MS/s IF/RF sampling pipelined ADC with 100 dB SFDR and 50 fs Jitter. *IEEE journal of Solid-State Circuits.* 2006, vol. 41, iss. 8, pp. 1846–1855. ISSN 0018-9200. DOI: 10.1109/JSSC.2006.875291.

[10] SINGER, L. A. and T. L. BROOKS. A 14-bit 10-MHz calibration-free CMOS pipelined A/D converter. In: *Symposium of VLSI Circuits, Digest of Technical Papers.* Honolulu: IEEE, 1996, pp. 94–95. ISBN 0-7803-3339-X. DOI: 10.1109/VLSIC.1996.507727.

[11] CHIU, Y., P. R. GRAY and B. NIKOLIC. A 14-b 12-MS/s CMOS pipeline ADC with over 100-dB SFDR. *IEEE Journal of Solid-State Circuits.* 2004, vol. 39, iss. 12, pp. 2139–2151. ISSN 0018-9200. DOI: 10.1109/JSSC.2004.836232.

[12] KELLY, D., W. YANG, I. MEHR, M. SAYUK, and L. SINGER. A 3V 340 mW 14 b 75 MSPS CMOS ADC with 85 dB SFDR at Nyquist. In: *IEEE International Solid-State Circuits Conference, Digest of Technical Papers.* San Francisco: ISSCC, 2001, pp. 134–135. ISBN 0-7803-6608-5. DOI: 10.1109/ISSCC.2001.912575.

[13] LEE B.-G., B. M. MIN, G. MANGANARO and J. W. VALVANO. A 14-b 100-MS/s pipelined ADC with a merged SHA and first MDAC. *IEEE Journal of Solid-State Circuits.* 2008, vol. 43, iss. 12, pp. 2613–2619. ISSN 0018-9200. DOI: 10.1109/JSSC.2008.2006309.

[14] YANG, W., D. KELLY, I. MEHR, M. SAYUK, and L. SINGER. A 3-V 340-mW 14-b 75-Msample/s CMOS ADC with 85-dB SFDR at Nyquist input. *IEEE Journal of Solid-State Circuits.* 2001, vol. 36, iss. 12, pp. 1931–1936. ISSN 0018-9200. DOI: 10.1109/4.972143.

[15] LEWIS, S. H. and P. R. GRAY. A pipelined 5-M sample/s 9-bit analog-to-digital converter. *IEEE Journal of Solid-State Circuits.* 1987, vol. 22, iss. 6, pp. 954–961. ISSN 0018-9200. DOI: 10.1109/JSSC.1987.1052843.

[16] KAPUSTA, R., H. SHINOZAKI, E. IBARAGI, K. NI, R. WANG, M. SAYUK, L. SINGER, and K. NAKAMURA. A 4-Channel 20-to-300 Mpixel/s analog front-end with sampled thermal noise below kT/C for digital SLR cameras. In: *IEEE international Solid-State Circuits Conference, Digest of Technical Papers, ISSCC.* San Francisco:

IEEE, 2009, pp. 42–43. ISBN 978-1-4244-3458-9. DOI: 10.1109/ISSCC.2009.4977298.

[17] MEHR, I. and L. SINGER. A 55-mW, 10-bit, 40-Msample/s Nyquist-rate CMOS ADC. *IEEE Journal of Solid-State Circuits.* 2000, vol. 35, iss. 3, pp. 318–325. ISSN 0018-9200. DOI: 10.1109/4.826813.

[18] ABOAND, A. M. and P. R. GRAY. A 1.5-V, 10-bit, 14.3-MS/s CMOS pipeline analog-to-digital converter. *IEEE Journal of Solid-State Circuits.* 1999, vol. 34, iss. 5, pp. 599–606. ISSN 0018-9200. DOI: 10.1109/4.760369.

[19] SINGER, L., S. HO, M. TIMKO and D. KELLY. A 12 b 65 MSample/s CMOS ADC with 82 dB SFDR at 120 MHz. In: *IEEE international Solid-State Circuits Conference, Digest of Technical Papers, ISSCC.* San Francisco: IEEE, 2000, pp. 38–39. ISBN 978-1-4244-3458-9. DOI: 10.1109/ISSCC.2000.839681.

[20] HU, J., N. DOLEV, and B. MURMANN. A 9.4-bit, 50-MS/s, 1.44-mW pipelined ADC using dynamic source follower residue amplification. *IEEE Journal of Solid-State Circuits.* 2009, vol. 44, iss. 4, pp. 1057–1066. ISSN 0018-9200. DOI: 10.1109/JSSC.2009.2014705.

[21] VAN DE VEL, H., B. A. J. BUTER, H. VAN DER PLOEG, M. VERTREGT, G. J. G. M. GEELEN and E. J. F. PAULUS. A 1.2-V 250-mW 14-b 100-MS/s Digitally Calibrated Pipeline ADC in 90-nm CMOS. *IEEE Journal of Solid-State Circuits.* 2009, vol. 44, iss. 4, pp. 1047–1056. ISSN 0018-9200. DOI: 10.1109/JSSC.2009.2014702.

[22] DEVARAJAN, S., L. SINGER, D. KELLY, S. DECKER, A. KAMATH and P. WILKINS. A 16-bit, 125 MS/s, 385 mW, 78.7 dB SNR CMOS Pipeline ADC. *IEEE Journal of Solid-State Circuits.* 2009, vol. 44, iss. 12, pp. 3305–3313. ISSN 0018-9200. DOI: 10.1109/JSSC.2009.2032636.

[23] MIYAHARA, Y. and M. SANO. A 14b 60 MS/s Pipelined ADC Adaptively Cancelling Opamp Gain and Nonlinearity. *IEEE Journal of Solid-State Circuits.* 2013, vol. 49, iss. 2, pp. 416–425. ISSN 0018-9200. DOI: 10.1109/JSSC.2013.2289902.

[24] PAYNE, R. and M. CORSI. A 16-Bit 100 to 160 MS/s SiGe BiCMOS Pipelined ADC with 100 dB SFDR. *IEEE Journal of Solid-State Circuits.* 2010, vol. 45, iss. 12, pp. 2613–2622. ISSN 0018-9200. DOI: 10.1109/JSSC.2010.2074650.

[25] YONGZHEN, C., C. CHIXIAO, F. ZEMIN, Y. FAN, and R. JUNYAN. 14-bit 100 MS/s 121 mW

pipelined ADC. *Journal of Semiconductors (IOP-Science)*. 2015, vol. 36, iss. 6, pp. 1–6. ISSN 1674-4926. DOI: 10.1088/1674-4926/36/6/065008.

[26] SHYLU, D. S. and D. JACKULINE MONI. A 1.8V 22mW 10 bit 165 MSPS Pipelined ADC for Video Applications. *Wseas Transactions on Circuits and Systems*. 2014, vol. 13, iss. 1, pp. 343–352. ISSN 2224-266X.

[27] KE, W., F. CHAOJIE, P. WENJIE, and Z. ZIAN-JUN. A 14-bit 100 MS/s SHA-less pipelined ADC with 89 dB SFDR and 74.5 dB SNR. *IEICE Electronics Express*. 2015, vol. 12, iss. 5, pp. 1–11. ISSN 1349–2543. DOI: 10.1587/elex.12.20150070.

About Authors

Swina NARULA was born in Punjab, India in 1978. She received her B.Sc. degree from M.D.U, Haryana, India in 2000 in electronics engineering, and the M.Sc. degree in VLSI designing from J.R.N Rajasthan, India in 2006. Currently she is pursuing her Ph.D. degree from Amity University, U.P, India. Her research interests include Pipelined Analog to Digital Converters, mixed-signal designing and high speed sampling circuits.

Munish VASHISHATH was born in Himachal Pradesh, India in 1976. He received his B.Sc. degree from Maharashtra, India 1997 in electronics engineering, the M.Sc. degree from BITS, Rajasthan, India in 1999, and Ph.D. degree from Thapar University, Punjab, India in 2009. He is currently working as a chairperson in electronics dept. at YMCA, Faridabad, India. His research interests include silicon carbide device, design of DIMOSFET with optimum value of breakdown voltage and Digital Signal Processing.

Sujata PADNEY did Master's in Electronics (VLSI) and Ph.D. in Electronics from the university of Delhi. Presently she is working as Professor in ECE Department at Amity School of Engineering and technology, Amity University and Professor in department of ECE. Her areas of research are microelectronics, Analog/digital VLSI Design, optical interconnects and Energy harvesting. She is a member of IEEE, USA, Electron Device Society, IET UK, founder member of VLSI and Semiconductor Society of India, ISTE, and life member of Indian Science Congress.

22

Disparity Line Utilization Factor Based Optimal Placement of IPFC for Congestion Management

Akanksha MISHRA, Gundavarapu Venkata Nagesh KUMAR

Department of Electrical and Electronics Engineering, Gandhi Institute of Technology and Management Institute of Technology, Gandhi Nagar, Rushikonda, 530045 Visakhapatnam, India

misakanksha@gmail.com, drgvnk@rediffmail.com

Abstract. *Recently, due to the adoption of power reforms, there is a marked increase of contracted power that flows in the transmission line and also the spontaneous power exchanges leading to complex power transmission congestion problems. The appearance of Flexible AC Transmission Systems (FACTS) devices specifically Interline Power Flow Controller (IPFC) has opened up new opportunities to overcome the congestion problem by increasing the possible system load. Hence, the optimal placement of FACTS devices is deservedly an issue of great importance. This paper proposes a Disparity Line Utilization Factor (DLUF) for the optimal placement of IPFC to control the congestion in transmission lines. DLUF determines the difference between the percentage MVA utilization of each line connected to the same bus. The proposed method is implemented for IEEE–14 and IEEE-57 bus test system. The IPFC is placed in all possible line combinations of IEEE-14 bus system to check the validity of the proposed methodology. To confirm the generality of the proposed method, the technique is also implemented and verified for IEEE-57 bus test system. An increased load of 110 % and 125 % is applied, and the results are presented and analysed in detail to establish the effectiveness of the proposed methodology.*

Keywords

Congestion, interline power flow controller, line utilization factor, optimal placement.

1. Introduction

Lately, the deregulated electric power industries have changed the way of operation, structure, ownership and management of the utilities. There is a huge enhance-ment of spontaneous power exchanges. More power is scheduled or flows across the transmission lines and transformers than the physical limits of those lines, which is the primary cause of congestion in transmission lines [1]. In the new competitive electric market, it is now mandatory for the electric utilities to operate such that it makes better utilization of the existing transmission facilities. It is, therefore, necessary to improve power delivery of system by reducing power loss in the interconnected electric power system. Many attempts have been made by researchers recently to improve the power transfer capability of the existing network [2], [3].

The concept of Flexible AC Transmission Systems (FACTS) devices was introduced by Hingorani [4] and they have been found very successful in solving various power system issues [5]. Several authors [6], [7] have proposed a sensitivity based approach for optimizing the location of FACTS devices for congestion management by controlling the device parameters. Kumar et al. [8] have used a sensitivity based approach for zonal/cluster-based congestion management. Acharya et al. [9] have proposed two new methodologies for the placement of series FACTS devices for congestion management. The overall objective of FACTS device placement can be minimization of the total congestion rent or maximization of social welfare.

Samimi et al. [10] have proposed a method to determine optimal location and best setting of Thyristor Controlled Series Compensator (TCSC). Seeking the best place is performed using the sensitivity analysis and optimum setting of TCSC is managed using the genetic algorithm. Yousufi et al. [11] have proposed a combination of Demand Response (DR) and Flexible Alternating Current Transmission System (FACTS) devices for congestion management. Esmaili et al. [12] have proposed optimization of total operating cost, voltage and transient stability margins for optimal placement and sizing of FACTS devices for congestion management. Hooshmand et al. [13] considered non-smooth fuel cost-function and penalty cost of emission for optimal placement of TCSC to manage congestion. Esmaili et al. [14] have used Real Power Performance Index (RPPI) and reduction of total system VAR power losses for optimal placement of TCSC. Ushasurendra et al. [15] have proposed Line Utilization Factor (LUF) for optimal placement of FACTS devices for congestion management. RPPI is based only on the real power flowing through a line, whereas, LUF takes into consideration the apparent power that flows in the line. Hence, LUF is chosen for the study. Line Utilization Factor has been used for the determination of congestion of a single transmission line. FACTS devices are placed on the transmission line with maximum LUF value. However, IPFC is a multiline series FACTS device [16]. In its simplest form it consists of at least two converters required to be placed on two transmission lines with a common bus. The 1st converter of IPFC can be placed on the line with maximum LUF. But the placement of the other converter is an issue which becomes more and more complex with the increase in the size of the system, number of IPFC's and the complexity of IPFC. Hence, LUF is not a sufficient index for obtaining the location for placement of IPFC.

In this paper, the difference of Line Utilization Factors between two lines has been used for determination of the optimal location of IPFC. The IPFC is placed in the lines with a maximum value of DLUF to reduce congestion. The effect of IPFC placement on the active and reactive power of the power system is also studied under different loading conditions. The proposed method is implemented and tested on IEEE-14 and IEEE-57 bus system for different loading conditions.

2.   Modelling of IPFC

An IPFC consists of at least two back to back DC-AC converters connected by a common DC link [17], [18]. V_i, V_j, V_k are complex voltages at bus-i, j, k respectively. $V_l = V_l\llcorner\Theta_l$ ($l = i, j, k$) and V_l, Θ_l are the magnitude and angle of V_l. Vse_{in} is the complex controllable, series injected voltage source. It shows the series compensation of the series converter. Vse_{in} is given by $Vse_{in} = Vse_{in}\llcorner\Theta se_{in}$ ($n = j, k$). Vse_{in} and Θse_{in} are the magnitude and angle of Vse_{in}.

The basic model of IPFC, as shown in Fig. 1 consisting of three buses-i, j and k. Two transmission lines are connected with the bus-i in common. The equivalent circuit of the IPFC with two converters is represented in Fig. 2. Zse_{in} is the series transformer impedance. Pse_{in} is the active power exchange of each converter via the common DC link. P_i and, Q_i as given in equa-

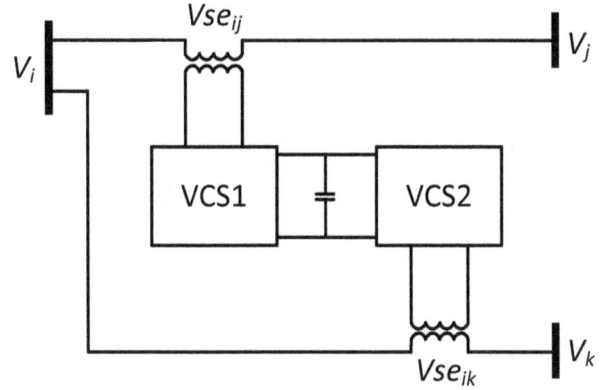

Fig. 1: Basic model of IPFC.

tions Eq. (1) and Eq. (2) are the sum of the active and reactive power flows leaving the bus-i. The IPFC branch active and reactive power flows leaving bus-n are P_{ni} and Q_{ni} and the expressions are given in equation Eq. (3) and Eq. (4). I_{ji}, I_{ki} are the IPFC branch currents of branch $j - i$ and $k - i$ leaving bus-j and k, respectively.

In Eq. (1), Eq. (2), Eq. (3) and Eq. (4) are:

- $n = j, k$,
- $g_{in} + jb_{in} = \dfrac{1}{zse_{in}} = yse_{in}$,
- $g_{nn} + jb_{nn} = \dfrac{1}{zse_{in}} = yse_{in}$,
- $g_{ii} = \sum\limits_{n=j,k} g_{in}$,
- $b_{ii} = \sum\limits_{n=j,k} b_{in}$.

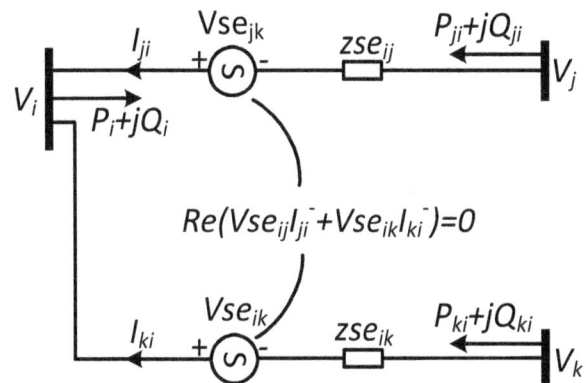

Fig. 2: Equivalent circuit of IPFC.

$$P_i = V_i^2 b_{ii} - \sum_n V_i V_n \left[g_{in} \cos\left(\Theta_i - \Theta_n\right) + b_{in} \sin\left(\Theta_i - \Theta_n\right) \right]$$
$$- \sum_n V_i V se_{in} \left[g_{in} \cos\left(\Theta_i - \Theta se_{in}\right) + b_{in} \sin\left(\Theta_i - \Theta se_{in}\right) \right]. \tag{1}$$

$$Q_i = -V_i^2 b_{ii} - \sum_{n=j,k} V_i V_n \left[g_{in} \sin\left(\Theta_i - \Theta_n\right) - b_{in} \cos\left(\Theta_i - \Theta_n\right) \right]$$
$$- \sum_{n=j,k} V_i V se_{in} \left[g_{in} \sin\left(\Theta_i - \Theta se_{in}\right) - b_{in} \cos\left(\Theta_i - \Theta se_{in}\right) \right]. \tag{2}$$

$$P_{ni} = V_n^2 g_{nm} - V_i V_n \left[g_{in} \cos\left(\Theta_n - \Theta_i\right) + b_{in} \sin\left(\Theta_n - \Theta_i\right) \right]$$
$$+ V_n V se_{in} \left[g_{in} \sin\left(\Theta_n - \Theta se_{in}\right) - b_{in} \cos\left(\Theta_n - \Theta se_{in}\right) \right]. \tag{3}$$

$$Q_{ni} = -V_n^2 b_{nm} - V_i V_n \left[g_{in} \sin\left(\Theta_n - \Theta_i\right) - b_{in} \cos\left(\Theta_n - \Theta_i\right) \right]$$
$$+ V_n V se_{in} \left[g_{in} \sin\left(\Theta_n - \Theta se_{in}\right) - b_{in} \cos\left(\Theta_n - \Theta se_{in}\right) \right]. \tag{4}$$

Assuming lossless converter, the active power supplied by one converter equals to the active power demanded by the other, if there are no underlying storage systems:

$$Re\left(V se_{ij} I_{ji}^* + V se_{ik} I_{ki}^*\right) = 0, \tag{5}$$

where the superscript $*$ denotes the complex conjugate.

3. Disparity Line Utilization Factor

Line Utilization Factor is an index used for determining the congestion of the transmission lines. It is given by Eq. (6):

$$LUF_{ij} = \frac{MVA_{ij}}{MVA_{ij\,\max}}, \tag{6}$$

where LUF_{ij} - Line Utilization Factor (LUF) of the line connected to bus-i and bus-j, $MVA_{ij\,max}$ - Maximum MVA rating of the line between bus-i and bus-j. MVA_{ij} - Actual MVA rating of the line between bus-i and bus-j.

LUF gives an estimate of the percentage of line being utilized and is an efficient method to estimate the congestion in a line. For placement of IPFC, there should be at least two lines connected to a common bus. Therefore, LUF is not sufficient for placement of IPFC.

Taking into consideration the fact that IPFC can directly transfer real power via the common DC link, it has the capability to transfer power demand from overloaded to under-loaded lines. Hence, a new index Disparity Line Utilization Factor is hereby proposed for the optimal placement of an IPFC. DLUF indicates the difference between the utilization of the lines. It gives an estimate of the difference of the percentage of line being used for the power flow. All the lines are first ranked in descending order of their line utilization factors. The line which has the first rank is considered to be the most congested line. DLUF is calculated for all the lines connected to the line with highest congestion. All the line pairs connected to the same bus are ranked on the basis of DLUF. The line set that has highest value of DLUF is considered to be the optimal location of IPFC for Congestion Management. Assuming both lines of same rating:

$$DLUF_{(ij)-(ik)} = \left| \frac{MVA_{ij} - MVA_{ik}}{MAV_{\max}} \right|, \tag{7}$$

where $DLUF_{(ij)-(ik)}$ - Disparity Line Utilization Factor (DLUF) of the line set ij and ik, MVA_{ij} - MVA rating ofthe line between bus-i and bus-j, $MVA_{\max}$ - maximum MVA rating of line, MVA_{ik} - actual MVA rating of the line between bus-i and bus-k.

Step by Step Procedure:

- STEP I – Read the line data and bus data.

- STEP II – Calculate the power flow and LUF of all lines

- STEP III – Calculate the DLUF values for all lines in pair wit the lines ranking highest in congestion.

- STEP IV – Place IPFC on the lines having highest value of DLUF.

Tab. 1: DLUF value calculation for line 4-5 of 14 bus test system.

Case	Line 1, SB No., RB No.	Line 2, SB No., RB No.	LUF Line1	LUF Line 2	DLUF	LUF of Line 1 with IPFC
1	4-5	4-7	0.5951	0.2851	0.1313	0.8374
2	4-5	4-9	0.5951	0.1849	0.3511	0.8370

- STEP V – Perform the load flow analysis and calculate the LUF of all lines.

4. Results and Discussion

4.1. IEEE-14 Bus System

An IEEE-14 bus test system has 4 generator buses, 9 load buses and 20 transmission lines. Bus 1 is the slack bus. Bus number 2, 3, 6, 8 are the generator buses as shown in Fig. 3. The remaining buses are the load buses. System base MVA is 100. An IPFC consisting of two converters has been used in the study. The inductive reactance and resistance of the coupling transformers are assumed to be 0.001 p.u. The voltage magnitude of the two converters of the IPFC is taken in the range $0 \leq V_{se} \leq 0.1$ and the angle is taken in the range $-\Pi \leq \Theta_{se} \leq \Pi$. Only load buses have been considered for IPFC placement.

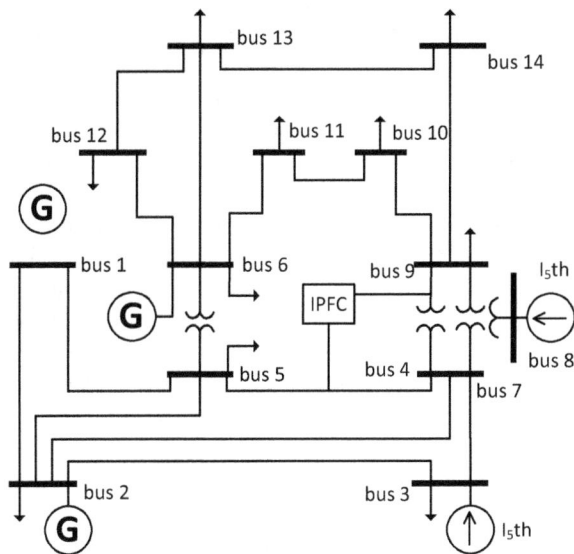

Fig. 3: IEEE-14 bus test system with IPFC installed at line connected between buses 4–5 and 4–9.

Table 2 displays the LUF values of all lines without IPFC and with IPFC. It is observed that the line connected between buses 4 and 5, with highest value of LUF is the most congested line. The line connected between buses 4–7 and buses 4–9 have been connected to the line 4-5 through a common bus. In Tab. 1 the value of DLUF has been calculated for the two possible

Tab. 2: LUF values of all lines without and with IPFC.

From Bus	To Bus	LUF without IPFC	LUF with IPFC at proposed location
2	5	0.4601	0.4417
3	4	0.3001	0.2623
4	5	0.5951	0.5787
4	7	0.2851	0.2802
4	9	0.1849	0.1598
5	6	0.5338	0.5713
6	11	0.2661	0.2908
6	12	0.2562	0.257
6	13	0.7323	0.7380
7	8	0.3724	0.4481
7	9	0.4853	0.4729
9	10	0.0763	0.0925
9	14	0.3556	0.3331
10	11	0.1326	0.1550
12	13	0.1159	0.1184
13	14	0.2405	0.2546

options of IPFC placement. It is observed from Tab. 1 that the line pair (4–5) and (4–9) have the maximum value of DLUF. Hence, the IPFC is proposed to be located at line 4–5 and 4–9. In order to prove that the proposed location is the best location for the placement of IPFC, the device is placed in all possible locations of the transmission system and the results have been presented in Tab. 3. It is observed from Tab. 3 and Fig. 4 that congestion in the line 4–5 is reduced most effectively when IPFC is placed on line 4–5 and 4–9 which is the location being proposed for optimal placement. Thus, it is verified that for reduction of congestion, the 1st converter of IPFC has to be placed on the most congested line (maximum LUF) while the 2nd converter has to be placed on the line that has maximum DLUF value with respect to the 1st line. The active power loss is also reduced at this location, although it may

Tab. 3: Placement of IPFC at all possible locations in the IEEE-14 bus test system.

S. No.	Location of IPFC on Line Pair	LUF of line 7	Total Active power loss
1.	7, 8	0.5820	22.313
2.	7, 9	0.5787	20.140
3.	8, 9	0.5980	22.368
4.	9, 16	0.8020	20.376
5.	9, 17	0.8020	20.383
6.	15, 16	0.8030	21.923
7.	15, 17	0.8030	21.929
8.	16, 18	0.7690	22.540
9.	17, 20	0.8050	21.929
10.	19, 20	0.8400	21.539

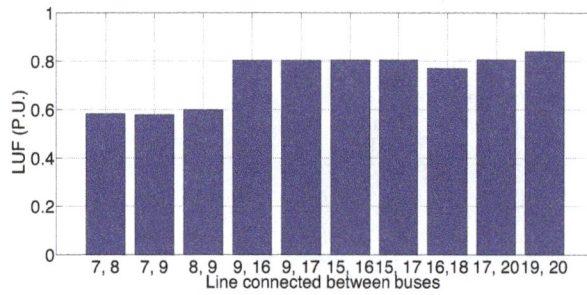

Fig. 4: LUF of line connected between buses 4–5 after placement of IPFC at all feasible locations of IEEE-14 bus test system.

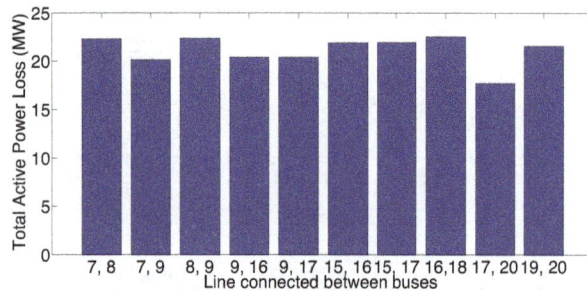

Fig. 6: Real power loss without and with IPFC for normal, 110 % and 125 % load.

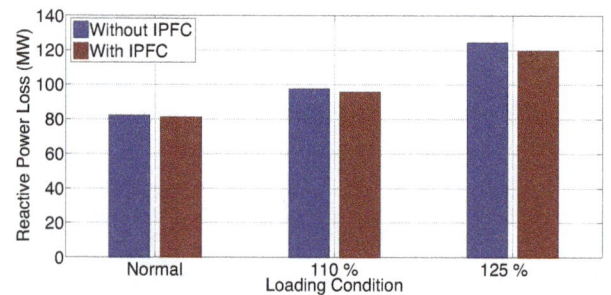

Fig. 5: Active power loss in IEEE-14 bus system after placement of IPFC at all feassible locations.

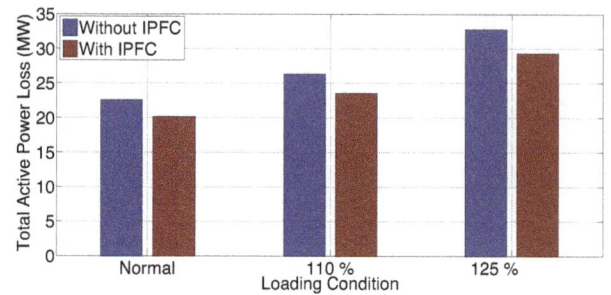

Fig. 7: Reactive power loss without and with IPFC for normal, 110 % and 125 %.

not achieve its minimum value as observed from Fig. 5. The next best location for IPFC placement, in terms of reduction of congestion, is line 4–5 and 4–7 where the value of DLUF is smaller in comparison to the proposed location.

Next, the load on the transmission system has been increased by 10 % and 25 % and the results have been presented in Tab. 4. It shows the improvement in active and reactive power loss with placement of IPFC at the proposed location at both normal and increased loading condition. The total active and reactive power loss for different loading conditions have been shown in Fig. 6 and Fig. 7 respectively.

Tab. 4: Active power loss and reactive power loss without and with optimally placed IPFC for normal, 110 % and 125 % loading condition.

Loading cond.	Real power loss		Reactive power loss	
	Without IPFC	With IPFC	Without IPFC	With IPFC
Normal	22.545	20.140	82.171	81.171
110 %	26.313	23.533	97.658	95.650
125 %	32.828	29.385	124.490	119.850

4.2. IEEE-57 Bus System

From the results obtained for IEEE-14 bus system, it is clear that the 1st converter of the IPFC has to be placed on the most congested line while the location of the 2nd

converter of the IPFC should be on the line with maximum DLUF with respect to the most congested line to obtain maximum LUF reduction. In order to confirm the validity and generality of the proposed method, the concept of optimal placement of IPFC using DLUF is verified again for an IEEE-57 bus test system. An IEEE-57 bus test system is considered. In an IEEE-57 bus system bus no. 1 is considered as a slack bus and bus nos. 2, 3, 6, 8, 9, 12 are considered as PV buses while all other buses are load buses. This system has 80 interconnected lines as shown in Fig. 8. LUF values of all the lines without and with Optimal placement of IPFC has been presented in Tab. 5. It is established that line connected between buses 14–46 (line 59) is the most congested line. All possible DLUF index calculations for line 14–46 have been shown in Tab. 6 as test cases. It is observed from Tab. 5, line 14–46 is the most congested line connected to load bus. In the 57 bus system, three lines have been connected to line 59. So, three test cases for IPFC placement have been considered, as shown in Tab. 6 DLUF has been calculated for each test case and it is observed that congestion in line 14–46 is reduced most when the line-2 used for IPFC placement is Line 13–14 where the DLUF value is maximum. Hence, lines 14–46 and 13–14 have been selected for optimal placement of IPFC. It is observed from Tab. 6 that placement of IPFC at the location where DLUF is maximum causes a maximum reduction in congestion in line 14–46.

It is observed from Fig. 9 that optimal placement of IPFC reduces congestion in line 14–46 (line no. 59)

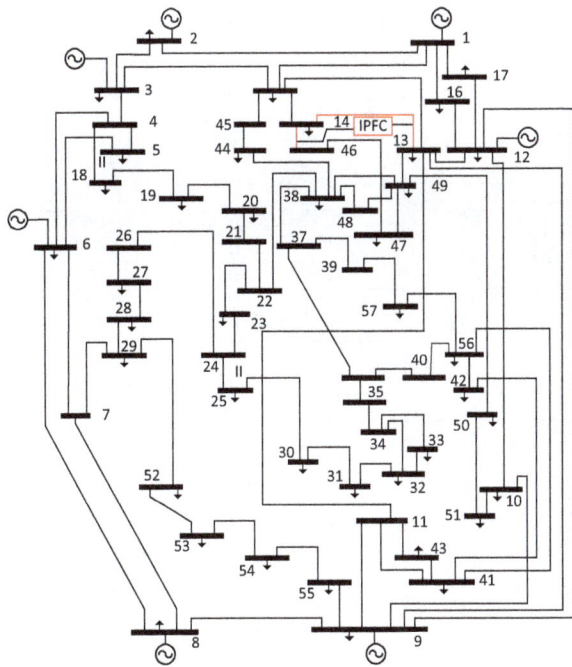

Fig. 8: IEEE-57 Bus test system with IPFC installed at line connected between buses 14–46 and 13–14.

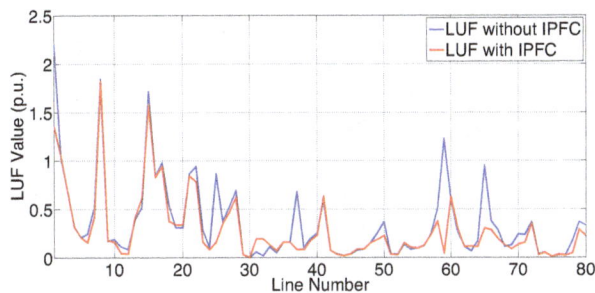

Fig. 9: Comparison of LUF values with and without IPFC with normal load.

and in the other lines in the system. Fig. 10 shows a marked improvement in voltage profile of the buses. It is observed from Tab. 7 that after the placement of IPFC using DLUF, line losses are considerably reduced.

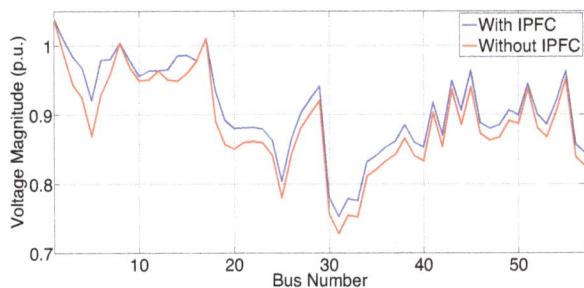

Fig. 10: Voltage profile without and with IPFC.

Simulation has been performed for 110 % and 125 % load on IEEE-14 and 57 bus test system. It is observed

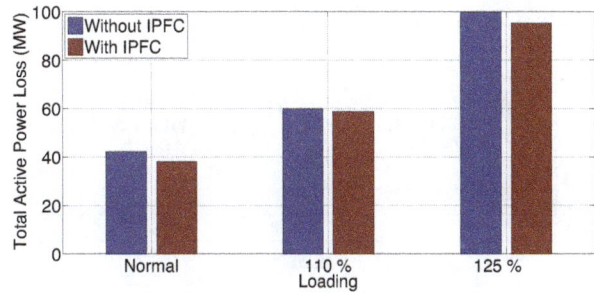

Fig. 11: Real power loss without and with IPFC for normal, 110 % and 125 % load.

from Tab. 7 and Fig. 11 and Fig. 12 that with increase in load the total real and reactive power loss increases. Placement of IPFC by the proposed method seems to be an effective method for reduction of the above parameters even in increased loading condition.

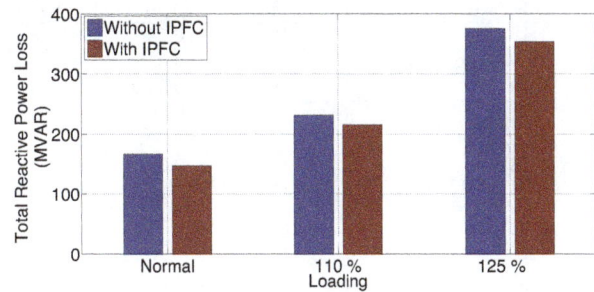

Fig. 12: Reactive power loss without and with IPFC for normal, 110 % and 125 % load.

5. Conclusions

In this paper, a Disparity Line Utilization Factor for the optimal placement of IPFC for congestion management has been implemented. The IPFC is being placed in the lines with highest DLUF value. It has been established that placement of IPFC using DLUF effectively reduces line congestion and active and reactive power loss simultaneously. The proposed method has been verified and implemented for IEEE-14 and IEEE-57 bus test system using MATLAB Software. It is observed that the placement of IPFC by the proposed methodology causes an effective reduction in congestion in the lines. The result of LUF value before and after placement of IPFC shows reduction of loading in congested line. Comparison of results with other locations ensures that placement of IPFC at the proposed location is a healthy location for the placement of IPFC in terms of reduction of congestion. A reduction in Real and reactive power loss has also been observed. Hence, the overall system performance has been studied under different loading conditions and the results are found to be favourable.

Tab. 5: Comparison of LUF values of all lines of 57 bus test system without and with optimally placed IPFC.

Line No.	From Bus, (SB No.)	To Bus, (RB No.)	LUF without IPFC	LUF with opt Placed IPFC	Line No.	From Bus, (SB No.)	To Bus, (RB No.)	LUF without IPFC	LUF with opt Placed IPFC
1.	1	2	2.2141	1.360	41.	7	29	0.6102	0.639
2.	2	3	1.0286	1.047	42.	25	30	0.0840	0.083
3.	3	4	0.6947	0.693	43.	30	31	0.0415	0.041
4.	4	5	0.3229	0.322	44	31	32	0.0257	0.026
5.	4	6	0.2115	0.214	45.	32	33	0.0430	0.043
6.	6	7	0.2522	0.164	46.	34	32	0.0834	0.094
7.	6	8	0.5275	0.427	47.	34	35	0.0913	0.093
8.	8	9	1.8483	1.830	48.	35	36	0.1591	0.159
9.	9	10	0.1792	0.186	49.	36	37	0.2597	0.193
10.	9	11	0.1904	0.164	50.	37	38	0.3742	0.235
11.	9	12	0.1205	0.050	51.	37	39	0.0395	0.041
12.	9	13	0.0897	0.046	52.	36	40	0.0389	0.035
13.	13	14	0.3954	0.415	53.	22	38	0.1433	0.157
14.	13	15	0.5226	0.624	54.	11	41	0.0930	0.109
15.	1	15	1.7232	1.585	55.	41	42	0.1034	0.104
16.	1	16	0.8385	0.834	56.	41	43	0.1302	0.133
17.	1	17	0.9834	0.950	57.	38	44	0.2488	0.241
18.	3	15	0.5463	0.382	58.	15	45	0.5156	0.381
19.	4	18	0.3160	0.344	59.	14	46	1.2301	0.053
20.	4	18	0.3160	0.344	60.	46	47	0.6038	0.636
21.	5	6	0.8674	0.852	61.	47	48	0.2786	0.316
22.	7	8	0.9475	0.787	62.	48	49	0.1220	0.119
23.	10	12	0.2954	0.168	63.	49	50	0.0755	0.123
24.	11	13	0.1092	0.091	64.	50	51	0.1785	0.120
25.	12	13	0.8694	0.169	65.	10	51	0.9562	0.312
26.	12	16	0.3787	0.358	66.	13	49	0.3884	0.293
27.	12	17	0.5314	0.477	67.	29	52	0.2874	0.202
28.	14	15	0.6985	0.629	68.	52	53	0.1184	0.14
29.	18	19	0.0391	0.040	69.	53	54	0.1310	0.097
30.	19	20	0.0079	0.006	70.	54	55	0.2471	0.141
31.	21	20	0.0655	0.200	71.	11	43	0.2405	0.162
32.	21	22	0.0249	0.200	72.	44	45	0.3737	0.362
33.	22	23	0.1188	0.138	73.	40	56	0.0415	0.035
34.	23	24	0.0565	0.078	74.	56	41	0.0595	0.061
35.	24	25	0.1653	0.165	75.	56	42	0.0148	0.017
36.	24	25	0.1653	0.165	76.	39	57	0.0376	0.041
37.	24	26	0.6851	0.089	77.	57	56	0.0320	0.030
38.	26	27	0.1095	0.089	78.	38	49	0.1835	0.054
39.	27	28	0.2021	0.183	79.	38	48	0.3734	0.298
40.	28	29	0.2578	0.237	80.	9	55	0.3367	0.227

Tab. 6: DLUF value calculation for line 14–46 of 57 bus test system.

Case	Line 1 SB No. RB No.	Line 2 SB No. RB No.	LUF Line 1	LUF Line 2	DLUF	LUF Of Line 1 with IPFC
1	14–46	46–47	1.230	0.603	0.627	0.166
2	14–46	14–15	1.230	0.698	0.532	0.160
3	14–46	13–14	1.230	0.395	0.834	0.053

Tab. 7: Active power loss and reactive power loss without and with optimally placed IPFC for normal, 110 % and 125 % loading condition.

Loading condition	Real power loss		Reactive power loss	
	Without IPFC	With IPFC	Without IPFC	With IPFC
Normal	42.258	38.110	166.112	146.724
110 %	59.989	58.736	231.139	215.918
125 %	99.721	95.216	375.397	353.829

References

[1] KUMAR, A., S. C. SRIVASTAVA and S. N. SINGH. Congestion management in competitive power market: A bibliographical survey. *Electric Power Systems Research*. 2005, vol. 76, iss. 1–3, pp. 153–164. ISSN 0378-7796. DOI: 10.1016/j.epsr.2005.05.001.

[2] NASRI, A. and B. GASBAOUI. Novel Power Flow Problem Solutions Method's Based on Genetic Algorithm Optimization for Banks Capacitor Compensation Using an Fuzzy Logic Rule Bases for Critical Nodal Detections. *Advances in Electrical and Electronic Engineering*. 2011, vol. 9, no. 4, pp. 150–156. ISSN 1336-1376. DOI: 10.15598/aeee.v9i4.548.

[3] SALHI, A., D. NAIMI and T. BOUKTIR. Fuzzy Multi-Objective Optimal Power Flow Using Genetic Algorithms Applied to Algerian Electrical Network. Power Engineering and Electrical Engineering. *Advances in Electrical and Electronic Engineering*. 2013, vol. 11, no. 6, pp. 443–454. ISSN 1336-1376. DOI: 10.15598/aeee.v11i6.832.

[4] HINGORANI, N. G. Fuzzy Multi-Objective Optimal Power Flow Using Genetic Algorithms Applied to Algerian Electrical Network. Power Engineering and Electrical Engineering. *IEEE Power Engineering Review*. 1988, vol. 8, iss. 7, pp. 3–4. ISSN 0272-1724. DOI: 10.1109/MPER.1988.590799.

[5] BHATTACHARYYA, B., V. K. GUPTA and S. KUMAR. Reactive Power Optimization with SVC & TCSC using Genetic Algorithm. *Advances in Electrical and Electronic Engineering*. 2014, vol. 12, no. 1, pp. 1–12. ISSN 1336-1376. DOI: 10.15598/aeee.v12i1.778.

[6] SINGH, S. N. and A. K. DAVID. Optimal location of FACTS devices for congestion management. *Electric Power Systems Research*. 2001, vol. 58, iss. 2, pp. 71–79. ISSN 0378-7796. DOI: 10.1016/S0378-7796(01)00087-6.

[7] VERMA, K. S., S. N. SINGH and H. O. GUPTA. Location of unified power flow controller for congestion management. *Electric Power Systems Research*. 2001, vol. 58, iss. 2, pp. 89–96. ISSN 0378-7796. DOI: 10.1016/S0378-7796(01)00123-7.

[8] KUMAR, A., S. C. SRIVASTAVA and S. N. SINGH. A Zonal Congestion Management Approach Using Real and Reactive Power Rescheduling. *IEEE Transactions on Power Systems*. 2004, vol. 19, iss. 1, pp. 554–562. ISSN 0885-8950. DOI: 10.1109/TPWRS.2003.821448.

[9] ACHARYA, N. and N. MITHULANANTHAN. Locating series FACTS devices for congestion management in deregulated electricity markets. *Electric Power Systems Research*. 2007, vol. 77, iss. 3–4, pp. 352–360. ISSN 0378-7796. DOI: 10.1016/j.epsr.2006.03.016.

[10] BESHARAT, H. and S. A. TAHER. Congestion management by determining optimal location of TCSC in deregulated power systems. *International Journal of Electrical Power and Energy Systems*. 2008, vol. 30, iss. 10, pp. 563–568. ISSN 0142-0615. DOI: 10.1016/j.ijepes.2008.08.007.

[11] MANDALA, M. and C. P. GUPTA. Congestion management by optimal placement of FACTS device. In: *Joint International Conference on Power Electronics, Drives and Energy Systems*. New Delhi: IEEE, 2010, pp. 1–7. ISBN 978-142447782-1. DOI: 10.1109/PEDES.2010.5712387.

[12] SAMIMI, A. and P. NADERI. A New Method for Optimal Placement of TCSC Based on Sensitivity Analysis for Congestion Management. *Smart Grid and Renewable Energy*. 2012, vol. 3, no. 1, pp. 10–16. ISSN 2151-481X. DOI: 10.4236/sgre.2012.31002.

[13] YOUSEFI, A., T. T. NGUYEN, H. ZAREIPOUR and O. P. MALIK. Congestion management using demand response and FACTS devices. *International Journal of Electrical Power and Energy Systems*. 2012, vol. 37, iss. 1, pp. 78–85. ISSN 0142-0615. DOI: 10.1016/j.ijepes.2011.12.008.

[14] ESMAILI, M., H. A. SHAYANFAR and R. MOSLEMI. Locating series FACTS devices for multi-objective congestion management improving voltage and transient stability. *European Journal of Operational Research*. 2014, vol. 236, iss. 2, pp. 763–773. ISSN 0377-2217. DOI: 10.1016/j.ejor.2014.01.017.

[15] USHASURENDRA, S. and S. PARATHASARTHY. Congestion management in deregulated power sector using fuzzy based optimal location technique for series flexible alternative current transmission system (FACTS) device. *Journal of Electrical and Electronics Engineering Research*. 2012, vol. 4, iss. 1, pp. 12–20. ISSN 2141-2367 DOI: 10.5897/JEEER11.143.

[16] HINGORANI, N. G. and L. GYUGYI. *Understanding FACTS: Concepts and Technology of Flexible AC Transmission Systems*. 1st ed. Piscataway: Wiley-IEEE Press, 1999. ISBN 978-0-7803-3455-7. DOI: 10.1002/9780470546802.

[17] ZHANG, X. P. Modelling of the interline power flow controller and the generalised unified power flow controller in Newton power flow. *Transmission and Distribution IEE Proceedings - Generation.* 2003, vol. 150, iss. 3, pp. 268–274. ISSN 1350-2360. DOI: 10.1049/ip-gtd:20030093.

[18] ACHA, E., C. R. FUERTE-ESQUIVEL, H. AMBRIZ-PEREZ and C. ANGELES-CAMACHO. *FACTS: Modelling and Simulation in Power Networks.* 1st ed. Chichester: John Wiley & Sons, 2004. ISBN 978-0-470-85271-2. DOI: 10.1002/0470020164.

About Authors

Akanksha MISHRA was born in Cuttack, India, in 1982. She received her bachelor degree in electrical engineering from Kalinga Institute of Industrial Technology, Bhubaneswar, India in 2004 and master's degree in power electronics and drives in 2006 from the same institute. She is presently pursuing her Ph.D. from Gandhi Institute of Technology and Management, Visakhapatnam, India. Her research interests are FACTS devices, power electronics and power system stability. She has published several research papers in national and international conferences.

Gundavarapu Venkata Nagesh KUMAR was born in Visakhapatnam, India in 1977. He received the B.E. degree from College of Engineering, Gandhi Institute of Technology and Management (GITAM), Visakhapatnam, India and M.E. degree from the College of Engineering, Andhra University, Visakhapatnam. He received his doctoral degree from Jawaharlal Nehru Technological University, Hyderabad. He is also working as an Professor in the Department of Electrical and Electronics Engineering, GITAM University. His research interests include gas insulated substations, fuzzy logic, high voltage testing, and wavelets and FACTS devices. He has published research papers in national and international conferences and journals.

23

REALIZATION OF OFCC BASED TRANSIMPEDANCE MODE INSTRUMENTATION AMPLIFIER

Neeta PANDEY, Deva NAND, V. Venkatesh KUMAR, Varun KUMAR AHALAWAT, Chetna MALHOTRA

Department of Electronics and Communication Engineering, Delhi Technological University, Shahbad Daulatpur, Main Bawana Road, Delhi-110042, India

n66pandey@rediffmail.com, devkamboj07@gmail.com, vvenkateshkumar.kumar@gmail.com, varun.ahlawat@gmail.com, chtmalhotra@gmail.com

Abstract. *The paper presents an instrumentation amplifier suitable for amplifying the current source transducer signals. It provides a voltage output. It has a high gain, common mode rejection ratio and gain independent bandwidth. It uses three Operational Floating Current Conveyors (OFCCs) and four resistors. The effect of nonidealities of OFCC on performance of proposed Transimpedance Instrumentation Amplifier (TIA) is also analyzed. The proposal has been verified through SPICE simulations using CMOS based schematicThe paper presents an instrumentation amplifier suitable for amplifying the current source transducer signals. It provides a voltage output. It has a high gain, common mode rejection ratio and gain independent bandwidth. It uses three operational floating current conveyors (OFCCs) and four resistors. The effect of nonidealities of OFCC on performance of proposed Transimpedance Instrumentation Amplifier (TIA) is also analyzed. The proposal has been verified through SPICE simulations using CMOS based schematic.*

Keywords

Difference amplifier, operational floating current conveyor, transimpedance instrumentation amplifier.

1. Introduction

The Operational Floating Current Conveyor (OFCC) [1], [2] is a versatile active building block which provides flexibility to the circuit designer. It inherits the features of current conveyor and the current feedback op-amp with additional current output terminal. The availability of both high and low impedance ports at input and output provides flexibility to circuit designer. The OFCC has been used for implementing instrumentation amplifier [3], read out circuit [4], logarithmic amplifier [5], rectifier [6], filters [7], [8], [9], [10], [11], variable gain amplifier [12], and wheatstone bridge [13].

An Instrumentation Amplifier (IA) is invariably used as an input block in applications such as automotive transducers [14], industrial process control [15], [16], [17], linear position sensing [18] and bio-potential acquisition systems [19], [20], [21], [22], [23], [24] to amplify differential signals and to suppress unwanted common mode signals. Generally the operational amplifier based IA are classified as Voltage Mode IA (VMIA) whereas current mode building block based IA are referred as Current Mode IA (CMIA). Another way to classification is based on the type of input and output signal on which IA is working. The Transimpedance IA (TIA) is one among such classification where the sensed current is amplified and converted into a voltage. There is limited literature available on TIAs [25], [26], [27], and no OFCC based TIA is available in open literature to the best of author's knowledge. The details of available TIAs are comprehended in Tab. 1 according to the number and type of active and passive elements used along with the impedance presented at input and output. The following points are observed from Tab. 1:

- The opamp based topology [26] uses excessive number of resistors.

- The input impedance is low for [27] which is ideal for current sensing whereas a high input impedance is presented for [25], [26].

- The output impedance of [26], [27] is proper i.e. low in contrast to the one provided by [25].

- Extra active elements are required for impedance matching at input [25], [26] and output [25].

Tab. 1: Characteristics of available instrumentation amplifiers.

Ref. no.	Active elements	No. of resistors	Input impedance	Output impedance
[25]	2 CCII+	3	High	High
[26]	3 opamps	10	High	Low
[27]	3 OTRA	5	Low	Low

It is clear from the above discussion that only topology [27] provides proper input and output impedance levels and does not require additional circuitry for impedance matching. The aim of this paper is to present an OFCC based TIA offering proper input/output interface. It uses three active blocks and four resistors i.e. same number of active blocks as [27] and the lesser passive components than [27]. Both input and output impedances of proposed topology are low.

The paper is organized in five sections as follows: Section 2. briefly discusses the basic characteristics of OFCC and detailed description of proposed TIA structure. Section 3. describes behavior of proposed TIA in presence of nonidealities. The simulation results are presented in Section 4. followed by conclusions in Section 5.

2.   Proposed Circuit

The key component of the proposed circuit is the OFCC block as shown in Fig. 1. It has two inputs (X, Y) and three outputs (W, Z+, Z-).

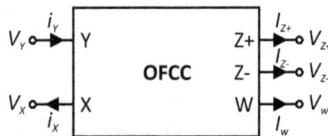

Fig. 1: OFCC block.

The port X is a low impedance current input while the port labeled Y is a high impedance voltage input. The ports Z+ and Z- are high impedance current outputs, where Z+ has positive polarity and Z- has negative polarity. The terminal marked W is the low impedance output voltage terminal. The terminal characteristics of the OFCC are characterized by the matrix given in Eq. (1):

$$\begin{bmatrix} I_Y \\ V_X \\ V_W \\ I_{Z+} \\ I_{Z-} \end{bmatrix} = \begin{bmatrix} 0 & 0 & 0 & 0 & 0 \\ 1 & 0 & 0 & 0 & 0 \\ 0 & Z_t & 0 & 0 & 0 \\ 0 & 0 & 1 & 0 & 0 \\ 0 & 0 & -1 & 0 & 0 \end{bmatrix} \begin{bmatrix} V_Y \\ I_X \\ I_W \\ V_{Z+} \\ V_{Z-} \end{bmatrix}, \quad (1)$$

where open loop transimpedance gain Z_t is impedance between the ports X and W and other symbols have their usual meanings.

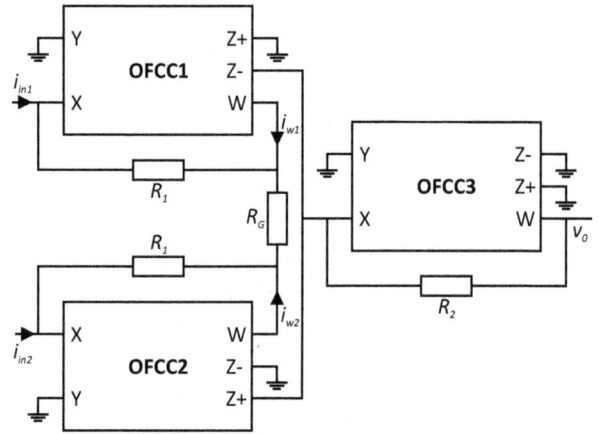

Fig. 2: Proposed OFCC based TIA.

Figure 2 shows the proposed TIA circuit. It consists of three OFCCs and four resistances. The third OFCC block in Fig. 2 is simply used as a current to voltage convertor which converts the amplified difference of currents that has been received as an input to OFCC1 and OFCC2, into voltage. The differential transimpedance gain of the instrumentation amplifier for an ideal case is computed as follows.

The currents (i_{w1}, i_{w2}) flowing out of W terminals of OFCC1 and OFCC2 respectively, are:

$$i_{w1} = -\left[i_{in1} + \frac{R_1}{R_G}(i_{in1} - i_{in2}) \right], \quad (2)$$

$$i_{w2} = -\left[i_{in2} + \frac{R_1}{R_G}(i_{in2} - i_{in1}) \right]. \quad (3)$$

The output voltage is computed as:

$$v_{out} = -R_2(-i_{w1} + i_{w2}) = R_2(i_{w1} - i_{w2}) =$$
$$= \left[(i_{in2} - i_{in1}) \left(1 + \frac{2R_1}{R_G} \right) \right]. \quad (4)$$

Using Eq. (4), the differential gain (A_d) is obtained as:

$$A_d = \frac{v_{out}}{i_{in2} - i_{in1}} = \left(1 + \frac{2R_1}{R_G} \right) R_2. \quad (5)$$

3. Non Ideal Analysis

Practically, there are two kinds of OFCC non-idealities. The first type of nonidealities comes from tracking errors between port voltages and currents and their effect depend strongly on topology. As the Y terminal in proposed topology is grounded the performance is unaffected due to voltage tracking errors. Considering the current tracking error, the currents at Z+ and Z-terminals are represented as:

$$i_{Z+} = \alpha i_{w2}, \tag{6}$$

$$i_{Z-} = -\gamma i_{w1}, \tag{7}$$

where α and γ are non ideality constants.

Therefore, Eq. (4) modifies to:

$$v_{out} = \left[(\alpha i_{in2} - \gamma i_{in1}) + \ldots \right.$$
$$\left. \ldots + \left((\alpha + \gamma) \frac{R_1}{R_G} (i_{in2} - i_{in1}) \right) \right] R_2. \tag{8}$$

Assuming $\alpha = \gamma = 1$ in Eq. (8) the differential gain (A_d) is written as:

$$A_d = \frac{v_{out}}{i_{in2} - i_{in1}} = \left(1 + \frac{2R_1}{R_G} \right) R_2. \tag{9}$$

Considering $i_{in1} = i_{in2} = i_{cm}$ in Eq. (8) for common mode operation, the common mode gain (A_{cm}) is obtained as:

$$A_{cm} = \frac{v_{out}}{i_{cm}} = (\alpha - \gamma) R_2. \tag{10}$$

Therefore the CMRR of the final circuit is:

$$\text{CMRR} = \frac{A_d}{A_{cm}} = \frac{1}{\alpha - \gamma} \left(1 + \frac{2R_1}{R_G} \right). \tag{11}$$

The second nonideality comes due to finite transimpedance gain Z_t and its frequency dependence which is approximated as $Z_t = 1/(sC_p)$ at high frequencies. The value of C_p is ($Z_{to} \omega_{tc}$), where Z_{to} and ω_{tc} represent open loop transimpedance gain and its cut off frequency respectively.

Considering finite Z_t, Eq. (4) is recalculated as

$$v_{out} = [\epsilon_1(s)i_{w1} - \epsilon_2(s)i_{w2}]\epsilon_3(s)R_2, \tag{12}$$

$$v_{out} = \left[\left((i_{in2} - i_{in1})(\epsilon_2(s) - \epsilon_1(s)) \frac{R_1}{R_G} \right) + \ldots \right.$$
$$\left. \ldots (\epsilon_2(s)i_{in2} - \epsilon_1(s)i_{in1}) \right] \epsilon_3(s)R_2, \tag{13}$$

where $\epsilon_i(s) = \dfrac{1}{1 + sC_{pi}R_1}$ for $i = 1, 2$ and $\epsilon_3(s) = \dfrac{1}{1 + sC_{p3}R_2}$.

Assuming $\epsilon_1(s) = \epsilon_2(s)$, differential gain is calculated as:

$$A_d = \left(1 + \frac{2R_1}{R_G} \right) R_2 \epsilon_{uc}, \tag{14}$$

where $\epsilon_{uc} = \dfrac{1}{1 + sC_{p3}R_2} \cdot \dfrac{1}{1 + sC_{p1}R_1}$ and is uncompensated error function.

Taking $i_{in1} = i_{in2} = i_{cm}$ the common mode gain is given by:

$$A_{cm} = \frac{sR_1R_2(C_{p1} - C_{p2})}{(1 + sC_{p1}R1)(1 + sC_{p2}R_1)(1 + sC_{p3}R_2)}. \tag{15}$$

Therefore the CMRR becomes:

$$\text{CMRR} = \frac{A_d}{A_{cm}} =$$
$$= \left(1 + \frac{R_G}{2R_1} \right) \left(\frac{2 + sR_1(C_{p1} + C_{p2})}{s(C_{p2} - C_{p1})R_G} \right). \tag{16}$$

4. Simulation Results

The CMOS based OFCC implementation [12] as shown in Fig. 3, is used for verifying functionality of proposed TIA. The transistor aspect ratios are given in Tab. 2. SPICE simulations are carried out using supply voltages of ±1.5 V and bias voltages of ±0.8 V. The simulated differential gain response of the proposed TIA is depicted in Fig. 4 for $R_G = 1$ kΩ, $R_1 = 5$ kΩ and R_2 is varied from 1 kΩ to 3 kΩ in step of 1 kΩ in order to obtain different gains. The CMRR frequency response is shown in Fig. 5. It may be noted that CMRR is independent of gain and has a bandwidth of 112 kHz. Figure 6 shows the noise spectral analysis of the proposed TIA using the same component values as those taken for obtaining differential gain response. It is observed that the output noise level has small magnitude. The power consumption of the proposed OFCC based TIA is found to be 1.5 mW.

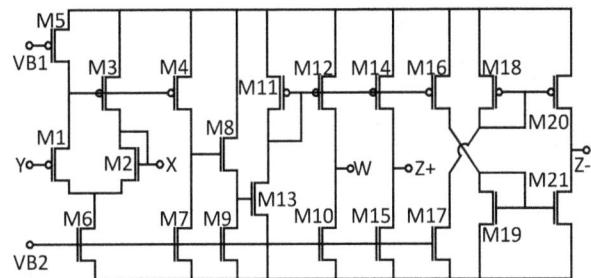

Fig. 3: CMOS schematic of OFCC [12].

5. Conclusion

In this paper, an OFCC based TIA is presented and simulated. The circuit requires only three OFCCs,

198 Electrical Systems Engineering

Tab. 2: Characteristics of available instrumentation amplifiers.

Transistor	W (µm)/L (µm)
M1, M2	50/1
M3, M4, M11, M12, M14, M16, M18, M20	50/2.5
M5, M7, M10, M15, M17, M19, M21	20/2.5
M6, M8	40/2.5
M9, M13	100/2.5

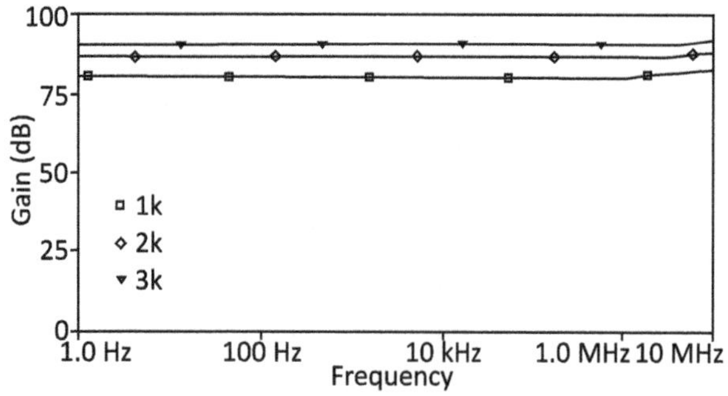

Fig. 4: Frequency response of the proposed TIA.

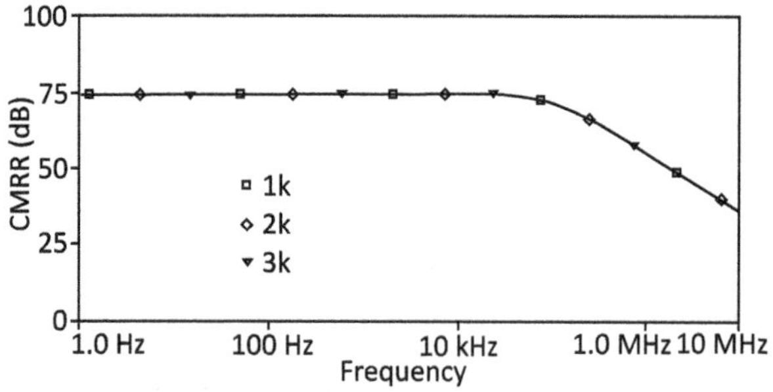

Fig. 5: CMRR of the proposed TIA for different gain values.

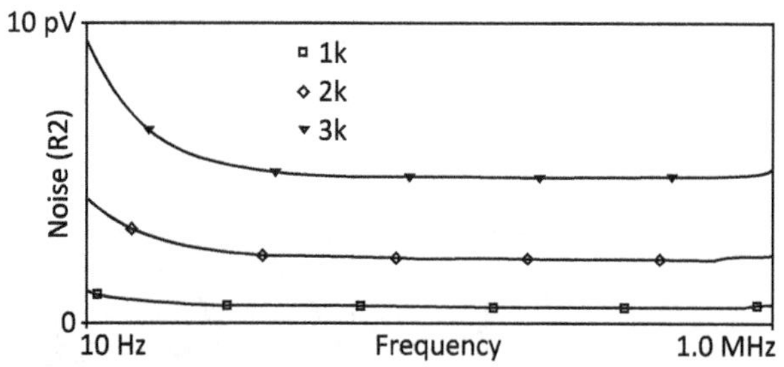

Fig. 6: Noise spectral density for different values of gain.

three feedback resistors and one grounded resistor. It works with current mode of input in order to produce an amplified output without using complex designs. The AC analysis proves the efficiency of this new circuit and the huge bandwidth it possesses. The proposed topology offers advantages over the existing operational amplifiers based TIAs, in terms of a wider bandwidth that stays independent of the finite open loop gain of the TIA. The proposed circuit also offers low component count as compared to the existing OTRA based design.

References

[1] KHAN, A. A., M. A. AL-TURAIGI and M. A. EL-ELA. Operational floating current conveyor: Characteristics, Modeling and applications. In: *IEEE Instrumentation and Measurement Technology Conference.* Hamamatsu: IEEE, 1994, pp. 788–790. ISBN 0-7803-1880-3. DOI: 10.1109/IMTC.1994.351887.

[2] GHALLAB, Y. H., M. A. EL-ELA and M. ELSAID. Operational floating current conveyor: Characteristics, Modeling and Experimental results. In: *Proceedings of International Conference on Microelectronics.* Kuwait: IEEE, 1999, pp. 59–62. ISBN 0-7803-6643-3. DOI: 10.1109/ICM.2000.884805.

[3] GHALLAB, Y. H., W. BADAWY, K. V. I. S. KALER and B. J. MAUNDY. A Novel Current-Mode Instrumentation Amplifier Based on Operational Floating Current Conveyor. *IEEE Transactions on Instrumentation and Measurement.* 2005, vol. 54, no. 5, pp. 1941–1949. ISSN 0018-9456. DOI: 10.1109/TIM.2005.854254.

[4] GHALLAB, Y. H. and W. BADAWY. A New Differential PH Sensor Current Mode Readout circuit using only two Operational Floating Current Conveyor. In: *IEEE International Workshop on Biomedical Circuits and Systems.* Singapore: IEEE, 2004, pp. 13–16. ISBN 0-7803-8665-5. DOI: 10.1109/BIOCAS.2004.1454145.

[5] PANDEY, N., P. TRIPATHI, R. PANDEY and R. BATRA. OFCC based Logarithmic Amplifier. In: *International Conference on Signal Processing and Integrated Networks (SPIN).* Noida: IEEE, 2014, pp. 522–525. ISBN 978-1-4799-2865-1. DOI: 10.1109/SPIN.2014.6777009.

[6] PANDEY, N., R. PANDEY, D. NAND and A. KUMAR. Current-mode rectifier configuration based on OFCC. In: *International Conference on Signal Propagation and Computer Technology (ICSPCT-2014).* Ajmer: IEEE, 2014, pp. 533–536. ISBN 978-1-4799-3139-2. DOI: 10.1109/ICSPCT.2014.6884960.

[7] GHALLAB, Y. H., W. BADAWY, K. V. I. S. KALER, M. A. EL-ELA and M. H. ELSAID. A New Second-order Active Universal Filter with Single Input and Three Outputs Using Operational Floating Current Conveyor. In: *14th International Conference on Microelectronics.* Beirut: IEEE, 2002, pp. 42–45. ISBN 0-7803-7573-4. DOI: 10.1109/ICM-02.2002.1161492.

[8] GHALLAB, Y. H., M. A. EL-ELA and M. EL-SAID. A novel universal voltage mode filter with

three inputs and single output using only operational floating current conveyor. In: *Proceedings of 12th International Conference on Microelectronics*. Tehran: IEEE, 2000, pp. 95–98. ISBN 964-360-057-2. DOI: 10.1109/ICM.2000.916422.

[9] GHALLAB, Y. H. and W. BADAWY. The Operational Floating Current Conveyor and its applications. *Journal of Circuits, Systems, and Computers*. 2006, vol. 15, iss. 3, pp. 351–372. ISSN 0218-1266. DOI: 10.1142/S0218126606003118.

[10] PANDEY, N., D. NAND and Z. KHAN. Single input four output current mode filter using operational floating current conveyor. *Active and Passive Electronic Components*. 2013, vol. 2013, iss. ID 318560, pp. 1–8. ISSN 1563-5031. DOI: 10.1155/2013/318560.

[11] PANDEY, N., D. NAND and Z. KHAN. Operational floating current conveyor based single input multiple output transadmittance mode filter. *Arabian Journal of Science and Engineering*. 2014, vol. 39, iss. 11, pp. 7991–8000. ISSN 2191-4281. DOI: 10.1007/s13369-014-1369-z.

[12] HASSAN, H. M. and A. M. SOLIMAN. Novel CMOS realizations of Operational Floating Current Conveyor and applications. *Journal of Circuits, Systems, and Computers*. 2005, vol. 14, iss. 6, pp. 1113–1143. ISSN 1793-6454. DOI: 10.1142/S0218126605002854.

[13] GHALLAB, Y. H. and W. BADAWY. A New Design of a Current-mode Wheatstone Bridge Using Operational Floating Current Conveyor. In: *International Conference on MEMS, NANO and Smart Systems*. Cairo: IEEE, 2006, pp. 41–44. ISBN 1-4244-0899-7. DOI: 10.1109/ICMENS.2006.348213.

[14] MILLER, B. D. and R. L. SAMPLE. Instrumentation amplifier IC designed for oxygen sensor interface requirements. *IEEE Journal of Solid State Circuits*. 1981, vol. 16, iss. 6, pp. 677–681. ISSN 0018-9200. DOI: 10.1109/JSSC.1981.1051661.

[15] SCHAFFER, V., M. F. SNOEIJ, M. V. IVANOV and D. T. TRIFONOV. A 36 V programmable instrumentation amplifier with sub-20 μV offset and CMRR in excess of 120 dB at all gain settings. *IEEE Journal of Solid State Circuits*. 2009, vol. 44, iss. 7, pp. 2036–2046. ISSN 0018-9200. DOI: 10.1109/JSSC.2009.2021921.

[16] REDOUTE, J.-M. and M. STEYAERT. An instrumentation amplifier input circuit with a high gain immunity to EMI. In: *2008 International Symposium Electromagnetic Compatibility - EMC Europe*. Hamburg: IEEE, 2008, pp. 1–6. ISBN 978-1-4244-2737-6. DOI: 10.1109/EMCEUROPE.2008.4786901.

[17] WITTE, J. F., J. H. HUIJSING and K. A. A. MAKINWA. A current feedback instrumentation amplifier with 5 μV offset for bidirectional high side current sensing. *IEEE Journal of Solid State Circuits*. 2008, vol. 43, iss. 12, pp. 2769–2775. ISSN 0018-9200. DOI: 10.1109/JSSC.2008.2005695.

[18] RAHAL, M. and A. DEMOSTHENOUS. A synchronous chopping demodulator and implementation for high frequency inductive position sensors. *IEEE Transactions on Instrumentation and Measurement*. 2009, vol. 58, iss. 10, pp. 3693–3701. ISSN 0018-9456. DOI: 10.1109/TIM.2009.2019314.

[19] STEYAERT, M. S. J. and W. M. C. SANSEN. A micropower low noise monolithic instrumentation amplifier for medical purposes. *IEEE Journal of Solid State Circuits*. 1987, vol. 22, iss. 6, pp. 1163–1168. ISSN 0018-9200. DOI: 10.1109/JSSC.1987.1052869.

[20] MARTINS, R., S. SELBERHERR and F. A. VAZ. A CMOS IC for portable EEG acquisition systems. *IEEE Transactions on Instrumentation and Measurement*. 1998, vol. 47, iss. 5, pp. 1191–1196. ISSN 0018-9456. DOI: 10.1109/19.746581.

[21] YEN, C.-J., W. Y. CHUNG and M. C. CHI. Micro-power low offset instrumentation amplifier IC design for biomedical system applications. *IEEE Transactions on Circuits Systems I: Regular papers*. 2004, vol. 51, iss. 4, pp. 691–699. ISSN 1549-8328. DOI: 10.1109/TCSI.2004.826208.

[22] NG, K. A. and P. K. CHAN. A CMOS analog front-end IC for portable EEG/ECG monitoring applications. *IEEE Transactions on Circuits Systems I: Regular papers*. 2005, vol. 52, iss. 11, pp. 2335–2347. ISSN 1549-8328. DOI: 10.1109/TCSI.2005.854141.

[23] YAZICIOGLU, R. F., P. MERKEN, R. PUERS and C. V. HOOF. A 60 μW 60 nV/√Hz readout front-end for portable biopotential acquisition systems. *IEEE Journal of Solid State Circuits*. 2007, vol. 42, iss. 5, pp. 1100–1110. ISSN 0018-9200. DOI: 10.1109/JSSC.2007.894804.

[24] WANG, C.-C., C.-C. HUANG, J.-S. LIOU, Y.-J. CIOU, I.-Y. HUANG, C.-P. LI, Y.-C. LEE, and W.-J. WU. A mini-invasive long term bladder urine pressure measurement ASIC and system. *IEEE Transactions on Biomedical Circuits*

Systems. 2008, vol. 2, iss. 1, pp. 44–49. ISSN 1932-4545. DOI: 10.1109/TBCAS.2008.921601.

[25] WILSON, B. Universal conveyor instrumentation amplifier. *Electronic Letters*. 1989, vol. 25, iss. 7, pp. 470–471. ISSN 0013-5194. DOI: 10.1049/el:19890323.

[26] AGOURIDIS, D. C. and R. J. FOX. Transresistance instrumentation amplifier. *Proceedings of the IEEE*. 1978, vol. 66, iss. 10, pp. 1286–1287. ISSN 0018-9219. DOI: 10.1109/PROC.1978.11125.

[27] PANDEY, R., N. PANDEY and S. K. PAUL. Electronically tunable transimpedance instrumentation amplifier based on OTRA. *Journal of Engineering*. 2013, vol. 2013, iss. ID 648540, pp. 1–5. ISSN 2051-3305. DOI: 10.1155/2013/648540.

About Authors

Neeta PANDEY did M.Tech. (Microelectronics) from Birla Institute of Technology and Sciences, Pilani, Rajasthan, India and Ph.D. from Guru Gobind Singh Indraprastha University, Delhi, India. At present she is Associate Professor in Department of Electronics and Communication Engineering, Delhi Technological University, Delhi. A life member of ISTE, and a member of IEEE, USA. She has published papers in International, National Journals of repute and conferences. Her research interests are in Analog and Digital VLSI Design.

Deva NAND did B.Tech (Electronics and Communication Engineering), M.Tech (Microelectronics and VLSI Design) from Kurukshetra University, Kurukshetra, India and he is pursuing Ph.D. from Delhi Technological University, Delhi, India. At present he is Assistant Professor in Department of Electronics and Communication Engineering, Delhi Technological University, Delhi, India. His research interests include Analog Mixed Signal VLSI Design and Digital VLSI Design.

V. Venkates KUMAR pursuing B.Tech (Electronics and Communication Engineering) from Delhi Technological University, Delhi, India. His research interests include Analog VLSI Design.

Varun KUMAR AHALAWAT pursuing B.Tech (Electronics and Communication Engineering) from Delhi Technological University, Delhi, India. His research interests include VLSI Design.

Chetna MALHOTRA pursuing B.Tech (Electronics and Communication Engineering) from Delhi Technological University, Delhi, India. Her research interests include Analog Electronics and VLSI Design.

Permissions

List of Contributors

Mohamed Anouar Ben Messaoud and Aicha Bouzid
Department of Electrical Engineering, National School of Engineers, University of Tunis El Manar, BP 37 Belvedere, Tunis, Tunisia

Ali Moshar Movahhed and Heydar Toosian Shandiz
School of Electrical & Robotic Engineering, Shahrood University of Technology, Shahrood, Iran

Syed Kamal Hosseini San
Department of Electrical Engineering, Faculty of Engineering, Ferdowsi University of Mashhad, Mashhad, Iran

Stefano Elia and Alessio Santantonio
Department DIAEE – Electrical Engineering Section, Sapienza University, Piazzale Aldo Moro 5, 00185 Rome, Italy

Djamel Benoudjit, Mohamed Said Nait-Said, Said Drid and Nasreddine Nait-Said
LSP-IE Laboratory, Electrical Engineering Department, Faculty of Technology, University of Batna 2, Rue Chahid Mohamed El-HadiBoukhlouf, 05000 Batna, Algeria

Ilias Ouachtouk, Soumia EL Hani, Khalid Dahi and Lahbib Sadiki
Electrical Laboratory Researche, Ecole Normale Superieure de Enseignement Technique, Mohammed V University, Avenue des Nations Unies, Agdal, Rabat, Morocco

Said Guedira
Higher National School of Mines, Ecole Nationale Superieure des Mines de Rabat, Avenue Hadj Ahmed Cherkaoui, Agdal, Rabat, Morocco

Hau Huu Vo, Chau Si Thien Dong and Thinh Cong Tran
Faculty of Electrical and Electronics Engineering, Ton Duc Thang University, 19 Nguyen Huu Tho, Dist. 7, Ho Chi Minh City, Vietnam

Pavel Brandstetter
Department of Electronics, Faculty of Electrical Engineering and Computer Science, VSB–Technical University of Ostrava, 17. listopadu 15, 708 33 Ostrava, Czech Republic

Lokeshwar Reddy Chintala
Department of Electrical and Electronics Engineering, CVR College of Engineering, Ibrahimpatnam, Hyderabad 501 510, Telangana, India

Satish Kumar Peddapelli
Department of Electrical Engineering, University College of Engineering, Osmania University, Hyderabad 500 007, Telangana, India

Sushama Malaji
Department of Electrical and Electronics Engineering, College of Engineering, Jawaharlal Nehru Technological University Hyderabad, Kukatpally, Hyderabad 500 085, Telangana, India

Pavel Stasa, Filip Benes, Jiri Svub, Jakub Unucka and Vladimir Kebo
Institute of Economics and Control Systems, Faculty of Mining and Geology, VSB–Technical University of Ostrava, 17. listopadu 15, 708 33 Ostrava, Czech Republic

Yong-Shin Kang
Engineering, College 2Department of Systems Management of Engineering, Sungkyunkwan University, 300 Cheoncheon-dong, Jangan-gu, Suwon, Republic of Korea

Lukas Vojtech
Department of Telecommunication Engineering, Faculty of Electrical Engineering, Czech Technical University in Prague, Technicka 2, 166 27 Prague 6, Czech Republic

Jong-Tae Rhee
Department of Industrial and Systems Engineering, College of Engineering, Dongguk University, 26 Pil-dong 3-ga, Jun-gu, Seoul, Republic of Korea

Jiri Jansa and Zdenek Hradilek
Department of Electrical Power Engineering, Faculty of Electrical Engineering and Computer Science, VSB–Technical University of Ostrava, 17. listopadu 15, 708 33, Ostrava, Czech Republic

Libor Gajdosik
Department of Telecommunications, Faculty of Electrotechnical Engineering and Computer Science, VSB–Technical University of Ostrava, 17. listopadu 15, 708 33 Ostrava-Poruba, Czech Republic

Iraida Kolcunova, Marek Pavlik and Lukas Lison
Department of Electric Power Engineering, Faculty of Electrical Engineering and Informatics, Technical University of Kosice, Letna 9, 040 01 Kosice, Slovak republic

Pavel Grigorievich Kolpakhchyan and Alexey Rifkatovich Shaikhiev
Science and Production Association "Don Technology", Mikhaylovskaya st. 164 a, 346414 Novocherkassk, Russian Federation

Alexander Evgenievich Kochin
Electrical Machines and Apparatus Department, Faculty of Power Engineering, Rostov State Transport University, Rostovskogo Strelkovogo Polka Narodnogo Opolchenya sq. 2, 344038 Rostov on Don, Russian Federation

Konstantin Stepanovich Perfiliev
OJSC "VNIKTI", Oktyabrskoy Revolutcii st. 410, 140400 Kolomna, Russian Federation

Jan Otypka
Department of Electrical Engineering, Faculty of Electrical Engineering and Computer Science, VSB–Technical University of Ostrava, 17. listopadu 15/2172, 708 33 Ostrava, Czech Republic

Andrey Valerievich Sukhanov
Electrical Machines and Apparatus Department, Faculty of Power Engineering, Rostov State Transport University, Rostovskogo Strelkovogo Polka Narodnogo Opolchenya sq. 2, 344038 Rostov on Don, Russian Federation
Department of Electrical Engineering, Faculty of Electrical Engineering and Computer Science, VSB–Technical University of Ostrava, 17. listopadu 15/2172, 708 33 Ostrava, Czech Republic

Zbynek Martinek and Ales Hromadka
Department of Electric Power Engineering and Ecology, Faculty of Electrical Engineering, University of West Bohemia, Univerzitni 2732/8, 306 14, Pilsen, Czech Republic

Jiri Hammerbauer
Department of Applied Electronics and Telecommunications, Faculty of Electrical Engineering, University of West Bohemia, Univerzitni 2732/8, 306 14, Pilsen, Czech Republic

Tomas Borovsky
Slovakia Steel Mills, a.s., Priemyselna 720, 072 22 Strazske, Slovak Republic

Karol Kyslan and Frantisek Durovsky
Department of Electrical Engineering and Mechatronics, Faectriculty of Elcal Engineering, Technical University of Kosice, Letna 9, 042 00 Kosice, Slovak Republic

Ilias Ouachtouk, Soumia El Hani and Khalid Dahi
Electrical Laboratory Researche, Ecole Normale Superieure de Enseignement Technique, Mohammed V University, Avenue des Nations Unies, Agdal, Rabat, Morocco

Radana Kahankova, Rene Jaros and Radek Martinek
Department of Cybernetics and Biomedical Engineering, Faculty of Electrical Engineering and Computer Science, VSB–Technical University of Ostrava, 17. listopadu 15, 708 33 Ostrava, Czech Republic

Janusz Jezewski
Institute of Medical Technology and Equipment ITAM, Roosevelt Street 118 , 41-800 Zabrze, Poland

He Wen
College of Electrical and Information Engineering, Hunan University, Lushan Road, Yuelu District, 410082 Changsha, Hunan Province, China

Michal Jezewski
Institute of Electronics, Faculty of Automatic Control, Electronics and Computer Science, Silesian University of Technology, Akademicka 2A, 44-100 Gliwice, Poland

Aleksandra Kawala-Janik
Automatic Control and Informatics, Faculty of Electrical Engineering, Opole University of Technology, Proszkowska 76/1, 45-758 Opole, Poland
Department of Biomedical Engineering, College of Engineering, University of Kentucky, 43 Graham Avenue, 40508 Lexington, United States of America

Amina Echchaachouai, Soumia El Hani and Imad Aboudrar
Department of Electrical Engineering, Ecole Normale Superieure de Enseignement Technique, Mohammed V University, Avenue des Nations Unies, Agdal, 10102 Rabat, Morocco

Ahmed Hammouch
National Center for Scientific and Technical Research, Angle avenues des FAR et Allal El Fassi, Hay Ryad, 10102 Rabat, Morocco

Abdulmalik Shehu Yaro
Department of Electronic and Computer Engineering, Faculty of Electrical Engineering, Universiti Teknologi Malaysia, UTM Johor Bahru, 81310 Johor, Malaysia

Department of Electrical and Computer Engineering, Faculty of Engineering, Ahmadu Bello University, Sokoto Road, PMB 06 Zaria, Nigeria

Ahmad Zuri Sha'ameri
Department of Electronic and Computer Engineering, Faculty of Electrical Engineering, Universiti Teknologi Malaysia, UTM Johor Bahru, 81310 Johor, Malaysia

Nidal Kamel
Department of Electrical and Electronic Engineering, Faculty of Engineering, Universiti Teknologi Petronas, 32610 Seri Iskandar, Malaysia

Rajesh Kumar Dubey
Department of Electronics and Communication Engineering, Jaypee Institute of Information Technology, Sector-62, 201309 Noida, Uttar Pradesh, India

Arun Kumar
Centre for Applied Research in Electronics, Indian Institute of Technology Delhi, Huaz Khas, 110016 New Delhi, India

Libor Gajdosik
Department of Telecommunications, Faculty of Electrotechnical Engineering and Computer Science, VSB–Technical university of Ostrava, 17. listopadu 15, 708 33 Ostrava, Czech Republic

Swina Narula and Sujata Pandey
Amity School of Engineering and Technology, Amity University Uttar Pradesh, Sector 125, Noida, Uttar Pradesh 201303, India

Munish Vashishath
YMCA University of Science and Technology Faridabad, Sector 6, Mathura Road, Opp. Sanjay Memorial Industrial Estate, Faridabad, Haryana 121006, India

Akanksha Mishra and Gundavarapu Venkata Nagesh Kumar
Department of Electrical and Electronics Engineering, Gandhi Institute of Technology and Management Institute of Technology, Gandhi Nagar, Rushikonda, 530045 Visakhapatnam, India

Neeta Pandey, Deva Nand, V. Venkatesh Kumar, Varun Kumar Ahalawat and Chetna Malhotra
Department of Electronics and Communication Engineering, Delhi Technological University, Shahbad Daulatpur, Main Bawana Road, Delhi-110042, India

Index

www.ingramcontent.com/pod-product-compliance
Lightning Source LLC
Chambersburg PA
CBHW080659200326
41458CB00013B/4913